The Quantum Nature of Things

This book offers readers an entirely original and unconventional view of quantum mechanics. It is a view that accepts quantum mechanics as the natural way to think about the way nature works, rather than the view commonly expressed, especially in books on quantum physics, that quantum theory is weird and counterintuitive. It is based on the concept of itemization.

From this simple premise, quantities like energy and momentum, both linear and angular, emerge naturally, as do configuration space, potentials, the electromagnetic field, many-body dynamics, special relativity, and relativistic wave mechanics. The many-body dynamics, because it is not tied to physics from the outset, can be applied to population dynamics outside physics as well as the usual physical situations.

From this emerges much of the basic physics that describes, mathematically, how the natural world behaves.

This accessible introduction does not require exotic maths and is aimed at inquisitive physics students and professionals who are interested in exploring unconventional approaches to physics. It may also be of interest to anyone studying quantum information theory or quantum computing.

Features

- Provides a unique, new approach to understanding quantum mechanics.
- Uses basic concepts and mathematical methods accessible at the undergraduate level.
- Presents applications outside physics, including a newly devised and original model of cell division that shows how cancer-cell population explosions occur.

T. R. Robinson is Emeritus Professor of Space Plasma Physics at the University of Leicester, where he obtained an M.Sc. in Experimental Space Physics and a PhD in Ionospheric Plasma Physics, before becoming a lecturer in 1982. He has had a research career at the University of Leicester of over 30 years in Space Plasma Physics, publishing over 100 papers in international refereed journals. He was awarded a personal chair in Space Plasma Physics in 1997. He has taught many undergraduate and postgraduate courses, including plasma physics, electromagnetism, quantum mechanics, and quantum field theory. He switched his research focus to quantum physics and the foundations of physics before retiring in 2016.

The Quantum Nature of Things

How Counting Leads to the Quantum World

T. R. Robinson

CRC Press
Taylor & Francis Group
Boca Raton London New York

CRC Press is an imprint of the
Taylor & Francis Group, an **Informa** business

First edition published 2023
by CRC Press
4 Park Square, Milton Park, Abingdon, Oxon, OX14 4RN

and by CRC Press
6000 Broken Sound Parkway NW, Suite 300, Boca Raton, FL 33487-2742

CRC Press is an imprint of Informa UK Limited

British Library Cataloguing-in-Publication Data
A catalogue record for this book is available from the British Library

ISBN: 978-1-032-44623-3 (hbk)
ISBN: 978-1-032-45546-4 (pbk)
ISBN: 978-1-003-37750-4 (ebk)

DOI: 10.1201/9781003377504

Typeset in Nimbus font
by KnowledgeWorks Global Ltd.

Publisher's note: This book has been prepared from camera-ready copy provided by the authors.

For Angelina, Adam, and Rachel,

without whom this book could not have been written.

Contents

Contents

Preface

This book was written, for the most part, during the COVID-19 pandemic between March 2020 and December 2021, although it has had a gestation period of more than 10 years. It is intended to be a radically different treatment from the usual one adopted in books that deal with quantum physics from a pedagogical standpoint. The choice of subject matter and approach adopted here have been motivated by a number of considerations, but the overriding one is a conviction that, given that quantum mechanics is regarded as the most accurate and successful theory of the physical universe yet devised, then it ought to be possible to find a straightforward explanation of the natural world that is fundamentally *quantum* in character and that does not rely on any preconceived notions associated with classical physics.

The usual way of introducing quantum mechanics at an undergraduate level is to follow Dirac's tried and tested prescription of *quantizing* the classical physics of Newton, particularly in its Hamiltonian guise. This is done by replacing the familiar algebra that represents numbers in classical physics with something a little less direct, the algebra that represents operators, in the quantum version. This *quantization*, as it is referred to, is a rather mysterious process that puzzles most physicists, even after a lifetime of familiarity with it, let alone undergraduates exposed to it for the first time.

It is arguable that we are doing things the wrong way round if we start with preconceived notions of what makes up the physical universe from the viewpoint of a classical theory that we know is at best incomplete, and get, what must be, a biased quantum theory that has all the baggage of the classical relationships between dynamical variables. After all, we know that classical physics is unable to explain many important phenomena on the atomic scale, hence the need for quantum physics in the first place. Dirac quantization seems to be very much a case of pulling a quantum mechanical rabbit out of a classical hat. This is the magic of quantization. Its main justification is that it works, but at what intellectual cost? It is arguably a key reason for quantum mechanics having a reputation of being weird and counterintuitive. It is the contention of this book that quantum mechanics is, on the contrary, a perfectly natural way to think about the universe, as long as we think about it in the right way and identify what is of primary importance.

It was with these long held reservations that I thought it worth trying an alternative approach, starting with quantum concepts alone, rather than with the observed behaviour of the macroscopic world, albeit one that has such a seductive appeal to our senses. However, we should not forget that we only perceive the world via senses that rely on optical, electrical and chemical processes that are themselves controlled by quantum physics.

With this in mind, I began to re-examine the ingredients of quantum theory with a view to finding a conceptually simple way into it, that avoided preconceived physical notions. Early ideas along these lines were presented, in embryonic form, in a talk I gave at a meeting organized by the Institute of Physics, at the University of Warwick in 2009. That meeting was aimed at improving the teaching of quantum mechanics to undergraduates. However, these ideas only reached maturity with the publication of a paper[1] in the European Journal of Physics in early 2021, entitled, *Natural number dynamics: reconstructing physics from generalized atomicity* [107]. The fundamental premise here is that the only information one needs about any system is that it comprises a number of items. Thus it is characterized by the natural numbers alone. The description of a system in this way is intrinsically quantum, since the natural numbers are discrete by their very nature. No structure needs to be attributed to the items in this primitive system, nor any properties other than that there are a countable number of them. The items themselves, rather like letters in an alphabet, need not even be material objects at all. Above all, this description of a system is independent of any scale, macroscopic or microscopic. Indeed, configuration space does not appear at all in the initial considerations, but emerges naturally. This primitive description of a system is conceptually simple and can surely be regarded as familiar, since we are surrounded by examples of the multiplicity of nature. This conjecture may possibly seem too familiar and too meagre for it to be able to facilitate the derivation of the laws that determine the behaviour of the physical universe. Remarkably, recognizable equations of physics do emerge. This book is built on these foundations.

Further thoughts and developments have lead to several new ideas in the ensuing months since publication of the EJP paper and have found their way into the present book, but the basic premise has remained. Put simply, this means that we identify quantization with itemization and counting. Historically, itemization and counting obviously precede both Planck's quantization of energy that emerged at the dawn of the twentieth century and also the ideas of Leucippus and Democritus, who proposed the quantization of matter in the form of atomism in the fifth century B.C. Counting was known for many millennia before either Planck's quanta or Greek atoms were thought of.

By a happy coincidence, as I was formulating this quantization as itemization idea, I came across a reference to the discovery in the 1950s of an artefact, *the Ishango bone*, that was at least 20,000 years old [93]. The bone had lines scored on it with a regularity that indicated it was a kind of tallying stick used for counting. The *Ishango bone*, which is now in the Royal Belgian Institute of Natural Sciences in Brussels, was unearthed at a site on the shores of Lake Rutanzige (Edward), close to the borders of the Democratic Republic of the Congo, Uganda, Rwanda and Tanzania, in the western arm of Africa's Great Rift Valley. That such a monumental achievement in human intellectual progress had occurred so long ago in the centre of Africa was not better known, is astonishing. Everyone knows that the Great Rift Valley and its surrounding regions are the cradle of humanity, but here was evidence that the region was also part of our early mathematical and scientific heritage too. After all, according to the great German mathematician, Gauss, mathematics is the

---

[1]The paper was written after work on the book had started and was a modified extract from an early draft of the book. The paper was really aimed at testing the water for its central idea.

Queen of the sciences and arithmetic (number theory according to some sources) is the Queen of mathematics. Counting is arguably the Empress of them all[2].

The book is set out as follows. We begin with a descriptive prologue that deals with a number of background issues that presage the unorthodox approach to quantum mechanics developed in the rest of the book. Next, the quantum mechanics proper begins with an extremely short chapter, the only purpose of which is to introduce the *universal quantum equation* from which emerges all of the physics covered in the rest of the book. In this fundamental equation, the natural numbers are associated with the eigenvalue spectrum of an operator, namely the (natural) number operator. This is a very natural way of explaining why operators enter the theory, something which remains puzzling in conventional approaches to quantum mechanics.

The essential mathematical properties of operators and Hilbert space representation are then introduced in Chapter 2. The mathematical core of the whole book is treated in Chapter 3. This shows how the universal quantum equation leads to the properties of the natural number operator and also automatically brings into play the creation and annihilation operators that have such a key role in developing the dynamical theory. This aspect is referred to as *natural number dynamics*. Chapter 4 extends the formalism to multi-category systems and explains why we need a universal variable that ultimately is associated with time. From this, in addition to bosons, fermions also emerge and their spinor properties are revealed without any preconceived notions about rotation or spin. Chapter 5 is concerned with the representations of the natural number operator in single category systems and shows how quantum mechanics emerges from natural number dynamics. Time and a dimension of configuration space also emerge as part of the dynamical picture. The importance of duality in the representation of the natural number operator by a pair of Hermitian operators is discussed, in the context of the wave-particle controversy. By generalizing the representation of the natural numbers, general potentials and bound states emerge, without any need for classical mechanical analogies.

Chapter 6 deals with two category systems and shows how interactions arise. This is the beginning of the idea of scattering of particles that leads to time dependent population numbers in many-body systems. Chapters 7 and 8 then examine the special cases of degenerate two- and three-category systems. From this analysis we discover a *Goldilocks principle* of three dimensions in which the properties of a self-consistent three-dimensional space emerge, with three components each of configuration space, linear momentum and angular momentum. Also, by generalizing the three sets of canonical commutation relations, the electromagnetic field together with the Maxwell equations that govern its behaviour emerge quite naturally, without any additional assumptions. Electromagnetic waves then provide a blueprint for the idea behind the itemization of a field and we encounter photons.

---

[2]After completing the manuscript of the present work, the author came across a recently published book [57] by one of the authors of ref. [93] that deals in greater detail with the archaeological history and analysis of the Ishango bone. This new book sheds further light on the view that the Ishango bone is the oldest of known mathematical artefacts and will hopefully lead to its being more widely known about in the future.

Chapter 9 deals with interactions in general multi-category systems and develops the equations of many-body interactions that lead to the theory of superconductivity and ultimately to an explanation of fermionic mass. Chapter 10 shows how by treating field amplitudes as the amplitudes of a density, then fields can be itemized. This essentially provides an introduction to the concept of a quantum field. Chapter 11 then takes invariant properties of the space-time dependent phase of quantum fields as the starting point for the development of both the theory of special relativity and relativistic quantum mechanics, that is completely independent of the light signal approach commonly found in elementary explanations of special relativity. This approach also yields Dirac's relativistic equation for the electron, without the need to quantize the relativistic theory of classical particle motion.

Chapter 12 deals with some examples of the use of natural number dynamics to model a variety of populations in ecology, microbiology as well as physics. Of particular interest here is the first quantum operator model of the cell division dynamics, which predicts the onset of unstable growth that mimics the behaviour of cancer cell populations. The chapter concludes with a remarkably simple but profoundly revealing model of *zitterbewegung* that is associated with the excitation of electron-positron pairs from the vacuum. The contents of this chapter are intended to illustrate something of the breadth and universality of *the quantum nature of things*. This in some ways mirrors the famous atomist poem, *On the nature of things*, written in the first century B.C. by the Roman, Lucretius, in which the poet attempted to explain a whole range of natural phenomena, based on the atomistic ideas of the earlier Greek philosophers, Democritus and Leucippus. The final chapter is an epilogue that serves as a brief summary of the ideas developed in the book and touches on what further physics might be developed. The final section contains a discussion on how classical physics arises by *de-quantizing* quantum theory. Clearly, the quantum starting point must lead to a correct description of the macroscopic classical world, but it is appropriate that this comes at the end of the journey rather than at the beginning.

This book is above all an exploration of an idea that attempts to encourage a way of thinking about quantum mechanics and physics in general, that is different from that usually taught to undergraduates in physics departments in most universities. It is aimed at inquisitive final year undergraduates and research students who have already had some exposure to a standard university course in quantum mechanics. It may even be of interest to open minded professionals who will not find the contents of this book too objectionable. In my experience, after several decades of teaching quantum physics, I have found that many students find it stimulating and exciting to be exposed to unconventional ideas. The intension is not to replace standard quantum mechanics, but to show that there are alternative ways of thinking about it, that may facilitate an acceptance of just how natural is *the quantum nature of things*. Indeed, the book might be deemed to have been successful if students return to the standard texts with renewed enthusiasm, but with many, many, more questions, after reading it.

Quantum mechanics is, after all, a controversial subject and I make no apologies for adding this little contribution to the mix. Peter Rowlands [108] has remarked that investigating the foundations of physics is a risky business, without status and career prospects. So, the endeavour is only likely to be attempted by the foolhardy and those of us who have already entered the *Grey Havens*.

T. R. Robinson
Leicester, 2022

Acknowledgments

I owe a huge debt of thanks to my colleagues in the Radio and Space Plasma Physics Research Group[3] in the Department of Physics and Astronomy at the University of Leicester. I have received tremendous support from them over the many years in which I have been a member of the group. They continued to listen patiently to my talks at our weekly research seminars, even when, in the past few years, the topic that I have presented has shifted from my longstanding research into waves in space plasmas to my more recent interest in fundamental physics. I know they had better things to do, with their own priorities in atmospheric, magnetospheric and planetary research, than to listen to my musings, but I much appreciate their interest and critical feedback.

Thanks are also due to Prof. Emmanuel Haven, who while he was on the staff of School of Management at Leicester, introduced me to the growing research field of *quantum-like* phenomena[4]. Through Emmanuel I was introduced to the work of Prof. Fabio Bagarello, who has pioneered the use of quantum operator methods to deal with population dynamics in social science and in ecological environments. Familiarity with these quantum-like studies, particularly through my collaborations with Prof. Haven, and personal contacts with Prof. Bagarello, have greatly enhanced my appreciation of the formal structures of quantum mechanics, that are not so strongly emphasized in the undergraduate physics material that I had been familiar with. Without, at least, an elementary familiarity with these formal structures, it would not have been possible to put this book together. I should add that I am entirely responsible for the contents of this book, so any errors or inadequacies in this regard are all mine. I would also like to thank Prof. Andrew Fry of the Department of Molecular and Cell Biology at Leicester for his patient explanation of cancer cell development. Thanks are also due to my colleague, Dr. Neil Arnold, who, during his time at Leicester, spent many a lunch hour trying to convince me of the value of ancient Greek philosophy. I was, for a long time, reluctant to acknowledge the relevance of Platonic thought to the modern scientific world view, however, a reading of this book will show how much I have now come to appreciate it.

---

[3] Now the Planetary Science Group.

[4] The term, *quantum-like*, refers to the use of the mathematical paraphernalia of quantum mechanics in situations outside physics.

Prologue

QUANTUM ORIGINS

This book is concerned with the central role that quantum mechanics plays in our understanding of the natural world. It is conventional to think of quantum theory as beginning around 1900 with Planck, who, in order to explain the black-body radiation spectrum, postulated that energy had to be considered as consisting of discrete chunks called *quanta* and so may be said to be *quantized*. The mathematical framework of the quantum theory, which has remained largely unchanged, was then formulated by Heisenberg, Schrödinger, Dirac and others[5] in the 1920s. This work was organized into definitive form in a classic text by von Neumann [84].

Prior to Planck's revolutionary idea of energy quanta, energy had been considered to be a continuum. However, this was not the first instance when something that had been considered as smoothly continuous was then suggested to be, in fact, made up of discrete units. Some two and a half millennia earlier than Planck's hypothesis, Leucippus and Democritus proposed that matter itself, which to our eyes and sense of touch seems so obviously continuous, was in fact made up of discrete units or atoms [90, 120]. This earlier conception is rightly regarded as one of the most momentous achievements of the human mind [129]. It was arguably even more revolutionary and outrageous than Planck's hypothesis, since it was made by the power of the intellect alone, without any direct experimental evidence, whereas Planck had detailed experimental data on which to base his hypothesis [72].

It is interesting to note that the idea of the atomic nature of matter was still being questioned as late as the 1890s. Planck himself had supposed that matter must be continuous. Here he was following Helmholz who had rejected atomism in favour of an elastic continuum picture of matter [72]. The newly developed electromagnetic theory of the time seemed to reinforce this view, since the electromagnetic field appeared to fill space without any discrete features. It was not until 1905, when Einstein showed that the electromagnetic field must have discrete features in order to explain the photoelectric effect, that the quantum nature of the physical world started to be taken seriously. Up until that time, Planck had not only doubted the reality of the atomic view of matter, but also of his own quantum theory of energy [37]. It was Einstein of course who, through his theory of special relativity, had shown the equivalence of matter and energy, so there should be no surprise that if one was discrete then so could be the other.

The two giants of the quantum revolution, Erwin Schrödinger [113] and Werner Heisenberg [51] both acknowledged the influence of early Greek thought on modern quantum ideas. Of course, what the ancient Greek atomists meant by *atoms* was not what we think of today as atoms, and indeed Schrödinger and Heisenberg were not trying to attribute any modern atomic physics to the earlier age. However, the idea

---

[5] Some of the key papers from this period may be found in ref. [124].

that matter consisted of discrete units, without any direct experimental evidence, inferred by shear force of intellect, stands even today as an incredible achievement. Moreover, the Roman poet, Lucretius [76], writing in the first century B.C., in his famous *De rerum natura* (*On the nature of things*), was able to offer remarkably accurate explanations, based on these early atomic ideas, for a wide variety of natural phenomena. The title of the present book is an obvious allusion to Lucretius's poem, and it too sets out to do something similar in explaining the behaviour of the physical Universe in terms of a kind of generalized atomicity [107] and completely accepting that after more than a 100 years of quantum theory, it is about time that we really did acknowledge the reality of *the quantum nature of things*.

PHYSICS BY NUMBERS

It is the fact that nature appears to favour the existence of energy and matter in discrete forms that raises the possibility that discreteness itself may be fundamental to the way the Universe works and that there is a kind of generalized atomism that characterizes the underlying nature of things. It is this underlying and universal discreteness that underpins the ideas explored in this book. More exactly, it is the idea that discreteness leads to countability and that this allows the discreteness to be encoded mathematically, as we shall see shortly. The fundamental premise here is that the only information one needs about any system is that it comprises a number of *atoms*. The description of a system in this way is intrinsically quantum, since the natural numbers are discrete by their very nature. No structure needs to be attributed to the atoms in this primitive universe, nor any properties other than that there are a countable number of them. Thus it is characterized by the natural numbers alone. The term *atom* here is meant in a very general way, i.e., not specifically in its modern usage, nor in the sense meant by ancient Greeks, but simply as the basic unit of a system. Then *item* might be a more appropriate term than *atom*.

It is interesting to note that the words, *atom* and *item*, although they sound very similar in English, and only differ in their vowels, have very different etymological roots. *Atom* is of Greek origin and means *cannot be cut* and hence indivisible. It came to mean the smallest part of matter and so became associated with the atomic theory of matter. *Item*, on the other hand is of Latin origin and is related to *iteration* and has to do with repetition and sequencing, or listing. It therefore has less of a material connotation than atom does and also, unlike atom, has no particular connection to size. However, given the idea that the properties of the physical world arise simply from this itemization, without any need to presume particular structure at the primitive level of items, we take a cue from the originators of atomism, Leuccipus and Democratus who famously said 'just atoms and the void' [120], in order to emphasis that the properties of material things arose not from any specific property at a macroscopic level, but rather from a simple, fundamental microstructure. So, here we begin with 'just items' and see where that takes us. This view is what will be referred to as *the quantum nature of things*, from which emerges much of the conventional physics that we are familiar with, as we shall see in later chapters of this book. The idea is that we remove all preconceived notions about the physical nature of the universe apart from the itemization conjecture, and so begin with a clean slate and then just see what emerges. This is very much an *Ockham's razor* approach.

The *items* themselves, rather like letters in an alphabet, need not even be material objects at all. The early Greek atomists themselves were aware of this more abstract definition of atomism [63]. Because this description of a system is independent of any scale, macroscopic or microscopic, configuration space does not appear at all in the initial considerations, but emerges naturally. This is contrary to the idea, adhered to in Newtonian physics and Kantian [65] philosophy alike, that configuration space is a prerequisite for an understanding of physical reality, a kind of theatre in which dynamical processes are played out. Here, this is regarded as an unnecessary presumption. As we shall see, configuration space emerges from the itemization premise, along with quantities like linear momentum, angular momentum and energy.

The idea that we might build a fundamental theory of physics on the basis of counting numbers alone takes the foundations of quantum physics further back in time, beyond Planck's discovery, beyond even ancient Greek atomism, to the discovery of counting itself and of the existence of what we now call the natural numbers. This remarkable stage in human intellectual development is known to have happened at least 20,000 year ago, as is evidenced by the discovery at Ishango, in the interlacustrine region of East Africa[6], of a bone [93, 57, 134] that was scored with well defined marks in the form of parallel lines. The neatness and distribution of these marks show an understanding of the natural numbers for counting and that the bone was probably used for tallying in a variety of number bases. There is also a suggestion that it might have been used as some kind of lunar calendar [57]. A photograph of the Ishango bone is illustrated in Fig. 0.1. There is no suggestion that, at this early stage, number and arithmetic were necessarily understood in any formal sense, but tallying seems to imply at least an awareness of one-to-one correspondence as a method of recording a number of items.

It is obviously highly unlikely that Ishango was the only place where early humans leaned to count. Indeed, artefacts have been found outside the immediate interlacustrine region, within Africa, and also in Europe [57], that indicate they were used at least as *aide-memoires* for recording a number of items. However, the degree of organization of the patterns of scored lines on the Ishango find mark it out as the earliest *mathematical* artefact. There is probably a large degree of uncertainty about the exact age of these artefacts, but what is certain is that, at a time long before Greek mathematicians developed geometry to the sophisticated level that we still use today, long even before Egyptian engineers made their preparatory measurements and calculations that enabled them to construct the pyramids, far away from the Mediterranean sea, in the centre of Africa, numerate Africans were counting the contents of their universe. Counting and number are the cornerstones of modern science and so here is an early example of the beginning of humanity's endeavour to understand the natural world in a quantitative way. It is the association between counting and the

---

[6]The interlacustrine region of East Africa is bounded by Lakes Kyoga, Albert and Rutanzige(Edward) in the west, and Lake Tanganyika in the east.

Figure 0.1 Four views of the Ishango bone. This artefact was unearthed at Ishango on the shores of Lake Rutanzige(Edward) in the western arm of the Great Rift valley in East Africa and is known to be at least 20,000 years old. This image was made available by the kind permission of the Royal Belgian Institute of Natural Sciences, Brussels.

natural numbers that provides the mathematical basis for the way in which physical theory is going to be reconstructed in the ensuing chapters of this book. Actually, counting strategy games continue to be an important element of modern African culture. This is typified by the game known as *bao* that is played widely throughout East Africa. Fig. 0.2 is part of a bao board in which the five cups contain respectively, $0, 1, 2, 3$ and 4 counters representing the first five natural numbers[7]. This is a hint to the way that the states of a system may be represented in *occupation number formalism* that we will meet later.

[7]Although the concept of nothing was understood by the ancient Greeks and Egyptians, zero as a numeric symbol like one, two or three, is thought to have been invented towards the end of the seventh century A.D., in India [114].

Figure 0.2 This figure illustrates part of a bao board. Bao is a counting strategy game that is played widely in East and Central Africa. The first five natural numbers (including 0) are represented by counters in the cups of the bao board.

THE TROUBLE WITH QUANTUM MECHANICS?

There are a number of reasons for considering an unconventional approach to developing what is the already well established body of knowledge that is quantum theory. One reason is that it is intrinsically interesting to look at something as familiar as quantum mechanics in a new light. Richard Feynman made a case for this in his Nobel prize acceptance lecture [32]. His view was that this could lead to new theoretical paths. However, he also cautioned that it could, and often did, lead up blind alleys. Even if this turns out to be so, sometimes, the trip may be interesting, instructive and worth taking.

Another important reason for trying to find a new perspective on quantum mechanics is to counter the common perception that quantum mechanics is weird, counterintuitive and beyond normal comprehension and experience. Even among physicists there is no universally accepted agreement about its meaning. What is universally agreed about quantum theory is that it is the most fundamental and accurate method we have of describing the natural world. However, it is remarkable that over a hundred years after the inception of modern quantum physics, it remains highly controversial. Famously, Richard Feynman observed that no-one really understands it [34].

In many ways, quantum theory remains the wilful and unruly child that first made an appearance at the beginning of the twentieth century with Planck's energy *quanta*. Rather like a Peter Pan, that child refuses to grow up and to behave itself in the physics classroom and do the bidding of the masters of classical physics like Newton, Faraday, and Maxwell. Even Einstein, who made one of the early important contributions to quantum theory, with his explanation of the photoelectric effect, could not fully condone the young child's behaviour, as is evidenced by his famous remark that *God doesn't play dice*. Indeed, Planck himself was reluctant to admit the reality of his discovery for several years, until persuaded to do so by eminent colleagues [72].

Part of the reason for the apparent incomprehensibility of quantum mechanics may be due to where we begin when introducing quantum ideas and that is typically by letting classical physics set the agenda. For example, quantum mechanics inherits the concepts of space, time, energy, and linear and angular momenta etc., from classical physics. Most students of physics spend their school years learning classical physics, understandably so. That is the physics that appeals directly to our senses. It grew out of the observations made by Galileo, among others, of the behaviour of tangible, visible objects in the macroscopic world around them. Detailed observations of less tangible, but still directly observable macroscopic bodies, such as the planets, made by Copernicus and Kepler, also contributed to this knowledge. These observations were given mathematical form as quantitative rules of behaviour by Newton, whose work was further developed in the following centuries by Lagrange, Hamilton, D'Alembert and others. It took some time after the early observations to decide what quantities were important. There was, for example, some confusion about the definitions of momentum and kinetic energy in terms of mass and velocity of moving bodies, but once that had been settled, Newton's famous laws of motion became the cornerstone of our understanding of the dynamical behaviour of the physical Universe, at all scales from tiny atoms to gigantic planets.

Newton himself unravelled the mysteries of planetary motion under the influence of gravity and somewhat later, Bernoulli, building on the work of Renaissance atomists like Gassendi [64, 90], applied Newton's laws to the motion of atoms and molecules in gases. Bernoulli's kinetic theory explained Boyle's law of gases in terms of molecular collisions and momentum exchange, precisely as Newton had decreed. Around the same time electrical phenomena were investigated. Coulomb gave mathematical rules for the behaviour of static charge, while Ampère and Ohm investigated electrical currents, and Øersted discovered the connection between current and magnetism. Faraday took matters further and discovered electromagnetism. This was put into a beautiful mathematical form by Maxwell. Eventually the source of the electric field, the electron, was discovered by J. J. Thomson.

The physics edifice looked unassailable in its phenomenological and mathematical certainties at the end of the nineteenth century. Then along came Planck and the problem of understanding the electromagnetic radiation that was emitted by a hot body. He found that the resulting *black-body* radiation spectrum could only be correctly understood if he assumed that the energy of the radiators was made up of discrete quanta. As a result of this momentous discovery, the physics world was thrown into turmoil and controversy that is still disturbing us today. The controversy concerning the interpretation of quantum mechanics still rages in the pages of international journals, in prestigious conferences and in physics departments in universities and research institutes the world over. Do we live in a multiverse? Does the wavefunction collapse when we measure something? Do wave functions actually represent physical waves? Is wave-particle duality a reality? Is Schrödinger's cat dead or alive? One can get the impression that this debate is akin to that in medieval Christendom over how many angels may be found on the head of a pin [47]. There is a plethora of popular science books that delve into these conjectures (see ref.[11], for a recent, informative example). Quantum mechanics, by its very nature, invites speculation about its interpretation. From its very beginning it has attracted philosophers interested in the nature of reality and what can be known about it [48, 51].

So why is quantum mechanics so controversial and apparently so weird? The problem cannot lie with quantum mechanics itself, since it is how the natural world actually works. Rather, the problem must lie with the way we think about quantum mechanics. It is arguable that physicists generally tend to over rely on ideas from classical physics to set the basic framework of the discussion about natural phenomena. Thus, we already have in mind the main parameters of physical theory, like energy, momentum, velocity, space and time, waves and particles etc., from the start. Not only that, when it comes to the basic ingredients of matter, the fundamental particles, such as the electron, which obviously are not directly accessible to our senses, then it is difficult not to think of them as miniature versions of the macroscopic objects that we see around us. It is no surprise then that when quantum mechanics tells a different story, that what we think of as little lumps of matter can be in two places at once, or that electrons can spread out like waves or be localized particles or both, it is quantum mechanics that we regard as pathological, when we should rather be questioning our mental image of what the fabric of the Universe might be. A full discussion concerning the ontological questions that arise from the quantum mechanical description of nature is beyond the scope of this book, but in the next section we will just dip into a few points that are connected to the way we think about and visualize the physical world, in particular, the role of images and the imagination.

THE TREACHERY OF IMAGES

Images are an essential part of the way we visualize the world around us. If someone we meet mentions a cat, then the image of a cat might well spring to mind. It does not have to be a particular cat we are thinking of, but the word *cat*, may conjure up a catlike image. Your acquaintance may show you a photograph of an actual cat and then the image in your mind's eye may be brought into co-incidence with the photograph. When it comes to macroscopic cats we have no problem distinguishing between an actual cat, a photograph of a cat and the thought of a cat. Educationalists often deliberately use images in the form of pictures to illustrate examples of what they are talking about. These images often take the form of diagrams and in many instances, especially when trying to illustrate conceptual points, these do not always correspond to actual objects. They may also be illustrations of what we might think of as actually being objects, but which are not directly accessible to our senses. The quantum world is necessarily an imagined world, although it is definitely not an imaginary one. Elkins [31] refers to quantum mechanics as being literally inconceivable. Arguably, it is important to acquire a *quantum imagination*, in order to think about quantum phenomena in a way that takes us beyond the classical surface.

Take, for example, the atom. One suspects that every student who has been taught about atoms in physics or chemistry lessons, in secondary school, has been shown a picture something like that in Fig. 0.3, but without the warning, *Ceci n'est pas un atome*, underneath it. The key point about this picture is that the viewer recognizes it as an atom. Of course the picture is not an atom, but is it a picture of an atom either? Before Rutherford's famous scattering experiments, the *picture* of an atom was very different. The atom was imagined to be a tiny pudding-like structure that was made up of a mixture of positive and negative charge, spread evenly inside the atomic volume. Rutherford's experiments, in which a gold foil was bombarded with alpha particles, showed that the atom mainly consisted of empty space with

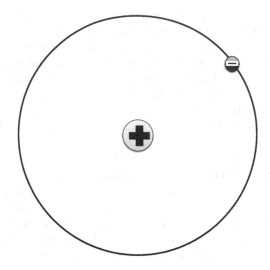

Ceci n'est pas un atome

Figure 0.3   This is not an atom.

a massive positively charged nucleus occupying a tiny fraction of its volume. The imagined pudding turned into an imagined miniature solar system. The illustration of the hydrogen atom in Fig. 0.3 is just such a model. However, this picture is still something imagined, since we cannot see it directly. What it looks like is inferred from experiment. This gives us only indirect information, from which we assemble the image. It looks like it does because we imagine the atom comprises Rutherford's positive nucleus and an equally small electron that somehow surrounds it. The orbiting possibility comes from incidental knowledge that a Coulomb force of electrical attraction is expected between two idealized points of charge. The Coulomb force has an inverse square law, just like gravity has in planetary motion, making the planetary model of the atom that much more plausible. However, classical physics tells us that the planetary model of the atom is unstable. Quantum mechanics contradicts the classical instability with quantum stability. Neither classical physics nor quantum physics actually tells us what an atom looks like. The image in Fig. 0.3 is literally a figment of our imaginations. Although this naive planetary model of the atom has been superseded by more elaborate constructions involving orbitals and electron clouds, which are also figments of the imagination, it still appears in compendium texts that are used in first year science and engineering courses in universities (e.g. [122]), typically in the context of the Bohr theory of the hydrogen atom. It can also be found in standard quantum mechanics text books (e.g.[3]), as a way of establishing the Coulomb potential in the Schrödinger equation for hydrogenic atoms. Every physicist knows that Coulomb's law describes the force between two point charges, just like the ones in Fig. 0.3. Coulomb's law then enters the Schrödinger equation in the form of a potential experienced by a negative point charge (the electron) as a function of the distance to a central positive nucleus. This reinforces the planetary picture of the atom, even if it is only by implication. The problem is that it is hard to get this image out of our minds when we think of the atom, even after working through the Schrödinger theory. The image is convincing; looks plausible; looks familiar; gives us a satisfactory feeling about what the atom is. This undermines our ability to accept quantum reality. An alternative treatment of hydrogenic atoms, that does not rely on assuming Coulomb's law for the force between point charges to get the Schrödinger equation, is developed in Section 8.3.

The *atom* in Fig. 0.3 is an obvious analogy with what the Belgian artist, René Magritte was alluding to in his famous painting, *The treachery of images*[8], which depicts a smoker's pipe that has underneath it the caption, *Ceci n'est pas une pipe*. This, of course, translates as, *This is not a pipe*. The easy interpretation is that of course it is not a pipe; it is a painting of a pipe. However, there is a deeper point, that the picture itself was imagined, a visualization by the brain after a complex chain of processes involving the eye, the brain, electromagnetic radiation, photoelectric phenomena and electrolytic charge exchange. The pipe itself gives up its image via electromagnetic scattering from an incredibly complex assemblage of atoms, so what was Magritte even looking at when he painted his famous pipe, assuming he actually had a pipe in front of him at all when he painted the picture? It is interesting to note that Magritte painted the pipe in 1928, at the very time that the foundations of quantum mechanics were being laid by Heisenberg, Dirac, Schrödinger and others [124].

Given that, even with some clues from scattering experiments about their notional structure, it is impossible to know what an atom really looks like, or even if contemplating such a notion makes any sense, the task of knowing anything about what we think of as individual particles like electrons may be, looks even more futile. Particle theorists believe that fundamental particles like the electron are singular points, despite their having intrinsic angular momentum. String theorists believe otherwise. It is even possible to find an entirely classical model of the electron as a singularity in a combination of Newtonian gravity and the Coulomb electric field [102]. Again these models are constructs of the imagination. Eddington, in his book, *Fundamental Theory*, that was published posthumously in 1946, made the following comment concerning fundamental particles:

The term 'particle' survives in modern physics, but very little of its classical meaning remains. A particle can best be defined as *the conceptual carrier of a set of variates* [29].

In other words, a particle is just an item to which one attaches some labels in order to categorize it, or, more accurately to define the state it happens to be in. An electron then should not be thought of as a little lump of matter, but only as an item that is labelled with, say, a mass, a momentum, a charge and a spin. We shall see that these labels emerge as part of the itemization process itself. However,

---

[8]The word, *trahison*, in the French title, *Le trahison des images*, is variously translated as *treason*[55] and *betrayal*[16]. Here, *treachery* is preferred.

there is no getting over the fact that despite the intellectual knowledge that electrons are best treated as Eddington suggests, it is almost impossible to rid oneself of the image of a miniaturized lump of matter whizzing through space under the influence of force fields and occasionally crashing into other little lumps of matter. It is hard to break the habit. Here again a quantum imagination is needed to overcome classical superficiality.

We can actually see this situation repeated in the macroscopic world. Take *the pound* for example. We often refer to the national currency of the UK as *The* pound. *The electron* is also referred to in a similar way. It is as if there were really only one of them, but that we make many copies for practical purposes. It's a bit like asking how many *ones* there are. Are there lots of *ones* or just a single, *one*, that we use many times over? This probably amounts to the same thing and our problem is a semantic one. In the case of electrons, their very properties depend on the fact that they are all identical and indistinguishable. It is hard to imagine that electrons can be *things* at all, if they are completely indistinguishable. It is easier to think of them like notes in a musical scale. We can repeat middle C on the piano as many times as we like, but there is, in a sense, only one middle C.

Fig. 0.4 is an image of three one pound coins. We may regard this as an image of three physical objects in the classical physics sense. It is just possible to distinguish between them by careful examination of the small blemishes and signs of wear on their surfaces. They occupy positions in configuration space. Transactions involving coins like these consist of movement along well-defined trajectories in space and time, in pocket or purse, from an initial location to where they are handed over in exchange for goods.

Fig. 0.5 is also an illustration of three pounds, but in a schematic way, using the pound symbol. These pounds are truly identical to one another, in a sense that those in Fig. 0.4 are not. There is really only one pound here, but the symbol appears three

Figure 0.4   Three pounds in the classical world. This illustrates three imperfect representations of the idealized perfection that is depicted schematically in Fig. 0.5.

times. Nor do the three pound symbols in Fig. 0.5 exist in any configuration space. It would be pointless to rearrange these symbols. Unlike the coins in Fig. 0.4, this would not lead to a new arrangement. We can imagine that Fig. 0.5 represents three pounds in a bank account. They are then clearly not material objects at all. However, they are no less real than the pound coins in their ability to allow us to purchase three pounds worth of goods, which is the point of pounds, after all. A transaction involving a pound like those in Fig. 0.5 consists of an electronic instruction to delete one from the total in one account and an instruction to add one to a second transacting account. The electronic instructions may travel along well-defined paths in wires or, more likely, in fibre optic cables, but the pounds themselves do not *travel*. This is one of the closest analogies we have in the macroscopic world of something akin to the nature of *objects* in the quantum world. The deletion and addition of each pound is analogous to the annihilation of a particle in a quantum state and the creation of one in another. This is carried out using so-called *annihilation* and *creation* operators in quantum theory. It is as pointless to ask if it is the same particle that is both annihilated in one state and then created in the other, as it is to ask if it is the same pound that is both deleted from one bank account and then added to the other.

Figure 0.5 Three pounds in the quantum world. This is a completely abstracted and idealized representation, bearing in mind the limitations due to the *treachery of images*.

The idea that things may not be what they seem, that we experience reality second hand, has long been a feature of how philosophers see the world. This notion, that there may be something beyond our senses is probably as old as human consciousness itself. It is famously epitomized by Plato's allegory of the cave. In his *Republic*, Plato envisaged humans as inhabitants of a cave, whose knowledge of the world was confined to shadows cast on the walls of their cave. Plato tried to develop concepts of reality based on what he called *Forms*. This is rather an illusive concept that Plato developed over a long period. The evolution of this concept over Plato's lifetime makes its exact meaning hard to pin down [99]. Platonic forms appear to be a kind of idealized essence, like the pound symbol in Fig. 0.5. However, for Plato these forms were the reality, the real world beyond the cave, whereas everyday experience was the illusion of the shadows on the cave walls, like the imperfect tokens in Fig. 0.4. There are parallels here with the relationship between the illusory nature of the everyday classical world and the fundamental reality of quantum nature, which we can only grasp through the idealized concepts of quantum forms. If we regard classical physics as the shadow on the wall of Plato's cave, then what we seek is the cause of the flickering images, in a reality beyond the cave [133]. This is quantum reality.

Given that classical physics is incapable of explaining key phenomena like the photoelectric effect, electron interference, atomic line spectra, the Compton effect and many other, essentially quantum, phenomena, it would seem unwise then, to use classical physics as a starting point when considering how to construct a quantum theory. However, using classical physics as a blueprint for developing quantum mechanics is more or less what physicists do. Mathematical theories of the classical world are formulated in terms of scalar valued functions that represent quantities like location coordinates, linear momentum, energy and so on. Quantum mechanics is then obtained by exchanging the scalar valued functions of classical physics for operators. These operators land like cuckoos in the classical nest. The relationships between these cuckoo operators is taken to be the same as those between their classical scalar counterparts. For example, the relationship between the momentum operator and the kinetic energy operator in quantum mechanics is taken to be the same as that between the classical momentum and classical kinetic energy that are familiar to physics students in secondary school [95]. The procedure is referred to as *quantization* of a classical theory. This *quantization* process is rather difficult to justify, apart from saying that it works. The quantization of a classical theory was strongly endorsed by Dirac [25], who was greatly influenced by Heisenberg's work. In one of his early papers on quantum mechanics that was also published in 1925, Dirac [24] made the following comment:

> "In a recent paper Heisenberg puts forward a new theory which suggests that it is not the equations of classical mechanics that are in any way at fault, but that the mathematical operations by which physical results are deduced from them require modification. *All* the information supplied by the classical theory can thus be made use of in the new theory."

The quantization of classical mechanics remains the most common way of developing quantum theory, particularly in a pedagogical context (see for example, [3, 19, 22, 80, 95, 126]), which is where physicists invariably encounter quantum mechanics for the first time. The problem with this approach is that it engenders a puzzled response among physics undergraduates, which may fail to resolve into acceptance and understanding, even after a whole career in physics. Drawing again on Plato's allegory, but in a modern context, one might say that using classical physics as a blueprint for the way the physical universe works is a bit like watching a video game in order to determine its rules. All one winds up with is the rules of the macroscopic image on the screen, when one should really be focussing on the computer programme that is driving the whole thing. The quantum world is that computer program and computer programs are not material objects. Hence the need for quantum imagination in order to appreciate it.

QUANTUM REALITY

The view adopted in this book is that the nature of reality is fundamentally quantum and that quantum mechanics is not some weird type of physics that it is often claimed to be, but is entirely natural. After all without quantum behaviour there could be no

Figure 0.6   Our quantum Sun at dawn, viewed from the western shores of lake Nyanza (Victoria), Bukoba, Tanzania. The surface temperature of the Sun is around 6000K, which corresponds to a black-body radiation spectrum that peaks at a wavelength of around 500 nm. Lake Nyanza is at the heart of the interlacustrine region, that is bounded by the lakes of the eastern and western arms of the Great Rift valley, where the Ishango bone was found.

natural world as we know it. We rely on our *quantum Sun* (Fig. 0.6), for example, for our very existence. If the electromagnetic radiation from the Sun obeyed classical rules we would end up with an ultraviolet catastrophe, the so-called *Rayleigh-Jeans catastrophe*, that would simply annihilate any living thing. That the Sun's radiation actually obeys the quantum rules discovered by Planck, that curtail the high frequency end of the black-body spectrum, is precisely what allows living organisms to survive. Our solar spectrum, which approximates to the black-body spectrum like that in Fig. 0.7, is dominated by visible light, but with enough infrared to keep us warm. It is this spectrum that has driven the whole of evolution on Earth.

   Quantum mechanics underpins the chemical and biological processes that are responsible for everything in our environment, from the simplest chemical elements to our complex brains. Quantum physics dominates our technological world too, from electronic communications to lasers and computers. A good example of the importance of applications of quantum science is modern medical physics. Almost from the beginning of quantum era, applications were found that had an impact on human health. Although X-ray technology had a shaky start, and reportedly killed its discoverer, Röntgen, it now underpins much of the key diagnostic armoury of modern medicine. So too with magnetic resonance imaging (MRI) which utilizes the nuclear magnetic resonance in the quantum states of hydrogen atoms in the water molecules of the human body to provided imaging capabilities for diagnosing the internal structure of the whole body. Treatments based on quantum processes are

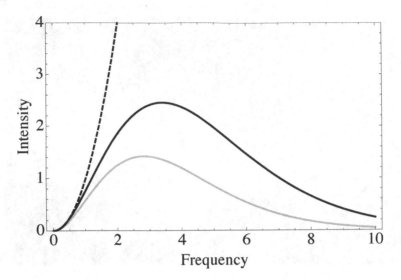

Figure 0.7 Black-body radiation spectrum in schematic form. The dark curve represents a temperature (in Kelvin) twice that of the grey curve. Higher temperatures increase the total energy output and also shift the peak intensity to higher frequencies. The dashed curve is the classical calculation, the so-called *Rayleigh-Jeans ultra-violet catastrophe.*

also important, such as the targeted destruction of cancer cells by gamma rays, that are generated by accelerating electrons. The interaction of the gamma rays with the cells can only be fully understood as quantum processes. Ultimately, the cancerous growth of cells, as well as general cell division process itself, are quantum processes.

All of the modern devices used in medical diagnostics and treatment rely on computer control and computer-aided imaging. The semiconductors at the heart of these computing devices can only be designed and manufactured through knowledge of quantum physics. Quantum physics is also entering directly into computing devices though the new science of quantum computing and quantum information [85]. It may soon be possible to build ultra-powerful computing devices, based on these principles. If anything can give us faith in quantum reality it is this understanding of the role that quantum mechanics plays in how life works and how we can intervene in its processes with the aid of quantum based devices.

The growing interest in applications of quantum methods to topics that have no direct connection to physics, such as quantum information and quantum computing, provide further incentives for the development of a more universal approach to quantum theory than that which is traditionally taught to physics undergraduates. As a first step towards this goal, we present, in the first chapter proper, a *universal quantum equation*, based on the itemization idea.

1 The universal quantum hypothesis: *Just items*

Haba na haba hujaza kibaba[1]

This book is based on the conjecture, first introduced in ref. [107], that the most fundamental description of a primitive universe is the number of items that it contains. Such a system is then characterized by a single natural number. As we shall see in the development of this idea, it is this that ultimately accounts for its physical behaviour. This idea can be thought of as expressing the *graininess* or *atomicity* of nature, but in a very general way. Initially, we will consider a system in which the items have no structure associated with them. They need not be thought of as material objects at all. They should be thought of as intrinsically indistinguishable, since distinguishable items would require further information about what distinguishes them. That would require, at least, a second label, which may or may not be a natural number. We will consider such more structured systems later, but for now our only information about this primitive system is the natural number we can associate with it. So we begin with no information concerning configuration space, time, mass, energy, momentum, neither linear or angular, nor any other characteristic commonly considered as an attribute of a physical system. Indeed, these qualities do not exist as prerequisites. If they are to exist, then they must emerge naturally. We must begin without any physical notions at all about our starkly primitive universe.

The first step towards a quantitative theory based on this *itemization* conjecture is to find a mathematical representation of natural numbers. One route would be to follow the Italian mathematician, Peano, who devised a set of axioms to provide a sound mathematical definition of natural numbers that are the basis of counting (see Appendix A). However, these axioms have very limited algebraic structure and do not allow the development of any dynamical features. Instead, we note that one can treat the natural numbers, including zero, as the spectrum of eigenvalues of an operator, the *natural number operator* [107]. This brings into consideration the rich mathematical structure of operators operating on a Hilbert space. It has the major advantage of immediately explaining why operators need to be involved in the theory, an assumption that is difficult to justify in conventional quantum mechanics that is based on quantizing an already extant classical physics, according to Dirac's prescription. By simply considering natural numbers as the eigenvalues of an operator that is free from preconceived physical notions allows us to explore the consequences without prejudice. We can then examine the emergent mathematical structures and see if we can identify any connection to what we see in nature.

---

[1]This is a well-known Kiswahili proverb that roughly translates as, *grain by grain fills the measure* [62].

DOI: 10.1201/9781003377504-1

The universal quantum hypothesis is encapsulated in the fundamental equation [107]

$$\hat{N}\Psi_n = n\Psi_n, \tag{1.1}$$

where $\hat{N}$ is an operator, Ψ_n is an eigenfunction and n is a scalar eigenvalue that is equal to any of the natural numbers, $0, 1, 2, 3, 4, \ldots$, and so on. Initially the only thing defined in Eq. (1.1) is n. What we intend to show is that a mathematical description of the physical universe emerges in quite a natural way from the fundamental equation, Eq. (1.1), which can be regarded as the *universal quantum equation*. This may seem an unlikely outcome from such a meagre starting point, but as we shall see, recognizable equations of physics do emerge. In this regard it is important to note that, because Eq. (1.1) just involves natural numbers, the equations that emerge are essentially in dimensionless form and do not automatically involve physical units. There is no particular problem in leaving the emergent equations in this dimensionless form, and this is in general how we will proceed. Dimensions and physical units will only be applied to the emergent equations when absolutely necessary to clarify their connection to physical systems.

Since the primitive system that is described by Eq. (1.1) is one in which the only information that characterizes it is the number of items it contains then, any experiment we could undertake to obtain information about it would simply involve counting. Then we can represent the states of such a system by tally marks like those on the Ishango bone in Fig. 0.1, or the number of counters in the cups in a bao board, like that in Fig. 0.2. It turns out that both of these representations can be used for quantum systems, as we shall see later. Our next task is to find a representation for the operator, $\hat{N}$, and its eigenfunctions, Ψ_n, in order to interpret Eq. (1.1). This will require us to learn something about operators on Hilbert spaces, which is what we will look at next.

2 An introduction to operators

2.1 LINEAR OPERATORS

The fundamental role of an operator is to effect a transformation. For example, an operator $\hat{F}$ may operate on a function Ψ and transform it into a new function, Φ, a process which is conventionally represented symbolically as

$$\hat{F}\Psi = \Phi. \tag{2.1}$$

The circumflex or *hat* symbol over a symbol like $\hat{F}$ will be generally used to indicate that it represents an operator. The algebraic expression in Eq. (2.1) is written like a product between the operator $\hat{F}$ and the function Ψ, the product of which is the function Φ. In practice, the operator in question is represented by a mathematical object whose properties are known from some algebraic context. A common feature of all of the operators that we will meet is that they will be linear. This means that if $\hat{F}$ is a linear operator then

$$\hat{F}(\Psi_1 + \Psi_2) = \hat{F}\Psi_1 + \hat{F}\Psi_2, \tag{2.2}$$

where Ψ_1 and Ψ_2 are functions. Also, if $\hat{F}_1$ and $\hat{F}_2$ are linear operators, then

$$(\hat{F}_1 + \hat{F}_2)\Psi = \hat{F}_1\Psi + \hat{F}_2\Psi. \tag{2.3}$$

An example of an operation on a function that transforms it into another function is differentiation. In this case we can define the operator $\hat{F}$ such that it operates on a function $\Psi(x)$, and transforms it into the derivative of $\Psi(x)$, then conventionally $\hat{F}$ is written symbolically as the *differential operator*, $\frac{d}{dx}$, such that

$$\frac{d}{dx}\Psi(x) = \Psi'(x), \tag{2.4}$$

where $\Psi'(x)$ is the derivative of $\Psi(x)$, with respect to x. We could have written $\frac{d\Psi(x)}{dx}$ instead of $\Psi'(x)$ in Eq. (2.4). Ψ is included in the numerator of the fractional form of the derivative to indicate that the symbols are not being treated as a stand alone differential operator, unlike the symbol on the lhs of Eq. (2.4). $\frac{d\Psi(x)}{dx}$ is a kind of hieroglyphic symbol that has its origins in the way calculus was developed as a method of determining the gradient of a curve. Then, the limiting ratio of a change in the function in response to changes in x was represented by $\frac{d\Psi(x)}{dx}$. The distinction between $\frac{d\Psi(x)}{dx}$ and $\frac{d}{dx}\Psi(x)$ becomes important when a differential operator operates on the product of two or more functions and the Leibniz rule is involved. An example of $\hat{F}$ being defined as *differentiate with respect to x* is

$$\hat{F}x^3 = 3x^2.$$

DOI: 10.1201/9781003377504-2

Functions themselves can trivially be thought of as operators, so that if $\hat{G} = x^2$, then

$$\hat{G}x^3 = x^5.$$

Actually we could stretch the definition of an operator even more trivially to include ordinary numbers as operators, since multiplying the function x^3 by the operator 2, say, results in the function $2x^3$.

The mathematical objects transformed by an operator do not have to be functions in the analytical sense. Another typical example that conforms to the rule in Eq. (2.1) involves matrices. Operators can be represented by square matrices and these operate on column vectors. For example, take the matrix product

$$\begin{pmatrix} 2 & -1 \\ 0 & 1 \end{pmatrix} \begin{pmatrix} x \\ y \end{pmatrix} = \begin{pmatrix} 2x - y \\ y \end{pmatrix}. \tag{2.5}$$

Here the operator is a 2×2 matrix and this transforms one 2-component column vector into another one. So in this example

$$\hat{F} = \begin{pmatrix} 2 & -1 \\ 0 & 1 \end{pmatrix}, \tag{2.6}$$

$$\Psi = \begin{pmatrix} x \\ y \end{pmatrix}, \tag{2.7}$$

and

$$\Phi = \begin{pmatrix} 2x - y \\ y \end{pmatrix}. \tag{2.8}$$

Differential operators and matrices are two important examples of how operators can be represented in a concrete way that makes quantitative calculations possible on systems that they represent. Both types of operator play important roles in quantum mechanics and we shall meet them again. While both differential operators and matrix operators are linear, they differ in their multiplicative properties, and this has important consequences when it comes issues of representation. Whereas, matrices obey the associative rule in their multiplication, differential operators do not, since they have to obey the Leibniz rule for the differentiation of a product and are mathematically referred to as *derivations* [36] (also see Section 2.9 and Appendix F).

2.2 EIGENVALUES AND EIGENFUNCTIONS

The operator equation, Eq. (1.1), that we used to introduce the concept of a natural number operator, constitutes an example of an important situation where an operator operates on a function and the result is the same function multiplied by a number. This special case represents a key concept where operators are used to investigate properties of systems of interest, whether the systems be physical or otherwise. When an operator operates on a function and leaves it unchanged, apart from multiplying it by a number, then the state of the system is said to be represented by the function that is operated on and the multiplying factor that results from the operation on it represents some quantitative information about that system while it is in that state. Thus, a special role is played by functions that are transformed into themselves, apart from a scaling factor. Such functions are termed *eigenfunctions*[1] of the operator involved and the numerical factor is called the *eigenvalue* of the operator. In a sense an operator interrogates a system represented by its eigenfunction, which results in a definite answer given by the eigenvalue. As we shall see, if the operator interrogates a function that is not among its eigenfunctions then no clear answer results and then there is a degree uncertainly about the state of the system of interest.

In many important situations, such as Eq. (1.1), a given operator may have more than one eigenfunction and then that results in a *spectrum of eigenvalues*. This situation is generally expressed as an eigenvalue equation of the form

$$\hat{F}\Psi_i = f_i\Psi_i, \qquad (2.9)$$

where Ψ_i is an eigenfunction of $\hat{F}$, with an eigenvalue of f_i. The index, i is a natural number which may or may not be limited to a finite value. It acts as a label of the state of the system represented by Ψ_i. To say the system is in a state represented by Ψ_i means that if we know Ψ_i then we know the eigenvalue associated with the operator. The eigenvalue can then be regarded as quantitative information about the system. There is usually some sequential connotation to these natural numbers which can be regarded as counting the eigenstates and eigenvalues. It sometimes occurs that two different eigenstates represented by different functions have the same eigenvalue. Such a state is called *degenerate*. Degenerate states are often associated with some kind of symmetry in a system, and play an important role in certain situations that occur in quantum theory.

The functions, Ψ_i, play a key role in representing a system. They are used to form an *orthonormal basis*,[2] $\mathcal{F}$, which is referred to as a Hilbert space. Then the set of Ψ_is can be thought of as basis vectors in the multi-dimensional Hilbert space. The orthonormal basis for $\mathcal{F}$ leads to the definition of an inner or scalar product

$$\langle\Psi_i|\Psi_j\rangle = \delta_{ij}. \qquad (2.10)$$

Eq. (2.10) is the mathematical expression of the orthonormality condition. There are several ways of representing the inner product, in Eq. (2.10), the two most common being an integral over a continuous real variable and as the scalar product of vectors in a Fock space [126]. Both of these representations are used in standard quantum mechanics and will be dealt with later.

The inner product in Eq. (2.10) is the basis of a notation, due to Dirac, in which, $\langle\Psi_i|\Psi_j\rangle$, is treated as the scalar product of a *ket vector*, $|\Psi_j\rangle$ and a *bra vector*, $\langle\Psi_i|$. This scalar product then constitutes the bracket or *bra-ket*, $\langle\Psi_i|\Psi_j\rangle$. The relationship between the bra and ket vectors will be developed further, after we have defined the properties of the inner product in a more formal way. First we must extend the orthonormal basis idea a little further.

---

[1] The German adjective word *eigen* has the meaning of *own* in the sense of belonging to.
[2] Orthonormal, means both orthogonal and normalized.

The orthonormal basis allows more general functions to be represented as linear superpositions of the basis functions. Thus a function, Ψ, may be represented as

$$\Psi = \sum_{i=1}^{S} \alpha_i \Psi_i, \tag{2.11}$$

where the α_is are scalar coefficients and S is the number of terms in the sum in Eq. (2.11), which may be infinite. Because Ψ is a linear superposition of the functions, Ψ_i, that act as basis vectors, then Ψ itself can be thought of as a vector in the Hilbert space.

Notice that, if we substitute Eq. (2.9) into Eq. (2.11), and recall Eq. (2.1), then we find

$$\hat{F}\Psi = \sum_{i=1}^{S} \alpha_i \hat{F}\Psi_i = \sum_{i=1}^{S} \alpha_i f_i \Psi_i = \Phi, \tag{2.12}$$

which implies that Φ is also a vector in $\mathscr{F}$. Note that we can only take $\hat{F}$ inside the summation sign in Eq. (2.12) because of its linear property.

The definition of the inner product in Eq. (2.10), can be extended to superposed vectors, such as, $\langle \Psi | \Phi \rangle$, but we need some formal rules to enable us to define this precisely. To do this, we treat the Hilbert space as a *normed space*, such that the inner product has the properties of linearity, conjugate symmetry and positivity, which are respectively [49, 56, 98, 126]

$$linearity : \langle \Phi | \alpha_1 \Psi_1 + \alpha_2 \Psi_2 \rangle = \alpha_1 \langle \Phi | \Psi_1 \rangle + \alpha_2 \langle \Phi | \Psi_2 \rangle,$$

$$conjugate\,symmetry : \langle \Psi | \Phi \rangle^* = \langle \Phi | \Psi \rangle,$$

$$positivity : \langle \Psi | \Psi \rangle \geq 0, \tag{2.13}$$

where α_1 and α_2 are scalars; Φ, Ψ, Ψ_1, and Ψ_2 are vectors in $\mathscr{F}$; and $\langle \Psi | \Phi \rangle^*$, is the complex conjugate of $\langle \Psi | \Phi \rangle$. The conjugate symmetry property above, is consistent with $(|\Theta\rangle)^* = \langle \Theta|$ and $((\Theta|)^* = |\Theta\rangle$, for any vector, Θ, in $\mathscr{F}$.

Using the positivity condition allows us to define a *norm*, $\| \, |\Psi\rangle \, \|$, of Ψ, by

$$\| \, |\Psi\rangle \, \|^2 = \langle \Psi | \Psi \rangle.$$

This norm is essentially the modulus of the vector, $|\Psi\rangle$.

A combination of the linearity condition and the complex conjugate definition means that, $\langle \Psi | \alpha \Phi \rangle = \alpha \langle \Psi | \Phi \rangle$, where α is a complex scalar, then

$$\langle \Psi | \alpha \Phi \rangle^* = (\alpha \langle \Psi | \Phi \rangle)^* = \alpha^* \langle \Phi | \Psi \rangle. \tag{2.14}$$

Also, substituting Eq. (2.11) into $\langle \Psi | \Psi \rangle$ and using the linearity condition and Eq. (2.14), we get

$$\langle \Psi | \Psi \rangle = \sum_{i,j} \alpha_i^* \alpha_j \delta_{ij} = \sum_i |\alpha_i|^2.$$

Notice that if $\sum_i |\alpha_i|^2 = 1$, then $\langle \Psi | \Psi \rangle = 1$, which is the condition for Ψ to be normalized. Since, in this case, the sum of all of the coefficients, $|\alpha_i|^2$, is 1, then it follows that $0 \leq |\alpha_i|^2 \leq 1$ for all i. Thus the $|\alpha_i|^2$s satisfy the condition for being interpretable as probabilities. It is this that allows the choice for a probabilistic interpretation of quantum mechanics. Furthermore, there is an important probabilistic interpretation to an inner product $\langle \Psi | \Phi \rangle$, since[3]

$$\langle \Psi | \Phi \rangle = \langle \Psi | \hat{F} \Psi \rangle = \sum_i |\alpha_i|^2 f_i. \tag{2.15}$$

Eq. (2.15) is interpreted as a sum of the eigenvalues f_i, weighted by the coefficients $|\alpha_i|^2$, which, since these can be interpreted as probabilities associated with the states, Ψ_i, may be considered as a weighted average or a mean value of the variable represented by the operator $\hat{F}$, for a system characterized by the superposed state vector, Ψ. Such a value is usually referred to as an *expectation value*. The construction in Eq. (2.15) provides quantum mechanics with its interpretative power in calculations on physical systems. The functions like Ψ, clearly determine the mean values obtained from operators and are regarded as indicators of the state of the system. For that reason they are commonly referred to as *state functions* or *state vectors*.

2.3 COMMUTATION PROPERTIES

It is often useful to operate with a pair of operators on the same state function, so for example we could write

$$\hat{F}\hat{G}\Psi.$$

This expression is well defined as long as $\hat{G}\Psi$ is also a vector in $\mathscr{F}$, but it does not have to be a normalized one, nor does $\hat{G}$ have to have the same set of eigenstates as $\hat{F}$. We have to be a little careful in interpreting the above expression. It could mean that we first operate on Ψ with $\hat{G}$ and then operate with $\hat{F}$ on the result of $\hat{G}\Psi$. We could have written the above expression as $\hat{F}(\hat{G}\Psi)$, where the bracket emphasizes this interpretation. However it could also mean that we first multiply the two operators together according to the algebraic rules that govern them to form a new operator and then operate with this on Ψ. In this case we could write the expression as $(\hat{F}\hat{G})\Psi$. If the algebra involved is associative then

$$(\hat{F}\hat{G})\Psi = \hat{F}(\hat{G}\Psi).$$

We shall encounter operators algebras that are associative as well as those which are not. Whether the operator algebras are associative or not, in general, the order in which two operators are applied to a function does affect the outcome, so if

$$\hat{F}\hat{G}\Psi \neq \hat{G}\hat{F}\Psi,$$

then $\hat{F}$ and $\hat{G}$ are said not to commute. Even in the associative case the operator multiplication itself may not be commutative, then

$$\hat{F}\hat{G} \neq \hat{G}\hat{F}.$$

---

[3] In standard quantum mechanics Eq. (2.15) is commonly written in Dirac's *bra* and *ket* notation [118], as $\langle \hat{F} \rangle = \langle \Psi | \hat{F} | \Psi \rangle$. Then, the operator $\hat{F}$ operates on the ket vector, $|\Psi\rangle$, rather than just Ψ. We will use the two forms interchangibly, but the form in Eq. (2.15) is sometimes clearer under certain circumstances. That is why we will retain it.

Non-commutation arises with differentiation. For example, if $\hat{F}$ is the differential operator defined in Eq. (2.4), and $\hat{G} = x^2$, then

$$\hat{F}\hat{G}\Psi(x) \neq \hat{G}\hat{F}\Psi(x),$$

since

$$\frac{d}{dx}x^2\Psi(x) \neq x^2\frac{d}{dx}\Psi(x).$$

Notice here that the associative rule does not apply in this case, since, in general, $\frac{d}{dx}(x^2\Psi(x)) \neq (\frac{d}{dx}x^2)\Psi(x)$.

Matrices also do not generally commute under multiplication, e.g.

$$\begin{pmatrix} 2 & -1 \\ 0 & 1 \end{pmatrix}\begin{pmatrix} 1 & 1 \\ -3 & 0 \end{pmatrix}\begin{pmatrix} x \\ y \end{pmatrix} \neq \begin{pmatrix} 1 & 1 \\ -3 & 0 \end{pmatrix}\begin{pmatrix} 2 & -1 \\ 0 & 1 \end{pmatrix}\begin{pmatrix} x \\ y \end{pmatrix}.$$

However, unlike differentiation, matrix multiplication is associative. Take for example

$$\begin{pmatrix} 2 & -1 \\ 0 & 1 \end{pmatrix}\begin{pmatrix} 1 & 1 \\ -3 & 0 \end{pmatrix}\begin{pmatrix} x \\ y \end{pmatrix}.$$

This may be evaluated either by first multiplying the column vector by the square matrix just to the left of it, which results in a new column vector

$$\begin{pmatrix} 1 & 1 \\ -3 & 0 \end{pmatrix}\begin{pmatrix} x \\ y \end{pmatrix} = \begin{pmatrix} x+y \\ -3x \end{pmatrix},$$

then

$$\begin{pmatrix} 2 & -1 \\ 0 & 1 \end{pmatrix}\begin{pmatrix} x+y \\ -3x \end{pmatrix} = \begin{pmatrix} 5x+2y \\ -3x \end{pmatrix}.$$

Alternatively, first multiplying the two matrix operators gives

$$\begin{pmatrix} 2 & -1 \\ 0 & 1 \end{pmatrix}\begin{pmatrix} 1 & 1 \\ -3 & 0 \end{pmatrix} = \begin{pmatrix} 5 & 2 \\ -3 & 0 \end{pmatrix}$$

and then

$$\begin{pmatrix} 5 & 2 \\ -3 & 0 \end{pmatrix}\begin{pmatrix} x \\ y \end{pmatrix} = \begin{pmatrix} 5x+2y \\ -3x \end{pmatrix}.$$

The result is the same either way. So if $\hat{F}$ and $\hat{G}$ are represented by square matrices, with Ψ a column vector, then $\hat{F}(\hat{G}\Psi) = (\hat{F}\hat{G})\Psi$, but, $\hat{F}\hat{G} \neq \hat{G}\hat{F}$.

It is useful to define a commutation bracket or *commutator* as

$$[\hat{F}, \hat{G}] = \hat{F}\hat{G} - \hat{G}\hat{F}. \tag{2.16}$$

This plays a key role in quantum mathematics, as will be seen later. We first note that, by definition, an operator commutes with any constant scalar, so

$$[\hat{F}, \lambda] = 0, \tag{2.17}$$

where λ is any constant scalar.

If two operators have the same eigenfunction then they must commute. To see this, suppose, in addition to Eq. (2.9), we also have an operator, $\hat{W}$ that shares the same eigenfunctions as $\hat{F}$, so, $\hat{W}\Psi_i = w_i\Psi_i$, where w_i is also an eigenvalue. Then with the aid of Eq. (2.11) [95]

$$[\hat{F},\hat{W}]\Psi = \sum_i \alpha_i(\hat{F}\hat{W} - \hat{W}\hat{F})\Psi_i$$
$$= \sum_i \alpha_i(f_iw_i - w_if_i)\Psi_i \qquad (2.18)$$
$$= 0,$$

since the numbers f_i and w_i always commute. Hence $[\hat{F},\hat{W}] = 0$, assuming $\Psi_i \neq 0$.

It is also straightforward to show the reverse is true, i.e., if two operators commute then they share the same eigenfunctions. Suppose an operator, $\hat{Z}$, commutes with $\hat{F}$, so that $[\hat{F},\hat{Z}] = 0$. Then

$$\hat{F}\hat{Z}\Psi_i = \hat{Z}\hat{F}\Psi_i = f_i\hat{Z}\Psi_i, \qquad (2.19)$$

where we have used Eq. (2.17) in the last step of Eq. (2.19). This implies that $\hat{Z}\Psi_i$ is an eigenfunction of $\hat{F}$ with an eigenvalue of f_i. So $\hat{Z}\Psi_i$ must be proportional to Ψ_i. Writing $\hat{Z}\Psi_i = z_i\Psi_i$, where z_i is a scalar shows that $\hat{Z}$ has the same eigenfunctions as $\hat{F}$.

Situations will arise where the commutator will contain products of operators and we will need to know how to simplify them. It is straightforward to show, by explicitly writing out the brackets that, for any three operators, $\hat{F}$, $\hat{G}$ and $\hat{H}$, then

$$[\hat{F},\hat{G}+\hat{H}] = [\hat{F},\hat{G}] + [\hat{F},\hat{H}]. \qquad (2.20)$$

Another useful result is

$$[\hat{F},\hat{G}\hat{H}] = \hat{G}[\hat{F},\hat{H}] + [\hat{F},\hat{G}]\hat{H}. \qquad (2.21)$$

Similarly

$$[\hat{F}\hat{G},\hat{H}] = \hat{F}[\hat{G},\hat{H}] + [\hat{F},\hat{H}]\hat{G}. \qquad (2.22)$$

These last two results can easily be checked by explicitly writing out the commutator brackets. In the case of the last two expansions the manipulation involves adding in some terms that sum to zero. For example in the case of Eq. (2.22), the trick is to write $\hat{F}\hat{G}\hat{H} - \hat{H}\hat{F}\hat{G}$ as

$$\hat{F}\hat{G}\hat{H} - \hat{F}\hat{H}\hat{G} + \hat{F}\hat{H}\hat{G} - \hat{H}\hat{F}\hat{G}.$$

Strictly speaking, the above explanation relies on the operators involved obeying the associative rule. However, the expansion rules still work with differential operators, which, of course, do not obey the associative rule (see Appendix E). This point will be touched on again in Sections 2.7 and 2.9.

2.4 ADJOINT AND HERMITIAN OPERATORS

Operators with real eigenvalues also play a key role in quantum theory. This is because the eigenvalues are interpreted as information about the state of a system and this information is universally regarded as requiring real values. Operators with real eigenvalues are conventionally represented by what are termed Hermitian or self-adjoint operators[4]. To understand how this works it is necessary to define the adjoint of an operator. This is done as follows.

Consider two vectors, Υ and Ψ, defined on the same orthonormal basis such as $\mathscr{F}$, so that their inner product, $\langle\Upsilon|\Psi\rangle$ is well defined. Then, from the conjugate symmetry condition in Eqs. (2.13), the complex conjugate $\langle\Upsilon|\Psi\rangle^*$ of the inner product, $\langle\Upsilon|\Psi\rangle$, is

$$\langle\Upsilon|\Psi\rangle^* = \langle\Psi|\Upsilon\rangle. \tag{2.23}$$

The adjoint, $\hat{F}^\dagger$, if it exists, of an operator $\hat{F}$, is then defined via the inner product[5], by

$$\langle\Upsilon|\hat{F}\Psi\rangle^* = \langle\hat{F}\Psi|\Upsilon\rangle = \langle\Psi|\hat{F}^\dagger\Upsilon\rangle. \tag{2.24}$$

From this definition we can also see that

$$\langle\Upsilon|\hat{F}\Psi\rangle^{**} = \langle\Upsilon|\hat{F}\Psi\rangle = \langle\Psi|\hat{F}^\dagger\Upsilon\rangle^*. \tag{2.25}$$

We can interpret the last equation as meaning $\hat{F}$ is the adjoint of $\hat{F}^\dagger$ or simply, $\hat{F} = \hat{F}^{\dagger\dagger}$. Further useful results that follow from the definition are $(\hat{F}+\hat{G})^\dagger = \hat{F}^\dagger + \hat{G}^\dagger$ and for a complex number α, $(\alpha\hat{F})^\dagger = \alpha^*\hat{F}^\dagger$.

For an Hermitian (self-adjoint) operator[6], $\hat{F}^\dagger = \hat{F}$ and so, if $\hat{F}$ has eigenvalues and eigenfunctions according to Eq. (2.9), then [126]

$$f_i\langle\Psi_j|\Psi_i\rangle = \langle\Psi_j|\hat{F}\Psi_i\rangle = \langle\hat{F}\Psi_j|\Psi_i\rangle = f_j^*\langle\Psi_j|\Psi_i\rangle. \tag{2.26}$$

If $i \neq j$ then first and fourth terms above are zero because of orthogonality. If $i = j$, then $f_i = f_i^*$ and hence, Hermitian operators have real eigenvalues. By the same token, if $\hat{F}^\dagger = -\hat{F}$, then $\hat{F}$ is termed anti-Hermitian and has imaginary eigenvalues. If $\hat{F}$ is Hermitian then it follows that $i\hat{F}$ is anti-Hermitian, since $(i\hat{F})^\dagger = -i\hat{F}^\dagger = -i\hat{F}$.

It is also very useful to note that a non-Hermitian linear operator may be written in the form $\hat{F} = \hat{X} + i\hat{Y}$, where $\hat{X}$ and $\hat{Y}$ are both Hermitian. One can check this by noting that if $\hat{X}$ and $\hat{Y}$ are Hermitian then $\hat{F}^\dagger = \hat{X} - i\hat{Y}$, so

$$\hat{F} + \hat{F}^\dagger = 2\hat{X} \tag{2.27}$$

and

$$\hat{F} - \hat{F}^\dagger = 2i\hat{Y}. \tag{2.28}$$

---

[4]It is possible to find non-Hermitian operators with real eigenvalues [12], but these require more complicated Hilbert spaces than those used in conventional quantum mechanics. These considerations do not contradict any of the results in this book and are not considered here.

[5]This definition is commonly written as $\langle\Psi_j|\hat{F}^\dagger|\Psi_i\rangle = \langle\Psi_i|\hat{F}|\Psi_j\rangle^*$, in Dirac notation, as noted in relation to Eq. (2.15). When adjoints are involved, the notation in Eq. (2.24) is somewhat clearer [126].

[6]Physicists take the definition of Hermitian and self-adjoint to be identical [19],whereas mathematicians make a distinction between them [45]. For more details see refs. [56, 98].

Thus

$$\hat{F} = \hat{X} + i\hat{Y} = \frac{\hat{F} + \hat{F}^\dagger}{2} + i\frac{\hat{F} - \hat{F}^\dagger}{2i}, \tag{2.29}$$

which is clearly the case.

A further useful result concerning adjoint operators that comes from the definition above is that $(\hat{G}\hat{F})^\dagger = \hat{F}^\dagger\hat{G}^\dagger$. To prove this from the definition, we first start with Eq. (2.1), $\hat{F}\Psi = \Phi$. Then we consider the expression, $\hat{G}\Phi = \hat{G}\hat{F}\Psi$. Then

$$\langle\Upsilon|\hat{G}\Phi\rangle^* = \langle\hat{G}\Phi|\Upsilon\rangle = \langle\Phi|\hat{G}^\dagger\Upsilon\rangle. \tag{2.30}$$

Now the first term in Eq. (2.30) is $\langle\Upsilon|\hat{G}\hat{F}\Psi\rangle^*$ and the last is $\langle\Phi|\hat{G}^\dagger\Upsilon\rangle = \langle\hat{F}\Psi|\hat{G}^\dagger\Upsilon\rangle$. So we get

$$\langle\Upsilon|\hat{G}\hat{F}\Psi\rangle^* = \langle\hat{F}\Psi|\hat{G}^\dagger\Upsilon\rangle = \langle\Psi|\hat{F}^\dagger\hat{G}^\dagger\Upsilon\rangle, \tag{2.31}$$

but the first term in Eq. (2.31) is just $\langle\Psi|(\hat{G}\hat{F})^\dagger\Upsilon\rangle$ by definition, so we can conclude that

$$(\hat{G}\hat{F})^\dagger = \hat{F}^\dagger\hat{G}^\dagger. \tag{2.32}$$

This result implies that even if $\hat{F}$ and $\hat{G}$ are both Hermitian, then their product is not necessarily Hermitian, since $(\hat{F}\hat{G})^\dagger = \hat{G}\hat{F}$, which is not the same as $\hat{F}\hat{G}$, unless $\hat{F}$ and $\hat{G}$ commute.

The definition of the adjoint operator facilitates the definition of the *norm* or square modulus of a vector like[7] $\hat{F}\Psi = \Phi$, as follows. Using the definition of the norm of a vector from Section 2.2, the norm of Φ is defined by $\| \,|\Phi\rangle\, \|^2 = \langle\Phi|\Phi\rangle$, then

$$\langle\Phi|\Phi\rangle = \langle\hat{F}\Psi|\hat{F}\Psi\rangle. \tag{2.33}$$

The key property of the norm is that it is a modulus squared and as such must be greater than or equal to zero. Consider a general operator, $\hat{F}$ which is not Hermitian, and its adjoint, $\hat{F}^\dagger$, then

$$\| \,|\hat{F}\Phi\rangle\, \|^2 = \langle\hat{F}\Phi|\hat{F}\Phi\rangle = \langle\Phi|\hat{F}^\dagger\hat{F}\Phi\rangle,$$

which demonstrates that the expectation values of $\hat{F}^\dagger\hat{F}$ must be non-negative.

Also, the eigenvalues of the combination, $\hat{F}^\dagger\hat{F}$ must be non-negative. To see this, we note that $\hat{F}^\dagger\hat{F}$ is Hermitian since, $(\hat{F}^\dagger\hat{F})^\dagger = \hat{F}^\dagger\hat{F}$, and so must have real eigenvalues. Now, if Θ is a normalized eigenfunction of $\hat{F}^\dagger\hat{F}$, with say, $\hat{F}^\dagger\hat{F}\Theta = \lambda\Theta$, then[8]

$$\lambda = \langle\Theta|\hat{F}^\dagger\hat{F}\Theta\rangle = \langle\hat{F}\Theta|\hat{F}\Theta\rangle = \| \,|\hat{F}\Theta\rangle\, \|^2 \tag{2.34}$$

and so λ must be non-negative, since the last term in Eq. (2.34) is a modulus squared. This result is crucial to the interpretation of Eq. (1.1) as we shall see in the next chapter.

The result in Eq. (2.32) is easily extended to a sequence of operators, e.g.

$$(\hat{A}\hat{B}\hat{C})^\dagger = \hat{C}^\dagger(\hat{A}\hat{B})^\dagger = \hat{C}^\dagger\hat{B}^\dagger\hat{A}^\dagger, \tag{2.35}$$

---

[7] Recall that Φ is not generally normalized.

[8] The fact that Θ is and eigenfunction of $\hat{F}^\dagger\hat{F}$, does not imply that it is an eigenfunction of either $\hat{F}$ or $\hat{F}^\dagger$.

so the adjoint of a sequence of operators is a sequence of their adjoint operators in reverse order. Notice also that in general, because of Eq. (2.32), then

$$[\hat{F},\hat{G}]^{\dagger} = [\hat{G}^{\dagger},\hat{F}^{\dagger}]. \tag{2.36}$$

Then if $\hat{F}$ and $\hat{G}$ are both Hermitian so that $\hat{F} = \hat{F}^{\dagger}$ and $\hat{G} = \hat{G}^{\dagger}$, then

$$[\hat{F},\hat{G}]^{\dagger} = [\hat{G},\hat{F}]. \tag{2.37}$$

However, we note that for any pair of operators, $[\hat{F},\hat{G}] = -[\hat{G},\hat{F}]$, which implies that for a pair of Hermitian operators

$$[\hat{F},\hat{G}]^{\dagger} = -[\hat{F},\hat{G}], \tag{2.38}$$

and so $[\hat{F},\hat{G}]$ is anti-Hermitian. So we could expect that, for Hermitian $\hat{F}$ and $\hat{G}$, then $[\hat{F},\hat{G}]$ has the form

$$[\hat{F},\hat{G}] = i\hat{J}, \tag{2.39}$$

where $\hat{J}$ is some Hermitian operator so that $i\hat{J}$ is anti-Hermitian. We will see this form plays an important role in the quantum theory in later chapters. It also has a key role in determining the level of uncertainty that is associated with the non-commutation of operators. This uncertainty is examined in the next section.

2.5 UNCERTAINTY

Non-commutation has important consequences when it comes to information about two different parameters associated with a system. We have already seen that if two operators do not commute, then they have no common eigenfunctions and so cannot simultaneously both have well defined eigenvalues. This gives rise to a level of uncertainty with regard to the state of a system when interrogated by the operators associated with the parameters in question. We can obtain a quantitative estimate of this mutual uncertainty as follows.

Consider the two non-commuting Hermitian operators, $\hat{F}$ and $\hat{G}$ that operate on a common Hilbert space, $\mathcal{H}$, and let Ψ represent a normalized vector in $\mathcal{H}$. The mathematical development that follows is greatly simplified if we take the expectation values of $\hat{F}$ and $\hat{G}$ to be zero, i.e. $\langle\Psi|\hat{F}\Psi\rangle = 0$ and $\langle\Psi|\hat{G}\Psi\rangle = 0$. If this is not the case, then we can simply take the operators, respectively, as $\hat{F} - \langle\Psi|\hat{F}\Psi\rangle$ and $\hat{G} - \langle\Psi|\hat{G}\Psi\rangle$, which will achieve the required property.

Now, let $\hat{F}\Psi = \Phi_f$ and $\hat{G}\Psi = \Phi_g$. As with Eq. (2.12), neither Φ_f nor Φ_g is normalized and indeed we can see that

$$\langle\Phi_f|\Phi_f\rangle = \langle\Psi|\hat{F}^{\dagger}\hat{F}\Psi\rangle = \| |\hat{F}\Psi\rangle \|^2,$$

so that $\langle\Phi_f|\Phi_f\rangle$ is a measure of the square of the modulus or length of the vector, $\hat{F}\Psi$. Similarly, we get for $\hat{G}$

$$\langle\Phi_g|\Phi_g\rangle = \langle\Psi|\hat{G}^{\dagger}\hat{G}\Psi\rangle = \| |\hat{G}\Psi\rangle \|^2.$$

Now, because $\hat{F}$ is Hermitian, then $\hat{F}^2 = \hat{F}^\dagger \hat{F}$ and so $\langle \Psi | \hat{F}^\dagger \hat{F} \Psi \rangle = \langle \Psi | \hat{F}^2 \Psi \rangle$. But $\langle \Psi | \hat{F}^2 \Psi \rangle$ is just the variance, V_f, of $\hat{F}$, since it has a zero mean. So, V_f is also the modulus squared of the vector, $\hat{F}\Psi = \Phi_f$. We can now define the variance, V_f of $\hat{F}$ as $V_f = \langle \Psi | \hat{F}^2 \Psi \rangle = \langle \Phi_f | \Phi_f \rangle$. Similarly, the variance, $V_g = \langle \Psi | \hat{G}^2 \Psi \rangle = \langle \Phi_g | \Phi_g \rangle$ is also the squared modulus of the vector, $\hat{G}\Psi = \Phi_g$.

Since the operators will both map Ψ to new vectors in the Hilbert space, and, since $\hat{F}$ and $\hat{G}$ are both Hermitian, we can define the inner product $\langle \Psi | \hat{F}\hat{G}\Psi \rangle = \langle \hat{F}\Psi | \hat{G}\Psi \rangle = \langle \Phi_f | \Phi_g \rangle$. From the Schwarz inequality [19] we know that

$$\| \langle \Phi_f | \Phi_g \rangle \| \leq \langle \Phi_f | \Phi_f \rangle^{\frac{1}{2}} \langle \Phi_g | \Phi_g \rangle^{\frac{1}{2}}. \tag{2.40}$$

Letting $[\hat{F}, \hat{G}] = i\hat{J}$ where $\hat{J}$ is Hermitian, as shown in Eq. (2.39), then, after a little manipulation, we can write

$$\hat{F}\hat{G} = \frac{1}{2}(\hat{K} + i\hat{J}), \tag{2.41}$$

where $\hat{K} = \hat{F}\hat{G} + \hat{G}\hat{F}$. It is easy to check that $\hat{K}$ is Hermitian, since

$$\hat{K}^\dagger = (\hat{F}\hat{G} + \hat{G}\hat{F})^\dagger = \hat{G}^\dagger \hat{F}^\dagger + \hat{F}^\dagger \hat{G}^\dagger = \hat{G}\hat{F} + \hat{F}\hat{G} = \hat{K}.$$

So,

$$\langle \Phi_f | \Phi_g \rangle = \langle \Psi | \hat{F}\hat{G}\Psi \rangle = \frac{1}{2}(\langle \Psi | \hat{K}\Psi \rangle + i\langle \Psi | \hat{J}\Psi \rangle). \tag{2.42}$$

Now because $\hat{K}$ and $\hat{J}$ are Hermitian, we know that their respective expectation values, $\langle \Psi | \hat{K}\Psi \rangle = \bar{k}$ and $\langle \Psi | \hat{J}\Psi \rangle = \bar{j}$, say, are real. So $\langle \Phi_f | \Phi_g \rangle = \frac{1}{2}(\bar{k} + i\bar{j})$ and hence that the modulus, $\| \langle \Phi_f | \Phi_g \rangle \| = \frac{1}{2}(\bar{k}^2 + \bar{j}^2)^{\frac{1}{2}}$, from which we can conclude that[9] $\| \langle \Phi_f | \Phi_g \rangle \| \geq \frac{1}{2}|\bar{j}|$. Thus

$$\| \langle \Phi_f | \Phi_g \rangle \| \geq \frac{1}{2} \| \langle \Psi | [\hat{F}, \hat{G}]\Psi \rangle \|. \tag{2.43}$$

From the two inequalities, Eqs. (2.40) and (2.43), we can conclude that

$$\Delta_f \Delta_g \geq \frac{1}{2} \| \langle \Psi | [\hat{F}, \hat{G}]\Psi \rangle \|, \tag{2.44}$$

where $\Delta_f = V_f^{\frac{1}{2}}$ and $\Delta_g = V_g^{\frac{1}{2}}$ can be regarded as the uncertainties in the values attributable, respectively, to the operators $\hat{F}$ and $\hat{G}$. The inequality in Eq. (2.44) shows that one can only obtain variance free information from two different operators only if they commute. Otherwise, the product of their variances will exceed some nonzero limit set by the commutation relation. Eq. (2.44) can be regarded as a general uncertainty principle for non-commuting operators. It is the basis of the famous *Heisenberg Uncertainty Principle* of standard quantum mechanics, as we shall see later in Chapter 5.

---

[9]The equality implies that $\bar{k} = 0$.

Notice that the limiting value of the product of the variances depends, in general of the state of the system as represented by Ψ, unless the commutator is a number. So, if $[\hat{F}, \hat{G}] = i\varepsilon$, where ε is a real number then

$$\Delta_f \Delta_g \geq \frac{1}{2} \parallel \langle \Psi | \varepsilon \Psi \rangle \parallel = \frac{|\varepsilon|}{2}, \tag{2.45}$$

which is independent of Ψ.

2.6 OPERATOR FUNCTIONS

We are familiar with the idea of a function of scalar variables like x and y. So the function x^2 takes the set of numbers represented by x and maps them into a new set of numbers obtained by calculating x^2. We want to be able to do something similar for operators. To see how to do this, we begin with an eigenvalue equation for an operator, $\hat{F}$, Eq. (2.9). If we operate with $\hat{F}$ a second time on this equation we get thus

$$\hat{F}^2 \Psi_i = f_i \hat{F} \Psi_i = f_i^2 \Psi_i. \tag{2.46}$$

Clearly, we could repeat this process to any order, p, and get

$$\hat{F}^p \Psi_i = f_i^p \Psi_i, \tag{2.47}$$

where p is a natural number. From the above result we can write the following series

$$(1 - \frac{\hat{F}^2}{2!} + \frac{\hat{F}^4}{4!} - \ldots)\Psi_i = (1 - \frac{f_i^2}{2!} + \frac{f_i^4}{4!} - \ldots)\Psi_i. \tag{2.48}$$

For the terms in the bracket on the rhs of Eq. (2.48), summed to infinity, we would have no hesitation in writing $\cos f_i$. By the same token we write the terms in the bracket on the lhs side as $\cos \hat{F}$. By this way of working we can define any analytic function of $\hat{F}$, $f(\hat{F})$ by

$$f(\hat{F})\Psi_i = f(f_i)\Psi_i. \tag{2.49}$$

Given the way the function of an operator is defined then

$$[\hat{F}, f(\hat{F})] = 0. \tag{2.50}$$

It is obvious that $[\hat{F}, \hat{F}] = \hat{F}\hat{F} - \hat{F}\hat{F} = 0$, but also $[\hat{F}, \hat{F}^2] = \hat{F}\hat{F}\hat{F} - \hat{F}\hat{F}\hat{F} = 0$, and so on. So, $[\hat{F}, \hat{F}^p] = 0$ for any natural number, p. Hence, any analytic function of $\hat{F}$, $f(\hat{F})$, commutes with $\hat{F}$ itself. It is also obvious that $f(\hat{F})$ has the same set of eigenfunctions as $\hat{F}$. This is consistent with the general result we found for two commuting operators in Eq. (2.18).

The above result has an important and powerful corollary. We can argue that if some operator $\hat{F}$ commutes with another operator $\hat{K}$, then we can infer the possibility that $\hat{F}$ is a function of $\hat{K}$, which also means, by inverting the functional relationship, that $\hat{K}$ is a function of $\hat{F}$. However this functional dependence is not guaranteed. It is perfectly possible that two commuting operators are independent of one another. So, for example, if

$$[\hat{K},\hat{F}] = 0, \tag{2.51}$$

then the most general relationship between $\hat{F}$ and $\hat{K}$ is $\hat{F} = f(\hat{K},\hat{L},\hat{M},\dots)$, where the argument contains other operators with which $\hat{F}$ commutes.

Finally, in this section, we note that for a function of an operator, $(f(F))^{\dagger} = f(\hat{F}^{\dagger})$. To see this we note that $(\hat{F}\hat{F})^{\dagger} = \hat{F}^{\dagger}\hat{F}^{\dagger}$ so $(\hat{F}^2)^{\dagger} = \hat{F}^{\dagger 2}$. This clearly can be extended to any power of $\hat{F}$ and to any function of $\hat{F}$ that can be expressed as a power series.

2.7 SYSTEM EVOLUTION

Changes in a system may be expressed in two superficially different but equivalent ways. First we assume that the eigenfunctions and hence any superposition of them are functions of some real continuous variable, ξ, so that as ξ changes, then $\Psi(\xi)$ evolves in some way. We can take $\Psi(\xi)$ as an analytic function of ξ so that it is differentiable with respect to ξ. Then we can define a linear operator, $\hat{\Lambda}$ that transforms $\Psi(\xi)$ into a new function of ξ, $\Phi(\xi)$, just as in Eq. (2.1), but here we want $\Phi(\xi) = \Psi'(\xi)$, where $\Psi'(\xi)$ is a function that is equal to the derivative of $\Psi(\xi)$. So, we write [104]

$$\frac{d\Psi(\xi)}{d\xi} = \hat{\Lambda}\Psi(\xi), \tag{2.52}$$

where $\hat{\Lambda}$ is a linear operator and we have used the traditional symbol $\frac{d\Psi(\xi)}{d\xi}$ to represent $\Psi'(\xi)$. At this stage, the operator, $\hat{\Lambda}$, is assumed not to depend on ξ.

The form of Eq. (2.52) is important. It is necessary that the differential of Ψ is equal to an expression that is linear in Ψ, since the operator $\hat{\Lambda}$ is linear, so if we operate on $\Psi_1 + \Psi_2$, then we get

$$\frac{d(\Psi_1 + \Psi_2)}{d\xi} = \hat{\Lambda}(\Psi_1 + \Psi_2) = \hat{\Lambda}\Psi_1 + \hat{\Lambda}\Psi_2. \tag{2.53}$$

Before proceeding, it is worth commenting on the form of the symbols used in Eq. (2.52). The rhs of Eq. (2.52) is straightforward. The operator $\hat{\Lambda}$ operates on a function of ξ, $\Psi(\xi)$, and maps it to the derivative of $\Psi(\xi)$ with respect to ξ, which is of course another function of ξ. This is represented on the lhs of Eq. (2.52) by the differential hieroglyph, $\frac{d\Psi(\xi)}{d\xi}$. We could equally well have used $\Psi'(\xi)$.

In principle the lhs of Eq. (2.52) could also be written with a differential operator as $\frac{d}{d\xi}\Psi(\xi)$. Both forms are found in standard text books on quantum mechanics, sometimes interchangeably. Here we choose not to use the stand-alone differential operator form, since not only is the meaning of Eq. (2.52) clear anyway, but also, it avoids the temptation of regarding $\frac{d}{d\xi}$ as a definition of $\hat{\Lambda}$ which it certainly is not, i.e.[10]

$$\hat{\Lambda} :\neq \frac{d}{d\xi}.$$

Indeed, $\hat{\Lambda}$ may not be a differential operator at all, in the conventional sense, in certain representations. In other representations it may involve scalar variables different from ξ and differentials with respect to these variables. Then the distinction between differentials represented by the hieroglyph form and differential operator forms, as described above, becomes important. These issues will become clearer when we deal with specific examples later on. They are also dealt with in more detail in Appendix E.

Proceeding now with the main issue of this section, i.e., system evolution, we note that Eq. (2.52) may be formally integrated to give

$$\Psi(\xi) = \exp(\xi\hat{\Lambda})\Psi(0). \tag{2.54}$$

To see how Eq. (2.54) comes about, suppose the eigenvalue equation for $\hat{\Lambda}$ is

$$\hat{\Lambda}\Psi_i = \lambda_i\Psi_i. \tag{2.55}$$

Then from Eq. (2.52) we have

$$\hat{\Lambda}\Psi_i(\xi) = \frac{d\Psi_i(\xi)}{d\xi} = \lambda_i\Psi_i(\xi), \tag{2.56}$$

from which we get

$$\Psi_i(\xi) = \exp(\lambda_i\xi)\Psi_i(0).$$

So if we take Ψ to be a vector in a Hilbert space constructed from the eigenfunctions, Ψ_i as in Eq. (2.11), then

$$\Psi(\xi) = \sum_{i=1}^{S} \alpha_i\Psi_i(\xi) = \sum_{i=1}^{S} \alpha_i\exp(\lambda_i\xi)\Psi_i(0). \tag{2.57}$$

Recalling the properties of functions of operators from the previous section, we get

$$\exp(\hat{\Lambda}\xi)\Psi(0) = \sum_{i=1}^{S} \alpha_i\exp(\hat{\Lambda}\xi)\Psi_i(0) = \sum_{i=1}^{S} \alpha_i\exp(\lambda_i\xi)\Psi_i(0), \tag{2.58}$$

and Eq. (2.54) follows immediately.

Even though $\Psi(\xi)$ varies with ξ, we insist that it remains normalized so $\langle\Psi(0)|\Psi(0)\rangle = \langle\Psi(\xi)|\Psi(\xi)\rangle = 1$. From the definition of adjoint in Eq. (2.24), this implies

$$\langle\exp(\xi\hat{\Lambda})\Psi(0)|\exp(\xi\hat{\Lambda})\Psi(0)\rangle = \langle\Psi(0)|\exp(\xi\hat{\Lambda}^\dagger)\exp(\xi\hat{\Lambda})\Psi(0)\rangle = 1, \tag{2.59}$$

---

[10]The symbol ':=' will be used to mean *is defined as*. By contrast, ':≠' will be used to mean *is not defined as*. These two symbols are not always necessary, but will be used where the symbols, =, and, ≡, are ambiguous.

where $\hat{\Lambda}^\dagger$ is the adjoint of $\hat{\Lambda}$. Eq. (2.59) then implies that

$$\exp(\xi\hat{\Lambda}^\dagger)\exp(\xi\hat{\Lambda}) = 1.$$

The Baker-Campbell-Hausdorff formula [130] implies that the product of $\exp(\xi\hat{\Lambda}^\dagger)$ and $\exp(\xi\hat{\Lambda})$ in this expression cannot immediately be equated to $\exp(\xi(\hat{\Lambda}^\dagger + \hat{\Lambda}))$, unless $[\hat{\Lambda}^\dagger, \hat{\Lambda}] = 0$, which at this stage is by no means obvious. To proceed, we simply multiply $\exp(\xi\hat{\Lambda}^\dagger)\exp(\xi\hat{\Lambda}) = 1$ by $\exp(-\xi\hat{\Lambda}^\dagger)$ throughout and get

$$\exp(\xi\hat{\Lambda}) = \exp(-\xi\hat{\Lambda}^\dagger). \tag{2.60}$$

By expanding each side of Eq. (2.60) and equating terms of the same order in ξ, we immediately get $\hat{\Lambda}^\dagger = -\hat{\Lambda}$. This result has profound implications for the nature of the changes to a system whose behaviour is governed by operators on a space of orthonormal vectors. It first of all implies that $\hat{\Lambda}$ must be anti-Hermitian and thus has the form $\hat{\Lambda} = -i\hat{K}$, where $\hat{K}$ is an Hermitian operator, which, like $\hat{\Lambda}$, does not depend on ξ. Then, from Eq. (2.54),

$$\Psi(\xi) = \exp(-i\xi\hat{K})\Psi(0). \tag{2.61}$$

Thus changes in the system are governed by an equation of the form[11]

$$i\frac{d\Psi}{d\xi} = \hat{K}\Psi. \tag{2.62}$$

It is important to emphasize again that Eq. (2.62) is not a definition of $\hat{K}$, i.e., $i\frac{d}{d\xi} :\neq \hat{K}$. Independent information is needed to define $\hat{K}$ as we shall see when we deal, below, with specific cases.

We will refer to Eq. (2.62) as an S-type equation since it is this form that is taken by the Schrödinger equation in standard quantum mechanics, although the S could also be interpreted as being related to *Shift*, since Eq. (2.61) implies that it transforms Ψ at $\xi = 0$ into Ψ at ξ. We can interpret $\Psi(\xi)$ as the state of the system for a given value of ξ. The S-type equation treats changes in the system as changes to $\Psi(\xi)$. This is the first way of dealing with changes in a system in operator algebra. The second, but equivalent, way of treating changes to a system is as follows. Suppose that $\Psi(\xi)$ is an eigenfunction of some operator $\hat{G}$, such that

$$\hat{G}\Psi(\xi) = g\Psi(\xi),$$

where g is an eigenvalue. Then,

$$\hat{G}\exp(-i\xi\hat{K})\Psi(0) = g\exp(-i\xi\hat{K})\Psi(0).$$

Multiplying this by $\exp(i\xi\hat{K})$ we get

$$\hat{G}(\xi)\Psi(0) = g\Psi(0),$$

---

[11]Notice that we could alternatively have chosen to make $\hat{\Lambda} = i\hat{K}$, which would have introduced a minus sign to one side of Eq. (2.62). However, the choice made above leads to the conventionally accepted form for Eq. (2.62).

where

$$\hat{G}(\xi) = \exp(i\xi\hat{K})\hat{G}\exp(-i\xi\hat{K}),$$

so $G(\xi)$ has exactly the same eigenvalue that G has[12]. Recalling that $\hat{K}$ is independent of ξ, explicitly differentiating $\exp(i\xi\hat{K})\hat{G}\exp(-i\xi\hat{K})$ with respect to ξ immediately leads to an H-type differential equation for the operator, $\hat{G}(\xi)$ of the form

$$i\frac{d\hat{G}}{d\xi} = \hat{G}\hat{K} - \hat{K}\hat{G} = [\hat{G}, \hat{K}]. \tag{2.63}$$

Here, H-type refers to the fact that Eq. (2.63) has the same form as the *Heisenberg equation of motion* in standard quantum mechanics and $\hat{G}(\xi) = \exp(i\xi\hat{K})\hat{G}\exp(-i\xi\hat{K})$ has the form of a Heisenberg (H-type) operator. Notice also that $\frac{d\hat{G}}{d\xi}$ in Eq. (2.63) represents an operator that is the differential of $\hat{G}$ with respect to ξ. So we could write

$$\frac{d\hat{G}(\xi)}{d\xi} \equiv \hat{G}'(\xi),$$

just to emphasis that $\frac{d}{d\xi}$ is not being used as a stand-alone operator symbol, as with Eq. (2.62). This point will be demonstrated in more detail in Appendix E.

Eqs. (2.62) and (2.63) represent two equivalent ways of expressing changes to the system represented either by $\Psi(\xi)$ and $\hat{G}(0)$ or equivalently by $\Psi(0)$ and $\hat{G}(\xi)$. These two representations are equivalent and to a certain extent interchangeable. In Appendix F it is shown that it is possible to derive the H-type equation entirely from the requirement for linearity and the Leibniz rule for the differentiation of a product.

An important property of the H-type and S-type representations is that products of operators obey the same rules as individual operators, since the product of two operators is itself an operator. For example, suppose an operator $\hat{F}$ has an S-type representation, $\hat{F}(0)$, then its H-type form is $\hat{F}(\xi) = \exp(i\xi\hat{K})\hat{F}\exp(-i\xi\hat{K})$. Suppose $\hat{E}(0) = \hat{F}(0)\hat{G}(0)$, then we find

$$\begin{aligned} \hat{E}(\xi) &= \exp(i\xi\hat{K})\hat{E}(0)\exp(-i\xi\hat{K}) = \exp(i\xi\hat{K})\hat{F}(0)\hat{G}(0)\exp(-i\xi\hat{K}) \\ &= \exp(i\xi\hat{K})\hat{F}(0)\exp(-i\xi\hat{K})\exp(i\xi\hat{K})\hat{G}(0)\exp(-i\xi\hat{K}) \\ &= \hat{F}(\xi)\hat{G}(\xi). \end{aligned} \tag{2.64}$$

An obvious extension of the result in Eq. (2.64) applies to commutator brackets, i.e.

$$[\hat{F}(0), \hat{G}(0)] = \hat{C}(0) \implies [\hat{F}(\xi), \hat{G}(\xi)] = \hat{C}(\xi). \tag{2.65}$$

Notice also, that, obviously, $[\hat{K}, \hat{K}] = 0$, so

$$i\frac{d\hat{K}}{d\xi} = [\hat{K}, \hat{K}] = 0,$$

which implies that, $\hat{K}$ is independent of ξ, as was originally assumed. Clearly, we also have, $\hat{K}(\xi) = \exp(i\xi\hat{K})\hat{K}\exp(-i\xi\hat{K}) = \hat{K}$, since $\hat{K}(0) \equiv \hat{K}$.

---

[12]Implicitly here, $\hat{G}$ means $\hat{G}(0)$.

Eq. (2.63) also implies

$$i\frac{\mathrm{d}\hat{F}\hat{G}}{\mathrm{d}\xi} = [\hat{F}\hat{G},\hat{K}] = [\hat{F},\hat{K}]\hat{G} + \hat{F}[\hat{G},\hat{K}]$$

$$= i\frac{\mathrm{d}\hat{F}}{\mathrm{d}\xi}\hat{G} + i\hat{F}\frac{\mathrm{d}\hat{G}}{\mathrm{d}\xi} \tag{2.66}$$

where the commutator expansion rule, Eq. (2.22) has been utilized. Eq. (2.66) is exactly what is needed for a self-consistent representation. It shows quite clearly that the expansion rule for the commutator brackets follows the Leibniz rule for differentiation of a product. Mathematically it means the commutator bracket is associated with the *derivation* property (see Section 2.9 for further comments on this point). This puts the emphasis for generating the Leibniz rule on the commutator bracket itself and not on the differential operator. This is essential in the differentiation of operators using H-type differential equations, since, as we shall see later, operators like $\hat{K}$, $\hat{F}$ and $\hat{G}$ may obey the associative rule under multiplication, so, for example $\hat{K}(\hat{F}\hat{G}) = (\hat{K}\hat{F})\hat{G}$. Indeed, under certain conditions, they can be represented by matrices, which of course, are associative under multiplication. Then they, by themselves, cannot satisfy the Leibniz rule for the differentiation of a product. However, the commutator, because it has the derivation property, ensures that operator products are differentiated correctly in H-type equations.

The connection between differentiation of a product, the Leibniz rule, and the commutator bracket, is indicative of the underlying algebraic structure of differential calculus. This structure does not depend on the traditional method of obtaining differential calculus by taking limits, in which form it is sometimes referred to as infinitesimal calculus. It is possible to generate the results of differential calculus without taking limits by beginning with a non-associative algebra based on the Leibniz rule (see Appendix F for a more detailed discussion of this method).

2.8 EVOLUTION OF EXPECTATION VALUES

We have seen that H-type operators are functions of some continuous variable, ξ, while the state vectors, $\Psi(0)$, are independent of time. From a practical predictive point of view we will be interested later in how the expectation values of the operators, vary with ξ. Take the example of $\hat{G}(\xi)$. Its expectation value, following the definition in Eq. (2.15), has the form[13]

$$\langle\hat{G}(\xi)\rangle = \langle\Psi(0)|\hat{G}(\xi)|\Psi(0)\rangle, \tag{2.67}$$

with respect to the state, $\Psi(0)$. Now, since the state vectors and their conjugates are invariant with respect to ξ, then, when we come to differentiate the expectation value of $\hat{G}(\xi)$, we get[14]

---

[13]Conventional Dirac notation works best here.

[14]Clearly we are treating $\langle\Psi(0)|\hat{G}(\xi)|\Psi(0)\rangle$ as a product of $\langle\Psi(0)|$, $\hat{G}(\xi)$, and $|\Psi(0)\rangle$.

$$\frac{d\langle \hat{G}(\xi) \rangle}{d\xi} = \frac{d\langle \Psi(0)|\hat{G}(\xi)|\Psi(0) \rangle}{d\xi} = \langle \Psi(0)|\frac{d\hat{G}(\xi)}{d\xi}|\Psi(0) \rangle. \tag{2.68}$$

This means that the rate of change of the expectation value of a H-type operator is equal to the expectation value of the rate of change of the operator, i.e.

$$\frac{d\langle \hat{G}(\xi) \rangle}{d\xi} = \langle \frac{d\hat{G}(\xi)}{d\xi} \rangle. \tag{2.69}$$

It is worth emphasizing the obvious here, that if an operator is a function of a continuous variable like ξ, then so, in general, is its expectation value. This is an important result that will be used with operator differential equations in the following sections.

2.9 COMMUTATORS AND DERIVATIONS

As we have seen in this chapter, there is a strong relationship between differentiation and commutators, particularly when it comes to the differentiation of operators that are functions of a continuous variable, in H-type equations. This is because the commutators themselves have the property of derivations. This property allows operators from an associative algebra, like matrices, to facilitate differentiation in both S-type and H-type equations. We can see this explicitly by defining a commutator bracket between non-commuting operators that are themselves from an associative algebra, as a product in its own right. To do this, consider three operators, $\hat{A}$, $\hat{B}$ and $\hat{C}$ that have the associative property in the sense that

$$\hat{A}(\hat{B}\hat{C}) = (\hat{A}\hat{B})\hat{C}, \tag{2.70}$$

for any permutations of $\hat{A}$, $\hat{B}$ and $\hat{C}$. Now we define a commutator as a product, using the multiplication symbol, $\bowtie$, by, for example

$$\hat{A} \bowtie \hat{B} = \hat{A}\hat{B} - \hat{B}\hat{A} = [\hat{A}, \hat{B}]. \tag{2.71}$$

Then we find that

$$\hat{A} \bowtie (\hat{B} \bowtie \hat{C}) \neq (\hat{A} \bowtie \hat{B}) \bowtie \hat{C}.$$

This inequality can be checked by noting that

$$\hat{A} \bowtie (\hat{B} \bowtie \hat{C}) = [\hat{A}, [\hat{B}, \hat{C}]] = \hat{A}\hat{B}\hat{C} - \hat{B}\hat{C}\hat{A} - \hat{A}\hat{C}\hat{B} + \hat{C}\hat{B}\hat{A},$$

whereas

$$(\hat{A} \bowtie \hat{B}) \bowtie \hat{C} = [[\hat{A}, \hat{B}], \hat{C}] = \hat{A}\hat{B}\hat{C} - \hat{C}\hat{A}\hat{B} - \hat{B}\hat{A}\hat{C} + \hat{C}\hat{B}\hat{A}.$$

It is also straightforward to show that

$$[\hat{A}, [\hat{B}, \hat{C}]] = [[\hat{A}, \hat{B}], \hat{C}] + [\hat{B}, [\hat{A}, \hat{C}]], \tag{2.72}$$

which implies that

$$\hat{A} \bowtie (\hat{B} \bowtie \hat{C}) = (\hat{A} \bowtie \hat{B}) \bowtie \hat{C} + \hat{B} \bowtie (\hat{A} \bowtie \hat{C}). \tag{2.73}$$

Eqs. (2.72) and (2.73) are equivalent statements of the rule for derivations [36]. What is striking about Eq. (2.73) is that the operator $\hat{A}$ behaves like a differential operator. We can compare Eq. (2.73) to a conventional statement of the Leibniz rule for the differentiation of a product in the form

$$\frac{\mathrm{d}(f(x)g(x))}{\mathrm{d}x} = \left(\frac{\mathrm{d}f(x)}{\mathrm{d}x}\right)g(x) + f(x)\left(\frac{\mathrm{d}g(x)}{\mathrm{d}x}\right). \tag{2.74}$$

However, it is clear that the derivation property in Eq. (2.73) lies with the product rule, $\bowtie$, not with the operators, $\hat{A}$, $\hat{B}$ and $\hat{C}$, which obey the associative rule in the sense implied by the multiplications in Eq. (2.70), whereas in the case of Eq. (2.74) the derivation property is associated with the differential, $\frac{\mathrm{d}}{\mathrm{d}x}$. So we can conclude that the derivation property in Eq. (2.72) lies with the algebraic properties of commutators and not with the operators, obeying the associative rule, themselves. This result explains why the operator, $\hat{K}$, in the S-type equation, Eq. (2.62) and the H-type equation, Eq. (2.63), does not have to be a differential operator in the conventional sense. We also note that rearranging Eq. (2.72) leads to

$$[\hat{A}, [\hat{B}, \hat{C}]] + [\hat{B}, [\hat{C}, \hat{A}]] + [\hat{C}, [\hat{A}, \hat{B}]] = 0, \tag{2.75}$$

which is a statement of the *Jacobi identity* [39].

3 Natural number dynamics I: The basic formulation

In this chapter we address the key goal of finding a representation of the natural numbers as a spectrum of eigenvalues of an operator. The set of natural numbers, including zero, $\mathbb{N}_0$,[1] is a subset of the set of integers. One of the obvious properties of this subset is that its members are all real and non-negative. This non-negative property provides an important starting point for developing a representation of the operator $\hat{N}$ in Eq. (1.1). So we begin our task by examining what kind of operator would guarantee non-negative, real eigenvalues.

3.1 NON-NEGATIVE OPERATORS AND THEIR FACTORIZATION

A possible representation of an operator that would be sure to have non-negative real eigenvalues would be one that has the form of a modulus, as was demonstrated in Section 2.4. Such an operator can be written as the product of an operator and its adjoint. This is analogous to seeing that a non-negative real number could be factorized into a complex number multiplied by its complex conjugate. To see that such an operator has eigenvalues that are real and also non-negative, consider the operator $\hat{A}^\dagger \hat{A}$. Clearly $\hat{A}^\dagger \hat{A}$ is Hermitian, since $(\hat{A}^\dagger \hat{A})^\dagger = \hat{A}^\dagger \hat{A}$. Now suppose that $\hat{A}^\dagger \hat{A}$ has an eigenfunction Φ_k such that $\hat{A}^\dagger \hat{A}\Phi_k = k\Phi_k$, where k is an eigenvalue.

Now $\langle \Phi_k | \hat{A}^\dagger \hat{A}\Phi_k \rangle = k$, so

$$k = \langle \Phi_k | \hat{A}^\dagger \hat{A}\Phi_k \rangle = \langle \hat{A}\Phi_k | \hat{A}\Phi_k \rangle = \| |\hat{A}\Phi_k \rangle \|^2, \tag{3.1}$$

which is a modulus squared, then $\hat{A}^\dagger \hat{A}$, necessarily has real, non-negative eigenvalues. Now we can make the operator $\hat{K}$ in Eq. (2.62) an operator with non-negative real eigenvalues by letting $\hat{K} = \hat{A}^\dagger \hat{A}$. We can think of $\hat{A}$ as a kind of operator amplitude [104] of the operator $\hat{K}$. Next we note that $\hat{K} = \hat{A}^\dagger \hat{A}$ is invariant to a phase shift in $\hat{A}$, so we can write $\hat{A}(\xi) = \exp(-i\kappa\xi)\hat{A}(0)$, where κ is an arbitrary constant and ξ is the real variable in Eq. (2.62). Treating $\hat{A}(\xi)$ as an H-type operator and differentiating it with the aid of Eq. (2.63), we get

$$i\frac{d\hat{A}}{d\xi} = [\hat{A}, \hat{K}] = \kappa\hat{A}. \tag{3.2}$$

Applying $[\hat{A}, \hat{K}] = \kappa\hat{A}$ from Eq. (3.2) to the eigenfunction Φ_k yields[2]

$$\hat{K}\hat{A}\Phi_k = (\hat{A}\hat{K} - \kappa\hat{A})\Phi_k = (k - \kappa)\hat{A}\Phi_k. \tag{3.3}$$

---

[1] The symbol $\mathbb{N}_0$ is used to indicate the set of natural numbers that includes zero, whereas the symbol $\mathbb{N}$ is used to mean the set of natural numbers, not including zero.

[2] Notice that associativity is implied since we have assumed that $(\hat{A}\hat{K})\Phi_k = \hat{A}(\hat{K}\Phi_k) = \hat{A}k\Phi_k = k\hat{A}\Phi_k$.

DOI: 10.1201/9781003377504-3

Eq. (3.3) tells us that $\hat{A}\Phi_k$ is an eigenfunction of $\hat{K}$ with an eigenvalue of $k - \kappa$, and so we can assume that

$$\hat{A}\Phi_k = \alpha(k)\Phi_{k-\kappa}, \tag{3.4}$$

where $\alpha(k)$ is a scalar factor that may depend on k. However, $\langle\Phi_k|\hat{A}^\dagger\hat{A}\Phi_k\rangle = \langle\hat{A}\Phi_k|\hat{A}\Phi_k\rangle = \| |\hat{A}\Phi_k\rangle \|^2$, so we must have $\alpha^2(k) = k$.

We can deduce what values k can take by applying $\hat{A}$ to Φ_k, p times. This yields

$$\hat{A}^p\Phi_k = \sqrt{k(k-\kappa)(k-2\kappa)\ldots(k-(p-1)\kappa)}\Phi_{k-p\kappa}, \tag{3.5}$$

where p is a natural number. Eq. (3.5) implies that if $p\kappa$ exceeds k in value, then the eigenfunction $\Phi_{k-p\kappa}$ will have a negative eigenvalue, which is not permissible. The only way we can avoid negative eigenvalues cropping up in the spectrum of eigenvalues is if we insist that $k = n\kappa$, where n is a natural number. This is because the value of the square root will be exactly zero for $p \geq n$, since it will then contain the factor, $n\kappa - n\kappa = 0$, whenever value of p reaches n. This ensures that there is a state, Φ_0, for which $\hat{A}\Phi_0 = 0$, beyond which the system cannot be taken by further applications of $\hat{A}$. We have now made sure that the system has non-negative eigenvalues. This result is crucially dependent on the existence of an eigenfunction that has an eigenvalue of zero. The system is then governed by the equations

$$\hat{K}\Phi_n = n\kappa\Phi_n \tag{3.6}$$

and

$$\hat{A}\Phi_n = \sqrt{n\kappa}\Phi_{n-1}. \tag{3.7}$$

Note that, from now on we will use n rather than k to label the eigenfunctions.

Next we can look at what $\hat{A}^\dagger$ does to Φ_n. We start with

$$i\frac{d\hat{A}^\dagger}{d\xi} = [\hat{A}^\dagger, \hat{K}] = -\kappa\hat{A}^\dagger. \tag{3.8}$$

With $k = n\kappa$, we then get

$$\hat{K}\hat{A}^\dagger\Phi_n = (\hat{A}^\dagger\hat{K} + \kappa\hat{A}^\dagger)\Phi_n = (n+1)\kappa\hat{A}^\dagger\Phi_n. \tag{3.9}$$

From Eq. (3.9) we can infer that $\hat{A}^\dagger\Phi_n$ is an eigenfunction of $\hat{K}$ with an eigenvalue of $(n+1)\kappa$. Thus we can assume that

$$\hat{A}^\dagger\Phi_n = \beta(n)\Phi_{n+1}, \tag{3.10}$$

where $\beta(n)$ is a scalar factor that can depend on n. However, we can also see that

$$\langle\Phi_n|\hat{A}^\dagger\hat{A}\Phi_n\rangle = \beta(n-1)\alpha(n) = n\kappa, \tag{3.11}$$

so we must have $\beta(n) = \sqrt{(n+1)\kappa}$. As a result we get

$$\hat{A}^\dagger\Phi_n = \sqrt{(n+1)\kappa}\Phi_{n+1}. \tag{3.12}$$

Thus

$$[\hat{A}, \hat{A}^\dagger] = \kappa. \tag{3.13}$$

The fact that the commutator in Eq. (3.13) is equal to a constant is significant. κ represents the step size in the eigenvalues that we obtain when we apply the operators $\hat{A}$ or $\hat{A}^\dagger$ to an eigenstate. This allows us to get to the natural number operator, as will be seen next.

3.2   THE NATURAL NUMBER OPERATOR

It is now a simple matter to obtain the properties of the natural number operator, $\hat{N}$. To do this we just set κ equal to 1 in Eqs. (3.11)-(3.13), so that the step in the eigenvalues becomes 1, and then $\hat{K}$ can be identified with $\hat{N}$, since from Eq. (3.6), with $\kappa = 1$, $\hat{K}$ has eigenvalues equal to the natural numbers. We may then write $\hat{N} = \hat{A}^\dagger \hat{A}$ and $\hat{A}^\dagger$ and $\hat{A}$ can be identified as *natural number operator amplitudes* [104]. We can summarize the properties of $\hat{N}$, $\hat{A}^\dagger$ and $\hat{A}$ as[3]

$$i\frac{d\Phi_n}{d\xi} = \hat{N}\Phi_n = n\Phi_n, \tag{3.14}$$

$$\hat{A}\Phi_n = \sqrt{n}\Phi_{n-1}, \tag{3.15}$$

$$\hat{A}^\dagger\Phi_n = \sqrt{(n+1)}\Phi_{n+1}, \tag{3.16}$$

and

$$[\hat{A}, \hat{A}^\dagger] = 1. \tag{3.17}$$

Furthermore, as a result of $\kappa = 1$, we now have $\hat{A}(\xi) = \exp(-i\xi)\hat{A}(0)$, which satisfies

$$i\frac{d\hat{A}}{d\xi} = [\hat{A}, \hat{N}] = \hat{A}. \tag{3.18}$$

Similarly, with $\hat{A}^\dagger(\xi) = \exp(i\xi)\hat{A}^\dagger(0)$, we get

$$i\frac{d\hat{A}^\dagger}{d\xi} = [\hat{A}^\dagger, \hat{N}] = -\hat{A}^\dagger. \tag{3.19}$$

It is important to remember that the differential symbol, $\frac{d}{d\xi}$ in Eqs. (3.14), (3.18) and (3.19) is not a stand-alone operator, and it certainly is not the definition of $\hat{N}$, i.e.

$$\hat{N} :\neq i\frac{d}{d\xi},$$

whereas we actually have

$$\hat{N} := \hat{A}^\dagger \hat{A}.$$

---

[3]The operators here form what is known as a *Heisenberg group* and obey an *associative Heisenberg algebra*.

Eqs. (3.14) to (3.19) constitute the equations which represent what we will refer to as *natural number dynamics*. These equations are the basis of the development of equations that govern the behaviour of the physical systems, including the equations of quantum mechanics that we will explore later in Chapter 5.

The eigenfunction, Φ_0 of $\hat{N}$ can be interpreted as corresponding to an empty system, i.e., one with no items in it. The fact that $\hat{A}\Phi_0 = 0$, from Eq. (3.15), provides the essential *backstop condition* that prevents the eigenvalues of $\hat{N}$ from falling below 0. Φ_0, together with multiple applications of the operator, $\hat{A}^\dagger$ in Eq. (3.16) can be used to create all of the eigenfunctions of $\hat{N}$, by noting that

$$\Phi_n = \frac{(\hat{A}^\dagger)^n}{\sqrt{n!}}\Phi_0. \tag{3.20}$$

These eigenfunctions give us the state of the system and when we operate with $\hat{N}$ on a particular Φ_n, the resulting eigenvalue gives us all the information we need, or indeed can have, about this simple system. So we can think of Φ_n as being the state function of the system. The set of state functions form an orthonormal set which constitutes the Hilbert space on which the operators operate and from which linear superposed states can be formed like those in Eq. (2.11).

Finally, in this section, a note of terminology. Because of their properties, defined by Eqs. (3.15) and (3.16), $\hat{A}$ is called a *lowering operator* and $\hat{A}^\dagger$ is called a *raising operator*. These terms are widely used in quantum mechanics, though they apply to more general forms, as we shall see later in Section 3.8. The terms raising and lowering are a little imprecise, since there is no restriction to whole numbers implied. Counting back and counting up operators may actually be more appropriate names, since $\hat{A}$ takes one item away from the n, and $\hat{A}^\dagger$ adds one. These operators may also be identified, respectively, with the *annihilation* and *creation* operators that play an important role in both quantum mechanics and quantum field theory.

3.3 UNIQUENESS OF $\hat{N} = \hat{A}^\dagger\hat{A}$.

At this point, it is worth considering the uniqueness of the representation, $\hat{N} = \hat{A}^\dagger\hat{A}$. Clearly, we could also have chosen to write $\hat{N} = \hat{A}\hat{A}^\dagger$ instead of $\hat{N} = \hat{A}^\dagger\hat{A}$, but this would not have affected the outcome as it would just have lead to the roles of $\hat{A}^\dagger$ and $\hat{A}$ being interchanged. However, we could more generally have written $\hat{N}$ as a linear combination of $\hat{A}\hat{A}^\dagger$ and $\hat{A}^\dagger\hat{A}$, so we will look at this possibility next. The most general form we need is

$$\hat{N} = \cos^2\theta\hat{A}^\dagger\hat{A} + \sin^2\theta\hat{A}\hat{A}^\dagger, \tag{3.21}$$

where θ is some fixed real angle. Generalizations of this type are referred to as *paraquantization* [86]. It is important to note that $\hat{N}$, as defined in Eq. (3.21), is still both an Hermitian and a non-negative operator.

Now $\hat{A}(\xi) = \exp(-i\xi)\hat{A}(0)$ still satisfies Eq. (3.21), so Eq. (3.18) still applies. This again leads to $\hat{A}\Phi_n = \beta(n)\Phi_{n-1}$, where $\beta(n)$ is a scalar factor that may depend on n, although now this is not generally equal to $\sqrt{n}$. Similarly, $\hat{A}^\dagger\Phi_n = \gamma(n)\Phi_{n+1}$, where $\gamma(n)$ is a scalar factor that may depend on n. From $\langle\Phi_n|\hat{A}^\dagger\hat{A}\Phi_n\rangle = \gamma(n-1)\beta(n)$ and $\langle\Phi_n|\hat{A}^\dagger\hat{A}\Phi_n\rangle = \langle\hat{A}\Phi_n|\hat{A}\Phi_n\rangle = \beta^2(n)$, we get $\gamma(n-1) = \beta(n)$. Thus, $\langle\Phi_n|\hat{A}\hat{A}^\dagger\Phi_n\rangle = \beta^2(n+1)$. Substituting these results into Eq. (3.21) and then into $\langle\Phi_n|\hat{N}\Phi_n\rangle$, we find, after a little manipulation

$$\beta^2(n+1) = n\csc^2\theta - \beta^2(n)\cot^2\theta. \tag{3.22}$$

Notice that we still need the backstop condition, $\hat{A}\Phi_0 = 0$, to prevent negative eigenvalues, so we must have $\beta(0) = 0$. From Eq. (3.22), with $n = 0$, we then get $\beta(1) = 0$. This is a very significant result because it means that $\hat{A}^\dagger\Phi_0 = 0$ and so this system remains empty when the *creation* operator is applied to the empty state. Thus a system governed by a number operator in Eq. (3.21) simply cannot be built up from its ground state of $n = 0$, unless $\sin\theta = 0$ (or $\cos\theta = 0$). So we can conclude that $\hat{N} = \hat{A}^\dagger\hat{A}$ (or equivalently $\hat{N} = \hat{A}\hat{A}^\dagger$) is a unique representation of the natural number operator, at least where linear dependence on the product of $\hat{A}^\dagger$ and $\hat{A}$ is concerned.

3.4   REPRESENTATION OF EIGENSTATES

In this section we look at two ways of representing the eigenfunctions of $\hat{N}$. These two representations are broadly characterized by their association with either S-form or H-form operators. Trivially, we find

$$i\frac{d\hat{N}}{d\xi} = [\hat{N},\hat{N}] = 0,$$

so that both the S-form and H-form of $\hat{N}$ are independent of ξ. Thus

$$\hat{N}\Phi_n(\xi) = n\Phi_n(\xi) \tag{3.23}$$

and

$$\hat{N}\Phi_n(0) = n\Phi_n(0). \tag{3.24}$$

Both the S-forms and H-forms of $\hat{A}$ satisfy Eq. (3.17), since

$$[\hat{A}(\xi),\hat{A}^\dagger(\xi)] = [\hat{A}(0),\hat{A}^\dagger(0)] = 1, \tag{3.25}$$

and we can formally write the S-form of Eq. (3.15) as

$$\hat{A}(0)\Phi_n(\xi) = \sqrt{n}\Phi_{n-1}(\xi). \tag{3.26}$$

The H-form then follows by writing $\Phi_n(\xi) = \exp(-i\xi\hat{N})\Phi_n(0)$ and multiplying Eq. (3.26) from the left by $\exp(i\xi\hat{N})$ to give

$$\exp(i\xi\hat{N})\hat{A}(0)\exp(-i\xi\hat{N})\Phi_n(0) = \hat{A}(\xi)\Phi_n(0) = \sqrt{n}\Phi_{n-1}(0). \tag{3.27}$$

Since, in general, $\hat{N}\Phi_n(0) = n\Phi_n(0)$ implies $f(\hat{N})\Phi_n(0) = f(n)\Phi_n(0)$, then it follows that

$$\Phi_n(\xi) = \exp(-i\xi\hat{N})\Phi_n(0) = \exp(-in\xi)\Phi_n(0), \tag{3.28}$$

from which

$$i\frac{\mathrm{d}\Phi_n(\xi)}{\mathrm{d}\xi} = n\Phi_n(\xi),$$

(3.29)

in agreement with Eq. (3.14).

3.4.1 FOCK SPACE REPRESENTATION

The representation based on H-form operators requires only a trivial modification to the notation. The orthonormal function, $\Phi_n(0)$, is treated as a unit vector in an infinite dimensional space, called a *Fock space*. This is just a special type of Hilbert space in which the state vector is represented by $|n\rangle$. This symbol is sufficient to characterize the system, since it does not depend on ξ and the only information it carries is the subscript, n. The symbol $|n\rangle$ is the simplest form of what is referred to as *occupation number representation*. It simply indicates the number of items occupying the system and this is then what entirely describes the state of the system. It becomes particularly useful when describing multi-category systems, as we shall see later. It is particularly relevant to a situation where the number n is specifically being used to count particles and is used in many-body quantum physics. In this notation the inner product in Eq. (2.10) becomes

$$\langle n|m\rangle = \delta_{nm}.$$

with $\langle n|m\rangle^* = \langle m|n\rangle$. Then

$$\hat{N}|n\rangle = n|n\rangle,$$

(3.30)

$$\hat{A}|n\rangle = \sqrt{n}|n-1\rangle,$$

(3.31)

and

$$\hat{A}^\dagger|n\rangle = \sqrt{n+1}|n+1\rangle.$$

(3.32)

Eqs. (3.30) to (3.32) can be given a concrete mathematical representation by noting that

$$\langle m|\hat{N}n\rangle = n\delta_{mn},$$

$$\langle m|\hat{A}n\rangle = \sqrt{n}\delta_{mn-1},$$

and

$$\langle m|\hat{A}^\dagger n\rangle = \sqrt{n+1}\delta_{mn+1},$$

where δ_{mn} is the Kronecker delta matrix. We note that $\langle m|\hat{N}n\rangle$ represents the elements of a diagonal square matrix of infinite dimensions. If we let $\langle m|\hat{N}n\rangle$ equal the matrix with elements[4], N_{nm}, then

$$N_{nm} = \begin{pmatrix} 0 & 0 & 0 & 0 & . & . \\ 0 & 1 & 0 & 0 & . & . \\ 0 & 0 & 2 & 0 & . & . \\ 0 & 0 & 0 & 3 & . & . \\ . & . & . & . & . & . \\ . & . & . & . & . & . \end{pmatrix}.$$

(3.33)

---

[4]Notice that n and m are numerical labels that run through the natural numbers, starting with 0.

Similarly with $A_{nm} = \langle m|\hat{A}n \rangle$ and $A^{\dagger}_{nm} = \langle m|\hat{A}^{\dagger}n \rangle$, then

$$A_{nm} = \begin{pmatrix} 0 & 1 & 0 & 0 & \cdot & \cdot \\ 0 & 0 & \sqrt{2} & 0 & \cdot & \cdot \\ 0 & 0 & 0 & \sqrt{3} & \cdot & \cdot \\ 0 & 0 & 0 & 0 & \cdot & \cdot \\ \cdot & \cdot & \cdot & \cdot & \cdot & \cdot \\ \cdot & \cdot & \cdot & \cdot & \cdot & \cdot \end{pmatrix} \tag{3.34}$$

and

$$A^{\dagger}_{nm} = \begin{pmatrix} 0 & 0 & 0 & 0 & \cdot & \cdot \\ 1 & 0 & 0 & 0 & \cdot & \cdot \\ 0 & \sqrt{2} & 0 & 0 & \cdot & \cdot \\ 0 & 0 & \sqrt{3} & 0 & \cdot & \cdot \\ \cdot & \cdot & \cdot & \cdot & \cdot & \cdot \\ \cdot & \cdot & \cdot & \cdot & \cdot & \cdot \end{pmatrix}. \tag{3.35}$$

It is a straightforward matter to check that the matrix N_{nm} is the matrix product of $A^{\dagger}_{nm}$ and A_{nm}, i.e.

$$N_{mn} = \sum_k A^{\dagger}_{nk} A_{km}. \tag{3.36}$$

The state vectors, $|n\rangle$ are represented by column vectors of unit length and infinite dimensions. Thus

$$|0\rangle = \begin{pmatrix} 1 \\ 0 \\ 0 \\ 0 \\ \cdot \\ \cdot \end{pmatrix}, |1\rangle = \begin{pmatrix} 0 \\ 1 \\ 0 \\ 0 \\ \cdot \\ \cdot \end{pmatrix}, |2\rangle = \begin{pmatrix} 0 \\ 0 \\ 1 \\ 0 \\ \cdot \\ \cdot \end{pmatrix}, |3\rangle = \begin{pmatrix} 0 \\ 0 \\ 0 \\ 1 \\ \cdot \\ \cdot \end{pmatrix}, \ldots \tag{3.37}$$

It is easy to check that these vectors are orthonormal by multiplying any one of them by the transposed (row vector) of itself or any other one.

Notice, the fact that $\hat{N}$, $\hat{A}$ and $\hat{A}^{\dagger}$ have a matrix representation emphasizes the fact that they form an associative algebra.

3.4.2 PHASOR OPERATOR REPRESENTATION

The first step to this approach involves writing the natural number operator amplitudes, $\hat{A}$ and $\hat{A}^{\dagger}$ in terms of a Hermitian pair, i.e., as

$$\hat{A} = \frac{1}{\sqrt{2}}(\hat{U} + i\hat{V}) \text{ and } \hat{A}^{\dagger} = \frac{1}{\sqrt{2}}(\hat{U} - i\hat{V}), \tag{3.38}$$

where $\hat{U}$ and $\hat{V}$ are Hermitian operators.

Substituting Eq. (3.38) into Eq. (3.17) yields

$$[i\hat{V}, \hat{U}] = 1. \tag{3.39}$$

Obviously Eq. (3.39) is equivalent to $[\hat{V},\hat{U}] = -i$, which shows that $\hat{U}$ and $\hat{V}$ can be interpreted as a pair of canonical variables as understood in standard quantum mechanics. As we shall see later when we develop quantum mechanics in Chapter 5, $\hat{U}$ is interpreted as the operator $\hat{X}$ representing the configuration space co-ordinate in a one-dimensional system, and $\hat{V}$ is interpreted as the corresponding linear momentum operator $\hat{P}$. Then Eq. (3.39) is referred to as the quantization condition. However, whereas in standard quantum mechanics this condition is treated as a postulate that is the key *assumption* which generates quantum behaviour, here no such postulate is needed as the condition comes naturally from the preceding analysis and no assumption is involved. The physical interpretation of these canonical variables will be left to a later chapter, Chapter 5, when we look at Hamiltonians and time dependence and see how standard quantum mechanics emerges naturally from natural number dynamics.

We are now in a position to interpret ξ by regarding $\hat{U}$ and $\hat{V}$ as rotating phasor operators. To see this we recall that $\hat{A}(\xi) = \exp(-i\xi)\hat{A}(0)$. So writing

$$\hat{A}(\xi) = \frac{1}{\sqrt{2}}(\hat{U}(\xi) + i\hat{V}(\xi)),$$

then substituting $\hat{A}(0) = \hat{U}(0) + i\hat{V}(0)$ into $\hat{A}(\xi) = \exp(-i\xi)\hat{A}(0)$ yields

$$\begin{pmatrix} \hat{U}(\xi) \\ \hat{V}(\xi) \end{pmatrix} = \begin{pmatrix} \cos\xi & \sin\xi \\ -\sin\xi & \cos\xi \end{pmatrix} \begin{pmatrix} \hat{U}(0) \\ \hat{V}(0) \end{pmatrix}. \tag{3.40}$$

Eq. (3.40) shows that ξ is a rotation angle and that $\hat{U}$ and $\hat{V}$ are a pair of phasors, but in the form of operators rather than the usual complex numbers. Notice also that Eq. (3.40) implies that $\hat{U}$ and $\hat{V}$ are separated by an angle, $\xi = \frac{\pi}{2}$. One can interpret this as $(\hat{U},\hat{V})$ acting like a two-dimensional phasor space, with $\hat{U}$ and $\hat{V}$ representing two orthogonal components.

Substituting $\hat{A} = \frac{1}{\sqrt{2}}(\hat{U} + i\hat{V})$ and its adjoint into $\hat{N} = \hat{A}^{\dagger}\hat{A}$, yields, with the aid of Eq. (3.17)

$$\hat{N} = \frac{1}{2}(\hat{U}^2 + \hat{V}^2 - 1), \tag{3.41}$$

and we can think of $\hat{U}^2 + \hat{V}^2 = 2\hat{N} + 1$ as the square of the radius of the phasor vector $(\hat{U},\hat{V})$. This operator thus has an eigenvalue of $2n + 1$ and we can treat $\sqrt{2n+1}$ as the radius of $(\hat{U},\hat{V})$. This result is depicted in Fig. 3.1. However, a little caution is needed with regard to Fig. 3.1, since strictly speaking $\hat{U}(0)$ and $\hat{V}(0)$ that are used to label the axes do not individually have well defined eigenvalues, with respect to $|n\rangle$. The consequences of this will be dealt with in more detail shortly. So, Fig. 3.1 is merely a schematic representation. With the aid of Eq. (3.41) we find

$$i\frac{d\hat{U}}{d\xi} = [\hat{U},\hat{N}] = \frac{1}{2}[\hat{U},\hat{V}^2] = i\hat{V}. \tag{3.42}$$

in the last step of Eq. (3.42) we used the expansion of $[\hat{U},\hat{V}^2]$ as

$$[\hat{U},\hat{V}^2] = \hat{V}[\hat{U},\hat{V}] = [\hat{U},\hat{V}]\hat{V} = 2i\hat{V},$$

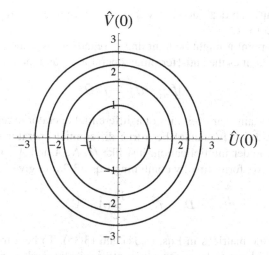

Figure 3.1 Schematic representation of $\hat{U}$ and $\hat{V}$ in phase space. The circles represent, $n = 0, 1, 2$ and 3, in increasing radii and indicate that $\hat{U}^2 + \hat{V}^2 = 2\hat{N} + 1$ is invariant as $\hat{U}$ and $\hat{V}$ rotate, as the phase space angle, ξ, changes.

together with Eq. (3.39). Similarly we get

$$i\frac{\mathrm{d}\hat{V}}{\mathrm{d}\xi} = [\hat{U}, \hat{N}] = \frac{1}{2}[\hat{V}, \hat{U}^2] = -i\hat{U}. \tag{3.43}$$

Substituting Eq. (3.43) into Eq. (3.42) and vice versa yields

$$\frac{\mathrm{d}^2\hat{U}}{\mathrm{d}\xi^2} + \hat{U} = \frac{\mathrm{d}^2\hat{V}}{\mathrm{d}\xi^2} + \hat{V} = 0. \tag{3.44}$$

Thus the operators $\hat{U}$ and $\hat{V}$ obey the same equations of motion as classical simple harmonic oscillators as functions of ξ. These *generalized harmonic oscillators* play an extremely important role in quantum physics, as we shall see in later chapters.

3.4.3 ANALYTIC REPRESENTATION

The phasor operator representation leads to second way of representing the states of the number operator, that relies on functional analysis. We first note that, iterating Eq. (3.39) implies that

$$[i\hat{V}, \hat{U}^p] = p\hat{U}^{p-1}, \tag{3.45}$$

for any natural number p. So, for any function that can be expressed as a power series, Eq. (3.45) can be generalized to

$$[i\hat{V}, f(\hat{U})] = f'(\hat{U}), \tag{3.46}$$

where $f(\hat{U})$ is an analytical function with operator arguments and $f'(\hat{U})$ is its derivative with respect to $\hat{U}$.

Now at this point it might be tempting to regard $i\hat{V}$ as some kind of differential operator and write it as the anti-Hermitian operator $\hat{D}_u$ and then

$$[\hat{D}_u, f(\hat{U})] = f'(\hat{U}), \tag{3.47}$$

so that $\hat{D}_u$ has certain characteristics of a differential operator in relation to functions of $\hat{U}$. However, $\hat{D}_u = i\hat{V}$ cannot be a true differential operator since it obeys the associative rule under multiplication, just like $\hat{U}$, $\hat{N}$, $\hat{A}$ and $\hat{A}^\dagger$. Indeed, $\hat{D}_u$ can be expressed in matrix form, first by combining Eqs. (3.38) to give

$$\hat{D}_u = i\hat{V} = \frac{1}{\sqrt{2}}(\hat{A} - \hat{A}^\dagger), \tag{3.48}$$

and then using the matrices in Eqs. (3.34) and (3.35). To be a true differential operator, $\hat{D}_u$ would have to be a *derivation* algebraically, and so could not have the associative property under multiplication, whereas

$$\hat{D}_u(f(\hat{U})g(\hat{U})) = (\hat{D}_u f(\hat{U}))g(\hat{U}),$$

showing that $\hat{D}_u$ by itself does obey the associative rule when it multiplies $f(\hat{U})g(\hat{U})$. However, using the rules for expanding the commutator we find

$$\begin{aligned}[\hat{D}_u, f(\hat{U})g(\hat{U})] &= f(\hat{U})[\hat{D}_u, g(\hat{U})] + [\hat{D}_u, f(\hat{U})]g(\hat{U}) \\ &= f'(\hat{U})g(\hat{U}) + f(\hat{U})g'(\hat{U}).\end{aligned} \tag{3.49}$$

What needs to be understood here is that the derivation rule is associated with the commutator, as was explained in Section 2.9, and not with the operator $\hat{D}_u$ alone. In spite of this apparent difficulty, it is possible to represent the system and the natural number operator in a powerful way using differential operators, but care is needed in setting this up.

One way of doing this is to replace the Hermitian pair $\hat{U}$ and $\hat{V}$ by two new Hermitian operators $\hat{u}$ and $\hat{v}$, where $\hat{u}$ is represented by a continuous scalar variable, u and $\hat{v}$ is defined as[5]

$$\hat{v} := -i\frac{\partial}{\partial u}. \tag{3.50}$$

In this case $\hat{v}$ is a derivation and is not associative when it multiplies functions of $\hat{u}$, so we must be careful how we proceed.

It is important to remember that these new operators still need to operate on the Hilbert space. The introduction of the continuous variable, u and its derivative suggests that the Hilbert space basis should comprise functions of the both u and ξ. So we shall write the Hilbert space eigenfunctions as $\Phi_n(\xi, u)$.

---

[5]Notice the difference here between the definition of $\hat{v}$ as a differential operator and the situation with regard to Eq. (3.14). Notice also that the definition, Eq. (3.50), implies that $\frac{\partial}{\partial u}$ in anti-Hermitian.

Notice that the differential operator is written with a partial derivative symbol because the functions involved depend on both ξ and u. Then, with regard to Eq. (3.39) we require

$$[\frac{\partial}{\partial u}, u]\Phi_n(\xi, u) = \frac{\partial}{\partial u}u\Phi_n(\xi, u) - u\frac{\partial}{\partial u}\Phi_n(\xi, u) = \Phi_n(\xi, u). \tag{3.51}$$

The question is, can we interpret $\frac{\partial}{\partial u}\Phi_n(\xi, u)$ as $\Phi'(\xi, u)$, where $\Phi'(\xi, u)$ represents the function obtained by differentiating $\Phi_n(\xi, u)$ with respect to u? Then the last equation in Eq. (3.51) would mean

$$(u\Phi_n(\xi, u))' - u\Phi'(\xi, u) = \Phi_n(\xi, u). \tag{3.52}$$

The issue here is that we still need to recall that both $\frac{\partial}{\partial u}$ and u are operators. So just as $\hat{v}$ does not operate on $\hat{u}$ then $\frac{\partial}{\partial u}$ does not, strictly speaking, operate on u. Rather the product $\frac{\partial}{\partial u}u$ operates on $\Phi_n(\xi, u)$, since we must have

$$\frac{\partial}{\partial u}u = u\frac{\partial}{\partial u} + 1, \tag{3.53}$$

so it must be the case that $\frac{\partial}{\partial u}u \neq 1$. Swanson [118] uses the term *phantom* Φ on which the operators in Eq. (3.53) need to operate to make sense. In some ways it would be safer always to include the function $\Phi(\xi, u)$ to the right of the operators, as in Eq. (3.51), but it is usual to see equations between stand-alone operators, without the functions that are operated on, as in Eq. (3.53), in the literature on quantum theory, and that convention will be utilized in certain situations in what follows. However, the phantom Φ must always be bourn in mind.

Notice that Eq. (3.51) is consistent with Eq. (3.52) as long as we insist on the convention that $\frac{\partial}{\partial u}u\Phi_n(\xi, u)$ means, first operate on $\Phi_n(\xi, u)$ with u and then operate on the result with $\frac{\partial}{\partial u}$. This amounts to a relaxation of the associativity condition that applies with Heisenberg algebra on the Fock space. We can test whether the representation involving the definition in Eq. (3.50) is valid in the following way.

The first step is to construct a pair of number amplitude operators to replace $\hat{A}$ and $\hat{A}^\dagger$, so we let

$$\hat{A}_u = \frac{1}{\sqrt{2}}(\hat{u} + i\hat{v}) = \frac{1}{\sqrt{2}}(u + \frac{\partial}{\partial u}) \tag{3.54}$$

and

$$\hat{A}_u^\dagger = \frac{1}{\sqrt{2}}(\hat{u} - i\hat{v}) = \frac{1}{\sqrt{2}}(u - \frac{\partial}{\partial u}). \tag{3.55}$$

The form of $\hat{A}_u^\dagger$ may be understood by recalling that $\hat{v}$ is Hermitian and noting that[6]

$$(\frac{\partial}{\partial u})^\dagger = (i\hat{v})^\dagger = -i\hat{v} = -\frac{\partial}{\partial u}, \tag{3.56}$$

which confirms that $\frac{\partial}{\partial u}$ is anti-Hermitian.

---

[6]See Appendix E.

We can now construct a new number operator, using

$$\hat{N}_u \Phi(\xi, u) = \hat{A}_u^\dagger \hat{A}_u \Phi(\xi, u). \tag{3.57}$$

Eqs. (3.54) and (3.55) then lead to

$$\hat{N}_u \Phi_n(\xi, u) = i\frac{\partial \Phi_n(\xi, u)}{\partial \xi} = \frac{1}{2}(-\frac{\partial^2}{\partial u^2} + u^2 - 1)\Phi_n(\xi, u). \tag{3.58}$$

The operators in Eq. (3.58) contains no products of u and $\frac{\partial}{\partial u}$, so issues of associativity do not arise. Now we can use Eq. (3.58) as a test to see if the assumption that the representation in Eq. (3.50) is valid. We are basically asking if

$$\frac{\partial^2}{\partial u^2}\Phi_n(\xi, u) = \Phi_n''(\xi, u).$$

We can do this by solving the eigenvalue equation

$$\frac{1}{2}(-\frac{\partial^2}{\partial u^2} + u^2 - 1)\Phi(\xi, u) = \lambda \Phi(\xi, u) \tag{3.59}$$

and checking that the eigenvalue spectrum is indeed represent by $\lambda = n$. The solutions to Eq. (3.59) are well known [95, 118, 126]. They have the form

$$\Phi_n(\xi, u) = \exp(-in\xi)\exp(-\frac{u^2}{2})H_n(u), \tag{3.60}$$

where $H_n(u)$ is a Hermite polynomial. Notice that $\Phi_n(\xi, u)$ in Eq. (3.60) is written as a product of a function of ξ times a function of u. This is to be expected, since the partial differential equation, Eq. (3.59) has implicitly been solved using the separation of variables. It is easy to check that the eigenvalues that result are indeed the natural numbers, n, i.e.

$$\hat{N}_u \Phi_n(\xi, u) = i\frac{\partial \Phi_n(\xi, u)}{\partial \xi} = n\Phi_n(\xi, u)$$

as required. So we are safe to assume that $\frac{\partial}{\partial u}\Phi_n(\xi, u) = \Phi'(\xi, u)$.

In this analytic representation, the orthonormality conditions for the eigenfunctions are now written in the concrete form

$$\langle \Phi_n | \Phi_m \rangle = \int_{-\infty}^{\infty} \Phi_n(\xi, u)^* \Phi_m(\xi, u) \mathrm{d}u = \delta_{nm}. \tag{3.61}$$

This orthonormality condition rests on the functions $\Phi_n(\xi, u)$ being square integrable with respect to u, i.e., that the integral

$$\int_{-\infty}^{\infty} \Phi_n(\xi, u)^* \Phi_n(\xi, u) \mathrm{d}u$$

is finite. Then the Schwarz inequality ensures that the integral in Eq. (3.61) is finite for all n and m.

Also we note that the orthonormality condition is independent of ξ as expected. In particular, $\Phi_n^*(\xi, u)\Phi_n(\xi, u) = \Phi_n^*(u)\Phi_n(u)$, where we have written $\Phi_n(u)$ for $\Phi_n(0, u)$. Then the normalization condition satisfies

$$\int_{-\infty}^{\infty} \Phi_n(\xi, u)^* \Phi_n(\xi, u) du = \int_{-\infty}^{\infty} \Phi_n(u)^* \Phi_n(u) du = 1,$$

as required.

We also note that the ground state, $\Phi_0(u)$, satisfies, $\hat{A}_u \Phi_0(u) = 0$, and may be obtained by solving

$$\frac{1}{\sqrt{2}}\left(\frac{\partial}{\partial u} + u\right)\Phi_0(u) = 0. \qquad (3.62)$$

The result is $\Phi_0(u) = \alpha_0 \exp(-u^2/2)$, where α_0 is an arbitrary constant that can be chosen to normalize $\phi_0(u)$. The normalization condition requires

$$1 = \int_{-\infty}^{\infty} \Phi_0(u)^* \Phi_0(u) du = \alpha_0^2 \int_{-\infty}^{\infty} \exp(-u^2) du = \sqrt{\pi}\alpha_0^2. \qquad (3.63)$$

So,

$$\Phi_0(u) = (\pi)^{-\frac{1}{4}} \exp(-\frac{u^2}{2}).$$

The set of Φ_ns may be generated by applying the creation operator thus

$$\Phi_{n+1}(u) = (\sqrt{n+1})^{-1}\hat{A}^\dagger \Phi_n(u) = \frac{1}{\sqrt{2(n+1)}}\left(-\frac{\partial}{\partial u} + u\right)\Phi_n(u). \qquad (3.64)$$

So, for example, from Φ_0 we obtain

$$\Phi_1(u) = \frac{1}{\sqrt{2}}\left(-\frac{\partial}{\partial u} + u\right)((\pi)^{-\frac{1}{4}} \exp(-\frac{u^2}{2})) = \sqrt{2}(\pi)^{-\frac{1}{4}} u \exp(-\frac{u^2}{2}), \qquad (3.65)$$

which is correctly normalized and in agreement with the results obtained in Eq. (3.60). The four eigenfunctions corresponding to $n = 0, 1, 2$ and 3 are depicted in Fig. 3.2. These results show that the Heisenberg algebra of operators on a Fock space are entirely equivalent to the partial differential equation representation of the number operator. These results have important consequences for quantum theory. The two representations of the natural number operator that we have explored in this section may be depicted schematically as in Fig. 3.3.

3.4.4 $\hat{N}$ AND u-v DUALITY

There is a complementary representation to that of the u-dependent representation of the number operator and the number amplitude operators. This may be obtained by

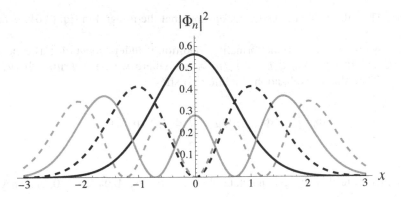

Figure 3.2 The squared moduli, $|\Phi_n|^2$, of the first four eigenfunctions of the natural number operator. The dark solid curve represents $n = 0$; the dark dashed curve, $n = 1$; the solid grey curve, $n = 2$; and the dashed grey curve, $n = 3$.

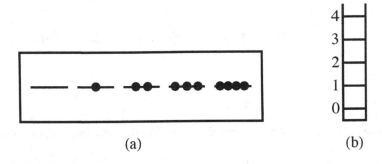

$$(a) \qquad\qquad\qquad\qquad\qquad (b)$$

Figure 3.3 Two schematic ways of representing the eigenvalues of the number operator. Panel (a) depicts n as an occupation number representing populations of items of a single category, that corresponds to H-type representation. Panel (b) depicts n as a labelled rung on a ladder which corresponds to S-type representation. The first five eigenvalues, $n = 0, 1, 2, 3$ and 4 are represented in each case. These two representations are mathematically entirely equivalent. The differences are entirely in the eye and mind of the beholder and can be attributed to *the treachery of images*. Notice also the similarities between panel (a) and Fig. 0.2, and panel (b) and Fig. 0.1.

noting the commutator in Eq. (3.39), implies $[\hat{V}, i\hat{U}] = 1$ and then, using the expansion rule for commutators one finds

$$[\hat{V}^2, i\hat{U}] = [\hat{V}, i\hat{U}]\hat{V} + \hat{V}[\hat{V}, i\hat{U}] = 2\hat{V}. \qquad (3.66)$$

So, following the same reasoning as in the previous section, we find

$$[f(\hat{V}), i\hat{U}] = f'(\hat{V}), \qquad (3.67)$$

which is a reverse of the roles of $\hat{U}$ and $\hat{V}$ that was found previously. So now, when we replace $\hat{U}$ and $\hat{V}$ by $\hat{u}$ and $\hat{v}$, this time we can represent $\hat{v}$ by a continuous variable v and $\hat{u}$ by a differential operator,

$$\hat{u} := i\frac{\partial}{\partial v}.$$

Notice that the sign of $i\frac{\partial}{\partial v}$ is reversed compared with that in Eq. (3.50). This has consequences for the definitions of new v-dependent number amplitude operators. Now $\hat{A}$ may be represented as

$$\hat{A}_v = \frac{i}{\sqrt{2}}\left(\frac{\partial}{\partial v} + v\right), \tag{3.68}$$

where the subscript, v, indicates the v-representation. $\hat{A}_v$ differs from $\hat{A}_u$ in the factor i, as well as the interchange of v for u. Notice that now the Hermitian part of $\hat{A}_v$ is $\frac{i}{\sqrt{2}}\frac{\partial}{\partial v}$, whereas its anti-Hermitian part is now $\frac{i}{\sqrt{2}}v$. Complementing Eq. (3.68) is its adjoint

$$\hat{A}_v^\dagger = \frac{i}{\sqrt{2}}\left(\frac{\partial}{\partial v} - v\right). \tag{3.69}$$

The Hilbert space for the system is now in the form of v-dependent functions, $\Psi(\xi, v)$ and then

$$\hat{N}_v\Psi(\xi, v) = \hat{A}_v^\dagger \hat{A}_v \Psi(\xi, v), \tag{3.70}$$

and we get, instead of Eq. (3.58)

$$\hat{N}_v\Psi_n(\xi, v) = i\frac{\partial \Psi_n(\xi, v)}{\partial \xi} = \frac{1}{2}\left(-\frac{\partial^2}{\partial v^2} + v^2 - 1\right)\Psi_n(\xi, v). \tag{3.71}$$

Remarkably, Eqs. (3.58) and (3.71) are identical in form and so we can expect the eigenfunctions of $\hat{N}_v$ to be identical to those of $\hat{N}_u$, apart from the interchange of u and v. There is an interesting way of connecting the two equations by noting that, if we define the relation between $\Phi_n(\xi, u)$ and $\Psi_n(\xi, v)$ via the Fourier transform [82]

$$\Phi_n(\xi, u) = \frac{1}{\sqrt{2\pi}}\int_{-\infty}^{\infty}\Psi_n(\xi, v)\exp(iuv)dv \tag{3.72}$$

and its inverse

$$\Psi_n(\xi, v) = \frac{1}{\sqrt{2\pi}}\int_{-\infty}^{\infty}\Phi_n(\xi, u)\exp(-iuv)du. \tag{3.73}$$

Then

$$-i\frac{\partial \Phi_n(\xi, u)}{\partial u} = \frac{1}{\sqrt{2\pi}}\int_{-\infty}^{\infty}v\Psi_n(\xi, v)\exp(iuv)dv \tag{3.74}$$

and

$$i\frac{\partial \Psi_n(\xi,v)}{\partial v} = \frac{1}{\sqrt{2\pi}} \int\limits_{-\infty}^{\infty} u\Psi_n(\xi,u)\exp(-iuv)\mathrm{d}u. \qquad (3.75)$$

Inverting Eqs. (3.74) and (3.75) gives

$$v\Psi_n(\xi,v) = \frac{1}{\sqrt{2\pi}} \int\limits_{-\infty}^{\infty} -i\frac{\partial \Phi_n(\xi,u)}{\partial u}\exp(-iuv)\mathrm{d}u \qquad (3.76)$$

and

$$u\Phi_n(\xi,u) = \frac{1}{\sqrt{2\pi}} \int\limits_{-\infty}^{\infty} i\frac{\partial \Psi_n(\xi,v)}{\partial v}\exp(iuv)\mathrm{d}v. \qquad (3.77)$$

Adding i times Eq. (3.74) to Eq. (3.77) then yields

$$(u+\frac{\partial}{\partial u})\Phi_n(\xi,u) = \frac{1}{\sqrt{2\pi}} \int\limits_{-\infty}^{\infty} i(v+\frac{\partial}{\partial v})\Psi_n(\xi,v)\exp(iuv)\mathrm{d}v, \qquad (3.78)$$

from which we can see that $\hat{A}_u\Phi_n(u)$ and $\hat{A}_v\Psi_n(v)$ are Fourier transforms of one another.

Furthermore, since

$$-\frac{\partial^2\Psi_n(\xi,v)}{\partial v^2} = \frac{1}{\sqrt{2\pi}} \int\limits_{-\infty}^{\infty} u^2\Phi_n(\xi,u)\exp(-iuv)\mathrm{d}u, \qquad (3.79)$$

and hence, from the inverse transform

$$v^2\Psi_n(\xi,v) = -\frac{1}{\sqrt{2\pi}} \int\limits_{-\infty}^{\infty} \frac{\partial^2\Phi_n(\xi,u)}{\partial u^2}\exp(-iuv)\mathrm{d}u, \qquad (3.80)$$

then adding Eqs. (3.79) and (3.80) makes it clear that Eq. (3.71) is just the Fourier transform of Eq. (3.58).

The function $\exp(iuv)$ is not only important in the definition of the Fourier transforms above, but also plays an important role in the interpretation of quantum mechanics, as we shall see later. However, we can already conclude that u- and v-representations of the system are entirely equivalent.

3.5 EIGENFUNCTIONS OF $\hat{A}, \hat{U}$ AND $\hat{V}$

As we have seen, the natural number operator, $\hat{N}$, has well defined eigenfunctions and eigenvalues, however, the operators, $\hat{A}, \hat{U}$ and $\hat{V}$, that are associated with it, do not commute with it and so do not share its eigenfunctions. It is of interest to find out if $\hat{A}, \hat{U}$ and $\hat{V}$ have any eigenfunctions and eigenvalues. We will first examine the operator $\hat{A}$.

3.5.1 EIGENFUNCTIONS OF $\hat{A}$

First we can seek an eigenstate in the form of a vector in a Fock space by solving

$$\hat{A}|\alpha\rangle = \alpha|\alpha\rangle, \tag{3.81}$$

where $|\alpha\rangle$ just means a vector in Fock space that has a eigenvalue of α. Now $|\alpha\rangle$ may be constructed from a superposition of the eigenvectors of $\hat{N}$, such that

$$|\alpha\rangle = \sum_n a_n|n\rangle, \tag{3.82}$$

where the a_n are constant coefficients. Substituting Eq. (3.82) into Eq. (3.81) then gives

$$\hat{A}|\alpha\rangle = \sum_n a_n\hat{A}|n\rangle = \sum_n a_n \sqrt{n}|n-1\rangle \tag{3.83}$$

and also

$$\hat{A}|\alpha\rangle = \sum_n \alpha a_n|n\rangle. \tag{3.84}$$

Equating the coefficients of $|n\rangle$ in Eqs. (3.83) and (3.84) gives

$$a_{n+1} = a_n\frac{\alpha}{\sqrt{n+1}}. \tag{3.85}$$

Iteration of Eq. (3.85) leads to

$$|\alpha\rangle = a_0 \sum_n \frac{\alpha^n}{\sqrt{n!}}|n\rangle. \tag{3.86}$$

Normalizing $|\alpha\rangle$, then leads to

$$\langle\alpha|\alpha\rangle = |a_0|^2 \sum_n \frac{(\alpha^*\alpha)^n}{n!} = |a_0|^2\exp\alpha^*\alpha = 1, \tag{3.87}$$

so $|a_0| = \exp(-\frac{\alpha^*\alpha}{2})$. It might be thought that a similar procedure could lead to an eigenvalue equation for $\hat{A}^\dagger$. However this leads to a sum over the Fock space vectors, $|n\rangle$ that does not converge, so there is no equivalent eigenstate for $\hat{A}^\dagger$.

We can further investigate the eigenstates of $\hat{A}$ by using its u-representation. The appropriate eigenvalue equation is then

$$\hat{A}_u\alpha(u) = \frac{1}{\sqrt{2}}(u + \frac{\mathrm{d}}{\mathrm{d}u})\alpha(u) = \lambda\alpha(u), \tag{3.88}$$

where $\alpha(u)$ and λ are the eigenfunction and eigenvalue, respectively. Notice that, because $\hat{A}_u$ is not Hermitian, then λ is in general a complex number. Eq. (3.88) is easily integrated to give

$$\alpha(u) = \alpha(0)\exp(\sqrt{2}\lambda u - \frac{u^2}{2}), \tag{3.89}$$

where $\alpha(0)$ is a constant of integration. This is a well-behaved function that is square integrable, and so can be normalized because the integral over the whole of u-space is finite. Recalling that λ is in general complex, then $|\alpha(0)|$ may be determined by normalization i.e.,

$$\int_{-\infty}^{\infty} \alpha^*(u)\alpha(u)\mathrm{d}u = |\alpha(0)|^2 \int_{-\infty}^{\infty} \exp(\sqrt{2}(\lambda^* + \lambda)u - u^2)\mathrm{d}u \qquad (3.90)$$

$$= |\alpha(0)|^2 \sqrt{\pi}\exp(\frac{1}{2}(\lambda^* + \lambda)^2) = 1, \qquad (3.91)$$

so $|\alpha(0)| = \pi^{-\frac{1}{4}} \exp(-\frac{1}{4}(\lambda^* + \lambda)^2)$.

As we found using the Fock space representation, if we try to find an eigenfunction for $\hat{A}_u^\dagger$ in the form in Eq. (3.88) we find that it is proportional to

$$\exp(-\sqrt{2}\lambda u + \frac{u^2}{2}),$$

which tends to infinity as $u \to \infty$ and is not square integrable, so cannot be normalized. As a consequence of the lack of an eigenvector for $\hat{A}^\dagger$ in either the Fock space or the u-representations, we cannot expect to find eigenvalue spectra for $\hat{U}$ or $\hat{V}$ since the depend on the sum and difference, respectively, of $\hat{A}^\dagger$ and $\hat{A}$. However, one can find eigenfunctions of sorts, for $\hat{u}$ and $\hat{v}$, in a rather ad hoc manner. We will examine these next, although the results will need to be treated with some caution.

3.5.2 EIGENFUNCTIONS OF $\hat{U}$ AND $\hat{V}$

The eigenvalue equation for u is

$$u\Phi(u) = \lambda\Phi(u), \qquad (3.92)$$

where λ is a scalar constant. There is no square integral function, $\Phi(u)$ that can be found to satisfy Eq. (3.92). However, by inspection one can see that [45]

$$u\delta(u - u_0) = u_0\delta(u - u_0), \qquad (3.93)$$

where u_0 is a constant and $\delta(u - u_0)$ is a Dirac delta function. Although $\delta(u - u_0)$ is not square integrable and so cannot be normalized, it does prove useful in certain contexts and is relatable mathematically to the eigenfunction of $\hat{v}$, which we will look at next.

In the u-representation, the eigenvalue equation for $\hat{v}$ is

$$\hat{v}\Psi(u) = -i\frac{\mathrm{d}\Psi(u)}{\mathrm{d}u} = v_0\Psi(u). \qquad (3.94)$$

The differential equation is easily integrated to give

$$\Psi(u) = \Psi(0)\exp(iv_0u), \qquad (3.95)$$

where $\Psi(0)$ is a constant of integration. Unlike the eigenfunction of $\hat{u}$, the eigen-function $\Psi(0)\exp(iv_0 u)$ is analytic, but it is still not square integrable. Indeed, $\Psi^*(u)\Psi(u) = \Psi^*(0)\Psi(0)$ is a constant. So once again, as with $\hat{u}$, the eigenfunction of $\hat{v}$ is somewhat problematic, from the point of view of its identity in the Hilbert space formalism. It may be unsurprising that the eigenfunctions for $\hat{u}$ and $\hat{v}$ are problematic in the analytic representation, since their Fock space representations involve the sum and differences of $\hat{A}$ and $\hat{A}^\dagger$, and we have seen that although $\hat{A}$ has an eigenstate, $\hat{A}^\dagger$ does not. This means that neither $\hat{U}$ nor $\hat{V}$ have eigenstates in Fock space. However, expectation values for functions of $\hat{A}^\dagger$, $\hat{U}$ and $\hat{V}$ may be calculated, so they still are important and useful operators as will be demonstrated in the next section. Similarly, the operators $\hat{u}$ and $\hat{v}$ play such an important role in standard quantum mechanics, as we shall see, that some allowance is made for their shortcomings with regard to their eigenfunctions.

One of the reasons why $\hat{u}$ and $\hat{v}$ are so useful is that there is an interesting connection between their eigenfunctions, in that they have the form of Fourier transforms of one another. So, for example

$$\frac{\exp(iuv_0)}{\sqrt{2\pi}} = \frac{1}{\sqrt{2\pi}} \int\limits_{-\infty}^{\infty} \delta(v - v_0) \exp(ivu) dv, \tag{3.96}$$

and

$$\delta(v - v_0) = \frac{1}{\sqrt{2\pi}} \int\limits_{-\infty}^{\infty} \frac{\exp(iuv_0)}{\sqrt{2\pi}} \exp(-ivu) du. \tag{3.97}$$

So, just as with the u and v representations of the number operator, the eigenfunctions of the operators $\hat{u}$ and $\hat{v}$ are related via Fourier transforms. This relationship turns out to be quite general.

3.6   EXPECTATION VALUES INVOLVING $\hat{A}, \hat{U}$ AND $\hat{V}$

3.6.1   FOCK SPACE REPRESENTATION

As pointed out in Section 2.2, the expectation value of an operator depends on the state it is in and so on the state function or state vector in the Hilbert space representation. We begin by examining expectation values of operators with respect to the basis vectors, $|n\rangle$ in a Fock space. Because, $\hat{A}|n\rangle = \sqrt{n}|n-1\rangle$ then the orthonormal properties of $|n\rangle$ lead to

$$\langle n|\hat{A}|n\rangle = 0. \tag{3.98}$$

Similarly one finds, for this space, that, $\langle n|\hat{A}^\dagger|n\rangle = 0$.

From Eqs. (3.38) we obtain

$$\hat{U} = \frac{1}{\sqrt{2}}(\hat{A}^\dagger + \hat{A}) \tag{3.99}$$

and

$$\hat{V} = \frac{i}{\sqrt{2}}(\hat{A}^\dagger - \hat{A}). \tag{3.100}$$

Then it is easy to show also that $\langle n|\hat{U}|n\rangle = \langle n|\hat{V}|n\rangle = 0$. Now although it follows that $\langle n|\hat{A}^p|n\rangle = \langle n|\hat{A}^{\dagger p}|n\rangle = 0$, for any natural number p, $\langle n|\hat{U}^2|n\rangle$ and $\langle n|\hat{V}^2|n\rangle$ are not zero. Actually

$$\langle n|\hat{U}^2|n\rangle = \frac{1}{2}\langle n|(\hat{A}+\hat{A}^{\dagger})^2|n\rangle = \frac{1}{2}\langle n|\hat{A}\hat{A}^{\dagger}+\hat{A}^{\dagger}\hat{A}|n\rangle = n+\frac{1}{2}. \qquad (3.101)$$

It is a straightforward matter to show that $\langle n|\hat{V}^2|n\rangle$ is equal to $\langle n|\hat{U}^2|n\rangle$.

It is also useful to examine expectation values on the state, $|\alpha\rangle$, the eigenstate of $\hat{A}$, in Eq. (3.81). Obviously $\langle \alpha|\hat{A}|\alpha\rangle = \alpha$, but more interestingly we can determine $\langle \alpha|\hat{N}|\alpha\rangle$. One finds

$$\langle \alpha|\hat{N}|\alpha\rangle = \langle \alpha|\hat{A}^{\dagger}\hat{A}|\alpha\rangle = \| \hat{A}|\alpha\rangle \|^2 = \| \alpha \|^2, \qquad (3.102)$$

so we can interpret $\| \alpha \|^2$ as the mean value of the number operator for the state α.

The state $|\alpha\rangle$ has another interesting property. It is easy to check that

$$\langle n|\hat{N}|n\rangle^2 = \langle n|\hat{N}^2|n\rangle = n^2.$$

However

$$
\begin{aligned}
\langle \alpha|\hat{N}^2|\alpha\rangle &= \langle \alpha|\hat{A}^{\dagger}\hat{A}\hat{A}^{\dagger}\hat{A}|\alpha\rangle \\
&= \langle \alpha|\hat{A}^{\dagger}\hat{A}^{\dagger}\hat{A}\hat{A}|\alpha\rangle + \langle \alpha|\hat{A}^{\dagger}\hat{A}|\alpha\rangle \\
&= \| \hat{A}^2|\alpha\rangle \|^2 + \| \hat{A}|\alpha\rangle \|^2 \\
&= \| \alpha \|^4 + \| \alpha \|^2 .
\end{aligned}
\qquad (3.103)
$$

So, in this case $\langle \alpha|\hat{N}|\alpha\rangle^2 \neq \langle \alpha|\hat{N}^2|\alpha\rangle$. In fact we can see that, from Eqs. (3.102) and (3.103), then the variance of $\hat{N}$ with respect to the state $|\alpha\rangle$ is given by

$$
\begin{aligned}
\Delta_n^2(\alpha) &= \langle \alpha|\hat{N}^2|\alpha\rangle - \langle \alpha|\hat{N}|\alpha\rangle^2 \\
&= \| \alpha \|^4 + \| \alpha \|^2 - \| \alpha \|^4 \\
\| \alpha \|^2 &= \langle \alpha|\hat{N}|\alpha\rangle.
\end{aligned}
\qquad (3.104)
$$

So we end up with $\Delta_n(\alpha) = \sqrt{\langle \alpha|\hat{N}|\alpha\rangle}$, which is effectively the statistical result that characterizes classical Poisson statistics. We will explore the consequences of this statistical behaviour in the context of u-v uncertainty later in the chapter.

Although the operator $\hat{A}^{\dagger}$ does not have an eigenstate and its expectation value with respect to $|n\rangle$ is zero, it does have a non-zero expectation value with respect to $|\alpha\rangle$, since, by Eq. (2.24)

$$\langle \alpha|\hat{A}^{\dagger}|\alpha\rangle = \langle \alpha|\hat{A}|\alpha\rangle^* = \alpha^*, \qquad (3.105)$$

from which we can also conclude that

$$\langle \alpha|\hat{U}|\alpha\rangle = \frac{1}{\sqrt{2}}\langle \alpha|(\hat{A}+\hat{A}^{\dagger})|\alpha\rangle = \frac{1}{\sqrt{2}}(\alpha+\alpha^*) \qquad (3.106)$$

and

$$\langle \alpha|\hat{V}|\alpha\rangle = \frac{i}{\sqrt{2}}\langle \alpha|(\hat{A}^{\dagger}-\hat{A})|\alpha\rangle = \frac{i}{\sqrt{2}}(\alpha^* - \alpha). \qquad (3.107)$$

3.6.2 u-REPRESENTATION

It is straightforward to obtain the results in Section 3.6.1 by using the u-representation or the v-representation. For example, Eq. (3.101) becomes

$$\langle u^2 \rangle = \int_{-\infty}^{\infty} \Phi_n(u)^* u^2 \Phi_n(u) du, \tag{3.108}$$

where the $\Phi_n(u)$ are the eigenfunctions defined in Eq. (3.60). Clearly, for values of n other than the lowest few, the integrals become increasingly tedious to evaluate.

Expectation values for states represented by $\alpha(u)$ in Eq. (3.89) are simpler to deal with. So for example,

$$\langle \alpha(u)|u|\alpha(u) \rangle = \pi^{-\frac{1}{2}} \exp(-\frac{(\lambda^* + \lambda)^2}{2}) \int_{-\infty}^{\infty} \exp(\sqrt{2}(\lambda^* + \lambda)u - u^2) du$$

$$\tag{3.109}$$

$$= \pi^{-\frac{1}{2}} \int_{-\infty}^{\infty} u \exp(-(u - \frac{(\lambda^* + \lambda)}{\sqrt{2}}) du.$$

A change of variable of $u - \frac{(\lambda^* + \lambda)}{\sqrt{2}} \to y$ then leads to

$$\langle \alpha(u)|u|\alpha(u) \rangle = \frac{1}{\sqrt{2}}(\alpha + \alpha^*),$$

which is identical to that obtained with the Fock space representation in Section 3.6.1.

The most general form that expectation values can take in the u-representation involves state functions, $\Phi(u)$ that are normalized linear combinations of the eigenstates as defined by Eq. (2.11). Then for example, the expectation value of some function, $f(u)$ is just

$$\langle f(u) \rangle = \int_{-\infty}^{\infty} \Phi(u)^* f(u) \Phi(u) du. \tag{3.110}$$

The special case of the expectation of u,

$$\langle u \rangle = \int_{-\infty}^{\infty} \Phi(u)^* u \Phi(u) du = \int_{-\infty}^{\infty} u \Phi(u)^* \Phi(u) du, \tag{3.111}$$

shows that the product $\Phi(u)^* \Phi(u)$ acts like a probability density and can be interpreted as the probability of finding the system in a state with a value of u between u and $u + du$. So $\Phi(u)^* \Phi(u)$ represents a *probability density distribution* of the system in terms of the variable u. Then $\Phi(u)$ is referred to as the corresponding *probability density amplitude*.

3.6.3 u-v FOURIER DUALITY

We have already seen that switching between the u- and v-representations in the case of the number operator, the number amplitude operators, and the eigenfunctions of $\hat{u}$ and $\hat{v}$ are connected via Fourier transforms. This connection may be generalized by construction the Fourier transform of the general normalized state function in Eq. (3.111), i.e.

$$\Psi(v) = \frac{1}{\sqrt{2\pi}} \int\limits_{-\infty}^{\infty} \Phi(u)\exp(-iuv)du. \tag{3.112}$$

We can check that $\Psi(v)$ is normalized, using the definition of the delta function from Eq. (3.97), and then, with the aid of Eq. (3.73) we get

$$\int\limits_{-\infty}^{\infty} \Psi^*(v)\Psi(v)dv = \frac{1}{2\pi} \int\limits_{-\infty}^{\infty}\int\limits_{-\infty}^{\infty}\int\limits_{-\infty}^{\infty} \Phi^*(u')\Phi(u)\exp(i(u'-u)v)du'dudv$$

$$= \int\limits_{-\infty}^{\infty}\int\limits_{-\infty}^{\infty} \Phi^*(u')\Phi(u)\delta(u'-u)du'du \tag{3.113}$$

$$= \int\limits_{-\infty}^{\infty} \Phi^*(u)\Phi(u)du = 1,$$

since $\Phi(u)$ has already been defined as being normalized.

We can take the Fourier inverse of Eq. (3.112) and apply the u-representation of operator $\hat{v}$ to give

$$\hat{v}\Phi(u) = -i\frac{\partial \Phi(u)}{\partial u} = \frac{1}{\sqrt{2\pi}} \int\limits_{-\infty}^{\infty} v\Psi(v)\exp(iuv)dv. \tag{3.114}$$

Using the same procedure as we did in the derivation in Eq. (3.111), we find

$$\langle \hat{v} \rangle = \int\limits_{-\infty}^{\infty} \Phi^*(u)\hat{v}\Phi(u)du = \int\limits_{-\infty}^{\infty} v\Psi^*(v)\Psi(v)dv. \tag{3.115}$$

So both integrals in Eq. (3.115) represent the expectation value of $\hat{v}$. This means that, from the integral over v, we can interpret $\Psi^*(v)\Psi(v)$ as the probability of finding the system with v between v and $v+dv$. So, just as $\Phi(u)^*\Phi(u)$ is the probability density distribution for the system in u-representation, $\Psi(v)^*\Psi(v)$ is the corresponding probability density distribution in v-representation. It means that for any probability distribution in u-representation, there is a probability distribution in v-representation and that these two distributions are related through a Fourier transform. This has profound implications for the interpretation of quantum mechanics as we shall see later.

3.6.4 $\hat{U}$-$\hat{V}$ UNCERTAINTY

The commutation relation between $\hat{U}$ and $\hat{V}$ in Eq. (3.39) has a consequence with regard to the expectation values for these operators. Clearly, $\hat{U}$ and $\hat{V}$ do not commute and so they have no common eigenstates. As was shown in Section 2.5, there is a relation between the expectation value of the commutator and the product of uncertainties associated with the expectation values of the two operators involved, with respect to a a given state. This is given by the inequality in Eq. (2.44), which in the present case can be written

$$\Delta_u \Delta_v \geq \frac{1}{2} |\langle \Psi | [\hat{U}, \hat{V}] \Psi \rangle|. \tag{3.116}$$

So in the present case, since Eq. (3.39) implies $[\hat{U}, \hat{V}] = i$, then

$$\Delta_u \Delta_v \geq \frac{1}{2}. \tag{3.117}$$

We will call the inequality in Eq. (3.117) *the u-v uncertainty principle*. Notice that since $[\hat{U}, \hat{V}]$ is just a number, then the product of the uncertainties is independent of the state, Ψ. However it is useful to evaluate the product $\Delta_u \Delta_v$ for some specific cases. For example, if the state is represented by the Fock space, $|n\rangle$, then, from Eq. (3.101) we get

$$\langle n | \hat{U}^2 | n \rangle = n + \frac{1}{2}$$

with the same result for $\langle n | \hat{V}^2 | n \rangle$. Now if we define the uncertainty in $\hat{U}$ with respect to the Fock state, $|n\rangle$ as $\Delta_u(n)$, then by definition

$$\Delta_u^2(n) = \langle n | \hat{U}^2 | n \rangle - \langle n | \hat{U} | n \rangle^2, \tag{3.118}$$

with a corresponding definition for $\hat{V}$. Then, recalling that $\langle n | \hat{U} | n \rangle = \langle n | \hat{V} | n \rangle = 0$, we find

$$\Delta_u(n) \Delta_v(n) = n + \frac{1}{2}. \tag{3.119}$$

We can also compare the result in Eq. (3.119) with the corresponding result for the state $|\alpha\rangle$ that was defined in Section 3.6.1. First we define

$$\Delta_u^2(\alpha) = \langle \alpha | \hat{U}^2 | \alpha \rangle - \langle \alpha | \hat{U} | \alpha \rangle^2, \tag{3.120}$$

with a corresponding definition for $\hat{V}$. The first term in Eq. (3.120) is

$$\begin{aligned}
\langle \alpha | \hat{U}^2 | \alpha \rangle &= \frac{1}{2} \langle \alpha | (\hat{A} + \hat{A}^\dagger)^2 | \alpha \rangle \\
&= \frac{1}{2} \langle \alpha | (\hat{A}^2 + \hat{A}^{\dagger 2} + \hat{A} \hat{A}^\dagger + \hat{A}^\dagger \hat{A}) | \alpha \rangle \\
&= \frac{1}{2} \langle \alpha | (\hat{A}^2 + \hat{A}^{\dagger 2} + 2\hat{A}^\dagger \hat{A} + 1) | \alpha \rangle \\
&= \frac{1}{2} (\alpha^2 + \alpha^{*2} + 2\alpha^* \alpha + 1) = \frac{1}{2} (\alpha^* + \alpha)^2 + \frac{1}{2},
\end{aligned} \tag{3.121}$$

where we have used $\langle\alpha|(\hat{A}^2|\alpha\rangle = \alpha^2$ and $\langle\alpha|(\hat{A}^{\dagger 2}|\alpha\rangle = \langle\alpha|(\hat{A}^2|\alpha\rangle^*$. So, with the result from Eqs. (3.121) and (3.106) we have

$$\Delta_u^2(\alpha) = \frac{1}{2}(\alpha^* + \alpha)^2 + \frac{1}{2} - \frac{1}{2}(\alpha^* + \alpha)^2 = \frac{1}{2}. \qquad (3.122)$$

Following a similar procedure for $\hat{V}$, one finds that $\Delta_v^2(\alpha)$ has exactly the same value as $\Delta_u^2(\alpha)$ so that

$$\Delta_u(\alpha)\Delta_v(\alpha) = \frac{1}{2}, \qquad (3.123)$$

which, remarkably is independent of α. It means that whatever value α has the product of the uncertainties in the expectation values of $\hat{U}$ and $\hat{V}$ are the minimum possible allowed by the u-v uncertainty principle. The state $|\alpha\rangle$ is referred to as the *coherent state* [2]. As we saw earlier, the expectation values of the number operator obey classical Poisson statistics and it is of importance in quantum optics [38, 78, 89].

3.7 IS THERE A $\hat{N}$-ξ DUALITY?

As we have seen, the commutation relation between the Hermitian operators, $\hat{u} = u$ and $\hat{v} = -i\frac{\partial}{\partial u}$ can be written in the form

$$[-i\frac{\partial}{\partial u}, u] = -i.$$

It is tempting to think that we could be used as a blueprint to construct an Hermitian phase operator, $\hat{\xi} = \xi$. Indeed, such proposals for the construction of a phase operator have occurred in the literature [17, 10, 38, 78, 89, 132]. These attempts are partly driven by the way quantum mechanics is often introduced as the quantization of classical systems by replacing scalar variables by operators in a rather ad hoc fashion. This was very much the view of some of the early developers of quantum mechanics like von Neumann [84]. We could certainly write

$$[i\frac{d}{d\xi}, \xi] = i. \qquad (3.124)$$

Eq. (3.124) might then be used to assume that $\hat{\xi}$ and $\hat{N}$ form a canonical pair like $\hat{u}$ and $\hat{v}$ such that

$$[\hat{N}, \hat{\xi}] = i ?$$

on the basis of Eq. (3.14). There is a telling argument, due to Pauli [3, 91][7], that if there were a commutation relation between $\hat{N}$ and $\hat{\xi}$ as that above, then

$$[\hat{N}, f(\hat{\xi})] = if'(\hat{\xi}).$$

This would then lead to, for any real number, p

$$\hat{N}\exp(ip\hat{\xi})|n\rangle = \exp(ip\hat{\xi})\hat{N}|n\rangle - p\exp(ip\hat{\xi})|n\rangle = (n-p)\exp(ip\hat{\xi})|n\rangle,$$

---

[7]Pauli used this argument in the context of speculation about a time operator (see Chapter 5).

the implication of which is that there exists an eigenstate, $\exp(ip\hat{\xi})|n\rangle$, of $\hat{N}$ with an eigenvalue of $n - p$. Since p is arbitrary this could lead to negative eigenvalue, which are inadmissible. Hence, there can be no phase operator with the commutation relation $[\hat{N}, \hat{\xi}] = i$.

However, as has been pointed out previously $\hat{N}$ is not identical to $i\frac{d}{d\xi}$, whereas, $\hat{v}$ is identical to $-i\frac{d}{du}$. This latter relation arises naturally from the Hermitian and anti-Hermitian components of $\hat{A}$, which form a canonical pair. $\hat{N}$ and ξ are not a canonical pair in this sense. Part of the problem in devising a phase operator on the above basis seems to be due to the fact that, unlike $\hat{v}$, $\hat{N}$ is bound from below due to its positivity property [38]. These issues also show the importance of taking great care when it comes to treating stand-alone differential operators, as was pointed out in Chapter 2.

The forging arguments do not entirely rule out the possibility that some form of phase operator could be constructed, but this does not appear to be possible on the basis of $\hat{N}$-ξ duality as a simple analogy with u-v duality. In principle, one could argue that $i\frac{d}{d\xi}$ and ξ do form such a dual pair, but then one could construct a new number amplitude, $\hat{A}_\xi$, such that

$$\hat{A}_\xi = \frac{1}{\sqrt{2}}(\xi + \frac{d}{d\xi}),$$

and get back to $\hat{A}_\xi^\dagger \hat{A}_\xi = \hat{N}_\xi$. All this would achieve would be to replace the dummy variable u by ξ and we could not then interpret ξ as a phase. This exercise shows the danger of assuming that any scalar variable can quantized by turning it into an operator in an ad hoc manner. We shall find a similar problem in relation to time as a variable in quantum mechanics.

3.8   GENERALIZATION OF RAISING AND LOWERING OPERATORS

Consider and operator $\hat{B}$ and its adjoint, $\hat{B}^\dagger$. Now the product, $\hat{B}^\dagger \hat{B}$ is invariant to the phase shift such that $\hat{B}(\xi) = \hat{B}(0)\exp(-i\xi)$, so we can conclude that

$$i\frac{d\hat{B}^\dagger \hat{B}}{d\xi} = [\hat{B}^\dagger \hat{B}, \hat{N}] = 0. \tag{3.125}$$

Bearing in mind the results in Section 2.8, we can infer, potentially, that Eq. (3.125) is consistent with $\hat{B}^\dagger \hat{B} = f(\hat{N})$, where f is some analytic function and hence that

$$\hat{B}^\dagger \hat{B}\Phi_n = f(\hat{N})\Phi_n = f(n)\Phi_n. \tag{3.126}$$

This supposition can be confirmed by noting that

$$i\frac{d\hat{B}}{d\xi} = \hat{B} = [\hat{B}, \hat{N}]. \tag{3.127}$$

As a consequence of this, then

$$\hat{N}\hat{B}\Phi_n = (\hat{B}\hat{N} - \hat{B})\Phi_n = (n-1)\hat{B}\Phi_n, \tag{3.128}$$

from which we can conclude that $\hat{B}\Phi_n = \beta(n)\Phi_{n-1}$ where $\beta(n)$ is some scalar coefficient that may depend on n. Similarly we find

$$i\frac{d\hat{B}^\dagger}{d\xi} = -\hat{B}^\dagger = [\hat{B}^\dagger, \hat{N}] \tag{3.129}$$

and so

$$\hat{N}\hat{B}^\dagger\Phi_n = (\hat{B}^\dagger\hat{N} + \hat{B}^\dagger)\Phi_n = (n+1)\hat{B}^\dagger\Phi_n, \tag{3.130}$$

from which we can conclude that $\hat{B}^\dagger\Phi_n = \gamma(n)\Phi_{n+1}$, where $\gamma(n)$ is some scalar coefficient that may depend on n. The connection between $\beta(n)$, $\gamma(n)$ and $f(n)$ then follows from

$$\langle\Phi_n|\hat{B}^\dagger\hat{B}|\Phi_n\rangle = \|\hat{B}\Phi_n\|^2. \tag{3.131}$$

Now the lhs of Eq. (3.131) is equal to $\gamma(n-1)\beta(n)$ while the rhs is $\beta^2(n)$, so we find $\gamma(n) = \beta(n+1)$. This, together with Eq. (3.126) lead to $\beta(n) = \sqrt{f(n)}$. Finally, we get

$$\hat{B}\Phi_n = \sqrt{f(n)}\Phi_{n-1} \text{ and } \hat{B}^\dagger\Phi_n = \sqrt{f(n+1)}\Phi_{n+1}. \tag{3.132}$$

These results show that $\hat{B}$ acts as a lowering operator and $\hat{B}^\dagger$ acts as a raising operator, but of more general form than the pair $\hat{A}$ and $\hat{A}^\dagger$ in Eqs. (3.15) and (3.16).

Notice that we still need a backstop condition

$$\hat{B}\Phi_0 = 0,$$

so that we do not generate eigenstates with $n < 0$. This means that we must have $f(0) = 0$. From the results above we can see that

$$[\hat{B}, \hat{B}^\dagger]\Phi_n = (f(n+1) - f(n))\Phi_n, \tag{3.133}$$

which means that $[\hat{B}, \hat{B}^\dagger] \neq 1$, unless $f(n) = n$.

So we can conclude that $\hat{B}^\dagger\hat{B}$ operates on the same Hilbert space as $\hat{N}$ and generates eigenvalues of $f(n)$ and that even in this general case the $\hat{B}$ and $\hat{B}^\dagger$ still operate respectively as lowering and raising operators for the eigenstates of $\hat{N}$. Clearly, when $f(\hat{N}) = \hat{N}$, then $\hat{B}$ reduces to $\hat{A}$.

Since $\hat{B}^\dagger\hat{B}$ is Hermitian, we can define a new Hermitian operator, $\hat{F} = \hat{B}^\dagger\hat{B}$, where $\hat{F} = f(\hat{N})$ is a *nonlinear natural number operator* and construct a new S-type system equation,

$$i\frac{d\Phi_n}{d\tau} = \hat{F}\Phi_n = f(n)\Phi_n, \tag{3.134}$$

where τ is a new real continuous system variable, whose meaning will become clear later. Eq. (3.134) may be integrated to give

$$\Phi_n(\tau) = \exp(-i\tau\hat{F})\Phi_n(0) = \exp(-i\tau f(n))\Phi_n(0). \tag{3.135}$$

It also follows that we can also construct an explicitly τ-dependent operator, $\hat{B}(\tau) = \exp(i\tau\hat{F})\hat{B}(0)\exp(-i\tau\hat{F})$, and then

$$i\frac{d\hat{B}}{d\tau} = [\hat{B},\hat{F}] = [\hat{B},\hat{B}^\dagger\hat{B}] = [\hat{B},\hat{B}^\dagger]\hat{B} = (f(n+1) - f(n))\hat{B}. \qquad (3.136)$$

Integrating the result in Eq. (3.136) then yields

$$\hat{B}(\tau) = \hat{B}(0)\exp(-i(f(n+1) - f(n))\tau).$$

We can now look for a representation for $\hat{B}$ and its adjoint in terms of a pair of Hermitian components as we did for $\hat{A}$. So, let us write

$$\hat{B} = \frac{1}{\sqrt{2}}(\hat{W} + i\hat{Z}), \qquad (3.137)$$

where $\hat{W}$ and $\hat{Z}$ are Hermitian. Then, from Eq. (3.135), we find

$$[\hat{B},\hat{B}^\dagger]\Phi_n = [i\hat{Z},\hat{W}]\Phi_n = (f(n+1) - f(n))\Phi_n, \qquad (3.138)$$

so, in general, $[i\hat{Z},\hat{W}] \neq 1$ and $\hat{W}$ and $\hat{Z}$ are not a pair of canonical variables. This means that if we seek a analytic representation in terms of a continuous variable and a differential operator, by switching to $\hat{w}$ and $\hat{z}$ in a similar way to that used for $\hat{A}$, we cannot represent $\hat{w}$ by w and $\hat{z}$ by $-i\frac{\partial}{\partial w}$. However, we can represent $\hat{z}$ by $-i\frac{\partial}{\partial w}$ as long as we represent $\hat{w}$ by some function of w, say $W(w)$.

Then we can define

$$\hat{B}_w = \frac{1}{\sqrt{2}}(\frac{\partial}{\partial w} + W(w)) \text{ and } \hat{B}_w^\dagger = \frac{1}{\sqrt{2}}(-\frac{\partial}{\partial w} + W(w)), \qquad (3.139)$$

where $W(w)$ is some real function of w. Then

$$\hat{B}_w^\dagger\hat{B}_w\Phi_n(w) = \frac{1}{2}(-\frac{\partial^2}{\partial w^2} + W^2(w) - W'(w))\Phi_n(w) = f(n)\Phi_n(w), \qquad (3.140)$$

where $W'(w) = \frac{\partial W(w)}{\partial w}$ and $\Phi_n(w)$ is some new eigenfunction. If we now let $W^2(w) - W'(w) = 2Y(w)$, then the eigenvalue equation that results is

$$(-\frac{1}{2}\frac{\partial^2}{\partial w^2} + Y(w))\Phi_n(w) = f(n)\Phi_n(w). \qquad (3.141)$$

Since we know that $\hat{B}_w^\dagger\hat{B}_w$ is Hermitian, then we can write the S-type equation, Eq. (3.134) as

$$i\frac{d\Phi_n}{d\tau} = \hat{F}_w\Phi_n = f(n)\Phi_n, \qquad (3.142)$$

where $\hat{F}_w\Phi_n = \hat{B}_w^\dagger\hat{B}_w\Phi_n$, and the eigenfunction Φ_n now depends on, τ, as well as w. Solving the differential equation with variable τ in Eq. (3.142) yields $\Phi_n(\tau) = \exp(-if(n)\tau)\Phi_n(0)$, where the 0 in $\Phi_n(0)$ applies to the value of τ.

It is interesting to note that there is a trivial form of Eq. (3.139) in which $W(w) = W'(w) = 0$, which means that $Y(w) = 0$. Then we find

$$\hat{B} = \frac{1}{\sqrt{2}} \frac{\partial}{\partial u} \text{ and } \hat{B}^\dagger = -\frac{1}{\sqrt{2}} \frac{\partial}{\partial u}.$$

Then we would get just

$$\hat{F} = -\frac{1}{2} \frac{\partial^2}{\partial u^2},$$

with $f(n)$ actually being independent of n. This trivial result has quite an interesting interpretation as far as standard quantum mechanics is concerned (see Chapter 5).

Now recall that Φ_n is an eigenfunction of both $\hat{N}_u$ and $\hat{F}_w$, so it must satisfy Eqs. (3.14), (3.58), (3.134) and (3.141), and hence it must be a function of ξ, u, τ and w. The eigenfunction must then have the form, $\Phi_n(\xi, u, \tau, w)$. This is no problem since it can be effective achieved by a separation of variables. Then we can write

$$\Phi_n(\xi, u, \tau, w) = \exp(-i(n\xi + f(n)\tau)\Psi_n(u)\Xi_n(w), \tag{3.143}$$

where $\Psi_n(u)$ is the solution to Eq. (3.58) and $\Xi_n(w)$ is a solution to Eq. (3.141). Equations like Eq. (3.141) that have been derived from the nonlinear natural number operators play a key role in standard quantum mechanics. As we shall see in Chapter 5, where we will deal with this topic in more detail, they appear in relation to the bound states of systems, such as a particle trapped in a potential well.

Finally, in this section, we note that it is quite possible that an operator $\hat{C}$ and its adjoint $\hat{C}^\dagger$ can act as a raising and lowering pair for some Hermitian operator, $\hat{G}$ say, if

$$[\hat{G}, \hat{C}] = -\hat{C}. \tag{3.144}$$

Suppose $\hat{G}$ has an eigenstate $|g\rangle$, with an eigenvalue g, i.e. $\hat{G}|g\rangle = g|g\rangle$. Then

$$\hat{G}\hat{C}|g\rangle = \hat{C}\hat{G}|g\rangle - \hat{C}|g\rangle = (g-1)\hat{C}|g\rangle, \tag{3.145}$$

from which we can conclude that $\hat{C}|g\rangle \to |g-1\rangle$ and so $\hat{C}$ lowers the eigenvalue of $\hat{G}$ by 1. Now we also find

$$[\hat{G}, \hat{C}]^\dagger = -\hat{C}^\dagger, \tag{3.146}$$

so, as long as $\hat{G}$ is Hermitian, then

$$[\hat{G}, \hat{C}]^\dagger = [\hat{C}^\dagger, \hat{G}] = -\hat{C}^\dagger \tag{3.147}$$

and so

$$[\hat{G}, \hat{C}^\dagger] = \hat{C}^\dagger. \tag{3.148}$$

It is then straightforward, following similar steps to those above, to show that $\hat{C}^\dagger|g\rangle \to |g+1\rangle$ and so $\hat{C}$ raises the eigenvalue of $\hat{G}$ by 1. Relations of this type will prove useful in multi-category case that will be treated later.

4 Multi-category systems: Bosons and fermions

4.1 A SUM OVER CATEGORIES

The results of Chapter 3 may be generalized to a multi-category system by noting that the sum of a set of natural numbers is itself a natural number, so the sum of natural number operators is just a natural number operator. So, let

$$\hat{N} = \sum_{i=1}^{S} \hat{N}_i, \tag{4.1}$$

where $\hat{N}_i$, is the number of items associated with a category with an index i, and we assume that there are a finite number, S, of categories. Then just like the single category case treated in Chapter 3, $\hat{N}_i$ can be represented by an S-type differential equation, like Eq. (3.14), as

$$i\frac{\mathrm{d}}{\mathrm{d}\xi_i}\Phi_{n_i} = \hat{N}_i\Phi_{n_i} = n_i\Phi_{n_i}, \tag{4.2}$$

where ξ_l is a phase angle in the phase space of the ith category and n_i is a natural number representing its population. Now these categories should simply be thought of as labels that can be attached to items in the system. A system with S categories has S different labels. One should not think of the items in the system as being intrinsically different. The differences that categorize them are only due to the labels we put on them. So we can, in principle take an item from one category and place it in another. It would simply then lose the label of the category it started with and acquire the label of the category it ended with. All that would change would be the value of n_i in the two categories involved. We will deal with dynamical systems in which this exchange of items takes place later, in Chapter 6. For now we assume that we have fixed numbers of items in the different categories. After all, we have not introduced any notion of time, so in that sense, change is not yet a meaningful concept.

Integrating Eq. (4.2) yields

$$\Phi_{n_i}(\xi_i) = \exp(-i\xi_i\hat{N}_i)\Phi_{n_i}(0) = \exp(-i\xi_i n_i)\Phi_{n_i}(0). \tag{4.3}$$

The eigenfunctions, $\Phi_{n_i}(\xi_i)$, can be individually normalized as before and then each will constitute a Hilbert space, $\mathcal{H}_i$. The whole system is then a product of the individual spaces that is written symbolically as

$$\mathcal{H} = \mathcal{H}_1 \otimes \mathcal{H}_2 \otimes \cdots \otimes \mathcal{H}_i \cdots \otimes \mathcal{H}_S.$$

DOI: 10.1201/9781003377504-4

The system as a whole can be represented as S sets of eigenfunctions. However, this is most conveniently represented by a generalization of the Fock state space introduced in Chapter 3. Its multi-category form is

$$|n_1, n_2, \ldots, n_i, \ldots, n_S\rangle,$$

where each of the S entries represents the population of a corresponding category. This is a general form of *occupation number representation*. The next step is to factorize $\hat{N}_i$ in a similar way to the single category case and we write

$$\hat{N}_i = \hat{A}_i^\dagger \hat{A}_i. \tag{4.4}$$

Then we note that a phase shift in $\hat{A}_i$ of the form $\hat{A}_i(\xi_i) = \hat{A}_i(0)\exp(-i\xi_i)$ leaves $\hat{N}_i$ unchanged, and then $\hat{A}_i$ obeys an H-type equation of the form

$$i\frac{d\hat{A}_i}{d\xi_i} = \hat{A}_i = [\hat{A}_i, \hat{N}_i]. \tag{4.5}$$

We can generalize Eq. (3.20) by applying the set of creation operators, $\hat{A}_i^\dagger$, to the state, $|0,0,0,0,\ldots,0\rangle$, in which all of the categories are empty. Then we can generate a state of the system in which the ith category contains n_i individuals by

$$|\{n_i\}\rangle = \prod_i \frac{\hat{A}_i^{\dagger n_i}}{\sqrt{n_i!}} |\{0\}\rangle, \tag{4.6}$$

where $|\{n_i\}\rangle \equiv |n_1, n_2, n_3, \ldots, n_i, \ldots\rangle$ is a shorthand way of representing the occupation number representation and obviously, $|\{0\}\rangle \equiv |0,0,0,0,\ldots,0\rangle$.

We can define an orthogonality condition for the state, $|\{n_i\}\rangle$, by

$$\langle\{m_i\}|\{n_i\}\rangle = \prod_i \delta_{m_i n_i}. \tag{4.7}$$

The creation and annihilation operators then have the following properties in relation to the multi-category state[1]

$$\hat{A}_i|\{n_i\}\rangle = \sqrt{n_i}|n_1, n_2, \ldots, n_i-1, \ldots\rangle,$$

$$\hat{A}_i^\dagger|\{n_i\}\rangle = \sqrt{n_i+1}|n_1, n_2, \ldots, n_i+1, \ldots\rangle,$$

$$\hat{A}_i^\dagger \hat{A}_i|\{n_i\}\rangle = n_i|\{n_i\}\rangle,$$

$$[\hat{A}_i, \hat{A}_j] = [\hat{A}_i^\dagger, \hat{A}_j^\dagger] = 0 \text{ and } [\hat{A}_i, \hat{A}_j^\dagger] = \delta_{ij}. \tag{4.8}$$

The last of Eqs. (4.8) can be deduced from the fact that

$$\hat{A}_i^\dagger \hat{A}_j|\ldots, n_i, \ldots, n_j, \ldots\rangle = \sqrt{(n_i+1)n_j}|\ldots, n_i+1, \ldots, n_j-1, \ldots\rangle,$$

[1]The explicit form of the occupation number representation is used only where necessary to avoid ambiguity.

and also

$$\hat{A}_j \hat{A}_i^\dagger | \ldots, n_i, \ldots, n_j, \ldots \rangle = \sqrt{(n_i+1)n_j} | \ldots, n_i+1, \ldots, n_j-1, \ldots \rangle$$

for $j \neq i$. Notice also that there is a backstop condition for each category of the form $\hat{A}_i | n_1, n_2, \ldots, n_i = 0, \ldots \rangle = 0$.

Although we now have a system of S categories, we do not really have any overall unification of this system. The reason is that the only differentials are those associated with the individual $\hat{N}_i$s and Eqs. (4.2) and (4.5). The global $\hat{N}$ may be associated with a continuous variable ξ, but this then cannot easily be related to the ξ_is, so we have neither an S-type nor an H-type equation that can be applied to the system as a whole. However, this situation can be remedied by construction a new operator for the whole system in the following way.

4.2   WHY WE NEED TIME

For the moment, if we want to investigate changes in our system of several categories, the only dynamical equation are Eqs. (4.2) and (4.5), which treat each category separately. It would be advantageous to have a single H-type equation of motion for the whole system, but, at the same time, also be able to to preserve the category labels that distinguish one category from another. In order to distinguish the different categories and at the same time provide a universal system variable we can parameterize each of the ξ_i with the same continuous variable, so, we let $\xi_i = \omega_i t$ and then we can write $\hat{A}_i(t) = \hat{A}_i(0) \exp(-i\omega_i t)$, where t is our continuous system-wide variable. The next step is to define a t-shift operator, $\hat{\Omega}$, by an S-type equation. Thus, we let

$$i \frac{\mathrm{d}}{\mathrm{d}t} \Psi = \hat{\Omega}\Psi, \tag{4.9}$$

where t is a continuous real variable that will be eventually identified with time. Here the function Ψ is a t-dependent function that characterizes the state of the system as a whole. This S-type equation is nothing but a scaled version of the Schrödinger equation, whose role will becomes clearer when we deal with single-category systems and the emergence of quantum mechanics, in the next chapter. For the moment time, t, is nothing more than a continuous real variable on which the function Ψ depends, in the sense that integrating Eq. (4.9) yields $\Psi(t) = \exp(-it\Omega)\Psi(0)$. From here we can construct t-dependent H-type operators by defining as an exemplar a t-dependent operator, $\hat{G}(t)$ in the same way we developed the ξ-dependent system in Chapter 3, such that

$$\hat{G}(t) = \exp(it\Omega)\hat{G}(0)\exp(-it\Omega), \tag{4.10}$$

which satisfies

$$i \frac{\mathrm{d}\hat{G}}{\mathrm{d}t} = [\hat{G}, \hat{\Omega}]. \tag{4.11}$$

This H-type equation above, for a t-dependent operator, is referred to as the *Heisenberg equation of motion*, albeit in a scaled form. If we now apply the Heisenberg equation of motion to $\hat{A}_i(t) = \exp(-i\omega_i t)\hat{A}_i(0)$, we get

$$i\frac{\mathrm{d}\hat{A}_i}{\mathrm{d}t} = \omega_i \hat{A}_i = [\hat{A}_i, \hat{\Omega}]. \tag{4.12}$$

It is now possible to deduce what form $\hat{\Omega}$ must have for the system as a whole, by noting that

$$\omega_i \hat{A}_i = [\hat{A}_i, \omega_i \hat{N}_i]. \tag{4.13}$$

So, subtracting Eq. (4.13) from Eq. (4.12), we get $[\hat{A}_i, \hat{\Omega} - \omega_i \hat{N}_i] = 0$, from which we can conclude that $\hat{\Omega} - \omega_i \hat{N}_i = K(\hat{A}_i, \hat{A}_j, \hat{A}_j^\dagger)$ for all i, with $j \neq i$, where $K(\hat{A}_i, \hat{A}_j, \hat{A}_j^\dagger)$ is an arbitrary function of its arguments. Similarly, from the time derivative of $\hat{A}_i^\dagger(t) = \hat{A}_i^\dagger(0)\exp(i\omega_i t)$, we find $\hat{\Omega} - \omega_i \hat{N}_i = L(\hat{A}_i^\dagger, \hat{A}_j, \hat{A}_j^\dagger)$, for all i, with $i \neq j$, where $L(\hat{A}_i^\dagger, \hat{A}_j, \hat{A}_j^\dagger)$ is a second arbitrary function of its arguments. The only way these relations can be satisfied for all i and j is if

$$\hat{\Omega} = \sum_{i=1}^{S} \omega_i \hat{N}_i + \lambda = \sum_{i=1}^{S} \omega_i \hat{A}_i^\dagger \hat{A}_i + \lambda, \tag{4.14}$$

where λ is an arbitrary scalar constant. Now we can see that we have an operator, $\hat{\Omega}$ that can be applied to the whole system, together with an equation of motion that applies to the system as a whole. From the above relations, one can show that $\frac{\mathrm{d}\hat{N}_i}{\mathrm{d}t} = 0$. This clearly implies that for the system as a whole, $\frac{\mathrm{d}\hat{N}}{\mathrm{d}t} = 0$. Obviously we can move to the single category case when $S = 1$ and this situation will be explored in the next chapter.

4.3 THE EMERGENCE OF FERMIONS

As noted in [105], it is also possible to deduce alternatives to the commutation relations for the creation and annihilation operators, that are still consistent with $\hat{N}_i = \hat{A}_i^\dagger \hat{A}_i$, the time shift operator, Eq. (4.9) and the Heisenberg equation for $\hat{A}_i(t) = \hat{A}_i(0)\exp(-i\omega_i t)$. We can treat $[\hat{A}_i, \hat{A}_j] = 0$ as an ordering rule that implies $\hat{A}_i \hat{A}_j = \hat{A}_j \hat{A}_i$, and then generalize this to $\hat{A}_i \hat{A}_j = \mu \hat{A}_j \hat{A}_i$, where μ is an as yet unknown constant scalar factor. Reversing the order once more gives $\hat{A}_i \hat{A}_j = \mu^2 \hat{A}_i \hat{A}_j$, so $\mu = \pm 1$. Similarly, if we assume that $\hat{A}_i \hat{A}_j^\dagger = \rho \hat{A}_j^\dagger \hat{A}_i + \sigma$, where ρ and σ are scalar constants, then we get

$$\omega_i \hat{A}_i = [\hat{A}_i, \sum_{j=1}^{S} \omega_j \hat{A}_j^\dagger \hat{A}_j] = \sum_{j=1}^{S} \omega_j (\hat{A}_j^\dagger \hat{A}_j \hat{A}_i(\mu\rho - 1) + \sigma \hat{A}_j). \tag{4.15}$$

Thus, since this expression has to be true for all i and j, we must have, $\mu = \rho = \pm 1$ and $\sigma = \delta_{ij}$. When $\mu = \rho = 1$ we recover our previous results in Eqs. (4.8), but when $\mu = \rho = -1$ we get something new, i.e.

$$\{\hat{A}_i, \hat{A}_j\} = 0, \{\hat{A}_i^\dagger, \hat{A}_j^\dagger\} = 0, \text{ and } \{\hat{A}_i, \hat{A}_j^\dagger\} = \delta_{ij}, \tag{4.16}$$

where $\{\hat{F},\hat{G}\} = \hat{F}\hat{G} + \hat{G}\hat{F}$, which are referred to as an *anti-commutator*. The first two relations in Eq. (4.16) imply that

$$\hat{A}_i^2 = \hat{A}_i^{\dagger 2} = 0.$$

From the last relation in Eq. (4.16), which is equivalent to $\hat{A}_i\hat{A}_i^{\dagger} = 1 - \hat{A}_i^{\dagger}\hat{A}_i$, we find

$$\begin{aligned} \hat{N}_i^2 &= \hat{A}_i^{\dagger}\hat{A}_i\hat{A}_i^{\dagger}\hat{A}_i \\ &= \hat{A}_i^{\dagger}(1 - \hat{A}_i^{\dagger}\hat{A}_i)\hat{A}_i \\ &= \hat{A}_i^{\dagger}\hat{A}_i = \hat{N}_i. \end{aligned} \tag{4.17}$$

So

$$\hat{N}_i^2 - \hat{N}_i = \hat{N}_i(\hat{N}_i - 1) = 0, \tag{4.18}$$

and then applying this result to the Fock space, $|\{n_i\}\rangle$, the only possible eigenvalues for $\hat{N}_i$ are, $n_i = 0$ or 1.

The first and third rules in Eqs. (4.8) still apply, but the second needs a slight modification to

$$\hat{A}_i^{\dagger}|n_i\rangle = \sqrt{1 - n_i}|n_i + 1\rangle.$$

The commutation relations in Eqs. (4.8) are referred to as *bosonic* and the items to which they apply are termed *bosons*. The new set of relations in Eqs. (4.16) are termed *fermionic* and apply to items called *fermions*. These have very different characteristics from those of bosons. These terms are of course those that are used in fundamental particle physics for the two categories of fundamental particles. Whereas an unlimited number of bosons may be accommodated in a single category, for fermions, the number of categories must exceed the number of items, since there can be either 0 or 1 item in any single category. This means a quantum state can either be empty or contain one fermion. This property of fermions can be recognized as the *Pauli exclusion principle*. Also note that the simple analysis above implies that bosons and fermions are the only two types of primitive systems that are possible. This has important implications when one considers interactions between different categories, as will be shown later.

4.4 FERMIONIC HERMITIAN OPERATORS

In the previous chapter we saw how it was possible the define what we now recognize as the bosonic number amplitude operators in terms of a pair of Hermitian operators. These proved to have an important and useful role as canonical variables, whose commutator led to a representation by differential operators. It is therefore of interest to see if the fermionic number amplitude operators that obey the anti-commutation rules, Eqs. (4.16) have a similar representation in terms of pairs of Hermitian operators. To check this, it is only necessary to look at a single category fermionic system, so we begin with the pair, $\hat{A}$ and $\hat{A}^{\dagger}$ that obey

$$\{\hat{A},\hat{A}\} = 0, \{\hat{A}^{\dagger},\hat{A}^{\dagger}\} = 0, \{\hat{A},\hat{A}^{\dagger}\} = 1, \tag{4.19}$$

which, following the results in Eqs. (4.17) and (4.18) lead to

$$\hat{A}^2 = 0, \hat{A}^{\dagger 2} = 0, \hat{A}^\dagger \hat{A} = \hat{N}, \tag{4.20}$$

where $\hat{N}$ is the fermionic number operator with eigenvalues of 0 or 1. Now we let

$$\hat{A} = \hat{U} + i\hat{V} \text{ and } \hat{A}^\dagger = \hat{U} - i\hat{V}, \tag{4.21}$$

where $\hat{U}$ and $\hat{V}$ are a pair of fermionic Hermitian operators. Substituting Eqs. (4.21) into the first two relation in Eqs. (4.20) and the last relation in Eqs. (4.19) yields, respectively

$$\hat{U}^2 - \hat{V}^2 + i\{\hat{U}, \hat{V}\} = 0, \tag{4.22}$$

$$\hat{U}^2 - \hat{V}^2 - i\{\hat{U}, \hat{V}\} = 0 \tag{4.23}$$

and

$$\hat{U}^2 + \hat{V}^2 = \frac{1}{2}. \tag{4.24}$$

Adding Eqs. (4.22) and (4.23), and the utilizing Eq. (4.24) leads to

$$\hat{U}^2 = \hat{V}^2 = \frac{1}{4}. \tag{4.25}$$

Subtracting Eq. (4.23) from Eq. (4.22) then gives

$$\{\hat{U}, \hat{V}\} = 0. \tag{4.26}$$

Next we can evaluate the commutator, $[\hat{U}, \hat{V}]$ by first rearranging Eqs. (4.21) to give

$$\hat{U} = \frac{1}{2}(\hat{A}^\dagger + \hat{A}) \text{ and } \hat{V} = \frac{i}{2}(\hat{A}^\dagger - \hat{A}). \tag{4.27}$$

Then

$$\begin{aligned}[\hat{U}, \hat{V}] &= \frac{i}{4}[\hat{A}^\dagger + \hat{A}, \hat{A}^\dagger - \hat{A}] \\ &= \frac{i}{2}[\hat{A}, \hat{A}^\dagger] \\ &= i(\frac{1}{2} - \hat{N}). \end{aligned} \tag{4.28}$$

We can immediately see from Eq. (4.28), that, unlike for the corresponding bosonic Hermitian operators in Section 3.4, their fermionic cousins here, are not canonical and therefore do not allow a representation in the form of differential operators. However, Eq. (4.28) does indicate that the fermionic $\hat{U}$ and $\hat{V}$, do have important symmetry properties. To explore this further we note that $\frac{1}{2} - \hat{N}$ is Hermitian, so we can let

$$\hat{W} = \frac{1}{2} - \hat{N}, \tag{4.29}$$

where $\hat{W}$ is an Hermitian operator and then

$$[\hat{U},\hat{V}] = i\hat{W}. \tag{4.30}$$

From Eq. (4.29) we can find

$$\hat{W}^2 = (\frac{1}{2} - \hat{N})^2 = \frac{1}{4} - \hat{N} + \hat{N}^2 = \frac{1}{4}, \tag{4.31}$$

which is the same as for $\hat{U}^2$ and $\hat{V}^2$. Furthermore, we find

$$\{\hat{V},\hat{W}\} = -\frac{i}{2}\{\hat{A}^\dagger - \hat{A}, \hat{A}^\dagger \hat{A}\}$$

$$= -\frac{i}{2}(\hat{A}^\dagger - \hat{A})\hat{A}^\dagger \hat{A} + \hat{A}^\dagger \hat{A}(\hat{A}^\dagger - \hat{A}) \tag{4.32}$$

$$= \frac{i}{2}(\hat{A}\hat{A}^\dagger \hat{A} - \hat{A}\hat{A}^\dagger \hat{A}) = 0.$$

Similarly, we get

$$\{\hat{W},\hat{U}\} = 0. \tag{4.33}$$

Eqs. (4.31) to (4.33) indicate that there is a great deal of symmetry between $\hat{U}$, $\hat{V}$ and $\hat{W}$. Just what this symmetry is can be understood by evaluating $[\hat{V},\hat{W}]$ and $[\hat{W},\hat{U}]$. We find, after a little manipulation, using Eqs. (4.21) and (4.27)

$$[\hat{V},\hat{W}] = -\frac{i}{2}[\hat{A}^\dagger - \hat{A}, \hat{A}^\dagger \hat{A}] = i\hat{U}, \tag{4.34}$$

and

$$[\hat{W},\hat{U}] = -\frac{1}{2}[\hat{A}^\dagger \hat{A}, \hat{A}^\dagger + \hat{A}] = i\hat{V}. \tag{4.35}$$

Eqs. (4.30), (4.34) and (4.35) indicate that $\hat{U}$, $\hat{V}$ and $\hat{W}$ form a closed Lie group of Hermitian operators (see Appendix D for further details). This is indicative of an important symmetry that we will explore in more detail in later chapters. For the moment it is worth determining their eigenvalues and eigenvectors.

The eigenvector of $\hat{W}$ is simple to identify since it depends only on $\hat{N}$, so it will have the single category Fock space vectors represented by $|n\rangle$ as its eigenvectors. This space is only two-dimensional for fermions and consists of $|0\rangle$ and $|1\rangle$. These are then the eigenvectors of $\hat{W}$. Clearly,

$$\hat{W}|n\rangle = (\frac{1}{2} - \hat{N})|n\rangle = (\frac{1}{2} - n)|n\rangle, \tag{4.36}$$

where $n = 0$ or 1. So the eigenvalues of $\hat{W}$ are $\pm\frac{1}{2}$.

Now neither $\hat{U}$ nor $\hat{V}$ commutes with $\hat{W}$ nor with $\hat{N}$ and so must have eigenvectors other than $|0\rangle$ or $|1\rangle$. However, they must have eigenvectors in the Fock space spanned by $|0\rangle$ and $|1\rangle$. These eigenvectors will be unit vectors that are linear combinations of $|0\rangle$ and $|1\rangle$. So let $\hat{U}$ have an eigenvector $c|0\rangle + s|1\rangle$, where c and s

are scalar coefficients and $|c|^2 + |s|^2 = 1$ is the normalization condition. Then the eigenvalue equation for $\hat{U}$ is

$$\hat{U}(c|0\rangle + s|1\rangle) = \frac{1}{2}(\hat{A}^\dagger + \hat{A})(c|0\rangle + s|1\rangle) = \lambda(c|0\rangle + s|1\rangle), \qquad (4.37)$$

where λ is an eigenvalue. This leads to

$$\frac{1}{2}(\hat{A}^\dagger + \hat{A})(c|0\rangle + s|1\rangle) = \frac{1}{2}(c|1\rangle + s|0\rangle) = \lambda(c|0\rangle + s|1\rangle). \qquad (4.38)$$

Equating the coefficients of $|0\rangle$ and separately of $|1\rangle$ then gives $c = \pm s$,

$$|c| = |s| = \frac{1}{\sqrt{2}},$$

and $\lambda = \pm\frac{1}{2}$. Then we can take the eigenvectors of $\hat{U}$ as

$$\frac{1}{\sqrt{2}}(|0\rangle \pm |1\rangle).$$

Similarly, for $\hat{V}$ one finds

$$\hat{V}(c|0\rangle + s|1\rangle) = \frac{i}{2}(\hat{A}^\dagger - \hat{A})(c|0\rangle + s|1\rangle) = \lambda(c|0\rangle + s|1\rangle), \qquad (4.39)$$

which leads to, $c = \pm is$,

$$|c| = |s| = \frac{1}{\sqrt{2}},$$

and $\lambda = \pm\frac{1}{2}$. We can take the eigenvectors for $\hat{V}$ as

$$\frac{1}{\sqrt{2}}(|0\rangle \pm i|1\rangle).$$

All three Hermitian operators, $\hat{U}$, $\hat{V}$ and $\hat{W}$ have eigenvalues of $\pm\frac{1}{2}$. Although $\hat{U}$ and $\hat{V}$ have eigenvalues and eigenvectors in this fermionic case, it is impossible to find a solution to the eigenvalue equation

$$\hat{A}(c|0\rangle + s|1\rangle) = \alpha(c|0\rangle + s|1\rangle)$$

where α is a number, apart from the trivial solution given by $s = \alpha = 0$. Thus we conclude that the fermionic annihilation operator, unlike its bosonic counterpart, has no non-trivial eigenstate.

As we have seen, the single category fermionic operators only require a two-dimensional Fock space to operate on, comprising unit basis vectors $|0\rangle$ and $|1\rangle$. This makes the representation of all of the operators we have met in this section take the form of 2×2 matrices. Indeed, it is a simple matter to adapt the matrix representation of the single category bosonic system we met in Chapter 3. We just take the first two rows and columns in each case, then we have

$$N_{nm} = \begin{pmatrix} 0 & 0 \\ 0 & 1 \end{pmatrix}, \tag{4.40}$$

$$A_{nm} = \begin{pmatrix} 0 & 1 \\ 0 & 0 \end{pmatrix}, \tag{4.41}$$

and

$$A_{nm}^\dagger = \begin{pmatrix} 0 & 0 \\ 1 & 0 \end{pmatrix}. \tag{4.42}$$

The state vectors, $|n\rangle$ are represented by two-dimensional column vectors of unit length as

$$|0\rangle = \begin{pmatrix} 1 \\ 0 \end{pmatrix}, |1\rangle = \begin{pmatrix} 0 \\ 1 \end{pmatrix}. \tag{4.43}$$

We can also use Eqs. (4.40) to (4.42) to construct 2×2 matrix representations of $\hat{U}$, $\hat{V}$ and $\hat{W}$, with the aid of Eqs. (4.27) and (4.29), then

$$U_{nm} = \frac{1}{2}(A_{nm}^\dagger + A_{nm}) = \frac{1}{2}\begin{pmatrix} 0 & 1 \\ 1 & 0 \end{pmatrix}, \tag{4.44}$$

$$V_{nm} = \frac{i}{2}(A_{nm}^\dagger - A_{nm}) = \frac{1}{2}\begin{pmatrix} 0 & -i \\ i & 0 \end{pmatrix} \tag{4.45}$$

and

$$W_{nm} = \frac{1}{2} - N_{nm} = \frac{1}{2}\begin{pmatrix} 1 & 0 \\ 0 & -1 \end{pmatrix}. \tag{4.46}$$

Now as with the bosonic case, the ξ dependence of $\hat{A}$ through $\hat{A}(\xi) = \exp(-i\xi)\hat{A}(0)$, leads to a phasor operator interpretation for the Hermitian components, $\hat{U}$ and $\hat{V}$. This again takes the form

$$\begin{pmatrix} \hat{U}(\xi) \\ \hat{V}(\xi) \end{pmatrix} = \begin{pmatrix} \cos\xi & \sin\xi \\ -\sin\xi & \cos\xi \end{pmatrix} \begin{pmatrix} \hat{U}(0) \\ \hat{V}(0) \end{pmatrix},$$

just as in the bosonic case. So in the fermionic case, ξ is again a rotation about the origin of the $(\hat{U},\hat{V})$, phasor space, where $\hat{U}$ and $\hat{V}$ are always $\frac{\pi}{2}$ out of phase and so can be thought of as being orthogonal to one another. The fermionic case differs from the bosonic one is that the is only one circle in $(\hat{U},\hat{V})$ space and that is defined by $\hat{U}^2 + \hat{V}^2 = \frac{1}{2}$.

In the bosonic case we essentially have just two phasors, namely the bosonic operators, $\hat{U}$ and $\hat{V}$ that we dealt with in Chapter 3. However in the fermionic case we have another related component, namely $\hat{W}$, but, unlike $\hat{U}$ and $\hat{V}$, this commutes with $\hat{N}$, and so

$$i\frac{d\hat{W}}{d\xi} = [\hat{W},\hat{N}] = 0, \tag{4.47}$$

which means that $\hat{W}$ is fixed as $\hat{U}$ and $\hat{V}$ rotate. If $\hat{W}$ does not rotate with $\hat{U}$ and $\hat{V}$, then we can take it as not being in the $(\hat{U},\hat{V})$ plane. This suggests we can view $\hat{W}$ as a component orthogonal in a phasor sense to $\hat{U}$ and $\hat{V}$.

So we have a triplet of mutually orthogonal components that are related through commutation in the form of a closed Lie group. This suggests that we can construct a three-dimensional vectorial operator, $\hat{\mathbf{S}} = (\hat{S}_1, \hat{S}_2, \hat{S}_3)$, where $\hat{S}_1 = \hat{U}$, $\hat{S}_2 = \hat{V}$ and $\hat{S}_3 = \hat{W}$. We can then construct the square modulus

$$\hat{\mathbf{S}}^2 = \hat{S}_1^2 + \hat{S}_2^2 + \hat{S}_3^2 = \frac{3}{4}. \tag{4.48}$$

Because $\hat{\mathbf{S}}^2$ is equal to a number, it must commute with all of the components of $\hat{\mathbf{S}}$, even though the components of $\hat{\mathbf{S}}$ do not commute with one another. Indeed, because of Eqs. (4.25) and (4.31), $\hat{S}_1^2$, $\hat{S}_2^2$ and $\hat{S}_3^2$, all individually commute with $\hat{S}_1$, $\hat{S}_2$ and $\hat{S}_3$. We shall meet this important fermionic Hermitian vector again, in some surprising contexts. We can anticipate what this might be by noting that the three components of $\hat{\mathbf{S}}$, which will specify as $\hat{S}_i$, where $i = 1, 2$, or 3 are

$$\hat{S}_i = \frac{1}{2}\hat{\sigma}_i, \tag{4.49}$$

where the three components $\hat{\sigma}_i$ are the three Pauli matrices. These are the three 2×2 matrices that appear in equations Eqs. (4.44) to (4.46). They are traditionally associated with the intrinsic spin of the electron and like the components $\hat{S}_i$ obey the mathematical rules for spinors. Notice however that here their properties have been derived solely from the properties of the fermionic creation and annihilation operators, without any reference to spin or rotation. So, in terms of $\hat{\mathbf{S}}$, we can interpret these results as $\hat{S}_1$ and $\hat{S}_2$ being a pair of orthogonal rotating components, that rotate about the axis defined by $\hat{S}_3$. It is important to note that this rotation involves the phase angle, ξ and is nothing to do with rotation in configuration space. We shall meet rotation in configuration space later with the emergence of angular momentum. As we shall see, there is a relationship, through symmetry, between the $\hat{S}_i$s and angular momentum. In this regard, the commutation relations between the components of $\hat{\mathbf{S}}$ play a crucial role. We can summarize these by noting that Eqs. (4.30), (4.34) and (4.35) can be written succinctly as

$$[\hat{S}_i, \hat{S}_j] = i\hat{S}_k, \tag{4.50}$$

where $i = 1, j = 2$ and $k = 3$ and their cyclic permutations. It is important to note that the spinor properties are associated with fermions but not at all with bosons.

It is important to recognize that the representation of the Pauli matrices by the specific 2×2 matrices in Eqs. (4.44) to (4.46) is to a degree arbitrary in the sense that we could exchange them by cyclically interchanging them as this would not change their interrelations through their mutual anti-commutators and commutators. This interchange preserves the *chirality*[2] of the set of triplet of Pauli matrices. As we shall

---

[2]This is just *handedness* in this case.

see, this allows a certain amount of flexibility when it comes to the representation of physical properties like electron spin that we will encounter later.

The key way that we have identified the importance of the three fermionic Hermitian components, $\hat{S}_i$, is that they form a closed Lie group. We will encounter a similar three component closed group of bosonic Hermitian operators later, that have a close connection to rotation, in the form of angular momentum. The two types of three component Lie groups can ultimately be brought together in relativistic wave mechanics. We will see this later when special relativity eventually emerges from natural number dynamics.

4.5 LINEARIZATION OF SQUARE ROOTS

The three 2×2 Pauli matrices that appear in Eqs. (4.44)-(4.46) represent an important triplet of operators that have some interesting algebraic properties that account for their ubiquity throughout quantum physics. These properties may be summarized as

$$\{\hat{\sigma}_i, \hat{\sigma}_j\} = 2\delta_{ij}, \tag{4.51}$$

where i and j can take on values of any of 1, 2 or 3, and

$$[\hat{\sigma}_i, \hat{\sigma}_j] = 2i\hat{\sigma}_k, \tag{4.52}$$

where $i = 1$, $j = 2$ and $k = 3$ or any cyclic permutation of these.

The property in Eq. (4.51) is particularly useful in linearizing square roots of moduli that takes on a meaning of profound importance in quantum mechanics. Take for example

$$\sqrt{p^2 + q^2},$$

where p and q are a pair of variables. We note that

$$(p\hat{\sigma}_i + q\hat{\sigma}_j)^2 = p^2 \hat{\sigma}_i^2 + q^2 \hat{\sigma}_j^2 + pq(\hat{\sigma}_i\hat{\sigma}_j + \hat{\sigma}_j\hat{\sigma}_i) = p^2 + q^2, \tag{4.53}$$

as long as $i \neq j$. Eq. (4.53) then implies that

$$\sqrt{p^2 + q^2} = p\hat{\sigma}_i + q\hat{\sigma}_j. \tag{4.54}$$

Thus we have succeeded in algebraically linearizing the square-root. This algebraic property of spinors plays a key role in situations as various as the Bardeen-Cooper-Schrieffer (BCS) theory of superconductivity [111], Dirac's relativistic equation for the electron [118] and Klein tunnelling in graphene [103].

5 The single-category system: The emergence of quantum mechanics

1 THE SCHRÖDINGER EQUATION

this chapter we are going to explicitly identify the parameter, t, from Chapter 4 as ontinuous variable that corresponds to time in a physical system. This will enable to show how natural number dynamics can be converted into Hamiltonian form d thus develop dynamical equations that can be applied to physical systems, that rm the basis for standard quantum mechanics. It is instructive to begin with a single tegory system with n items. We will take the system to be bosonic so that n can any natural number. We then simply parameterize the variable ξ with t, as we d for the multi-category case, by writing $\xi = \omega_0 t$, where ω_0, is a constant angular equency. Eq. (3.14) can then be transformed into a time evolution S-type equation plied to an eigenfunction, Φ_n of the operator $\hat{N}$. Comparing the result with Eq. .9), one then gets

$$i\frac{d\Phi_n}{dt} = \hat{\Omega}\Phi_n = \omega_0 \hat{N}\Phi_n, \tag{5.1}$$

d so

$$\hat{\Omega} = \omega_0 \hat{N} = \omega_0 \hat{A}^\dagger \hat{A}.$$

A complementary H-type equation to Eq. (5.1) is then

$$i\frac{d\hat{A}}{dt} = [\hat{A}, \hat{\Omega}] = \omega_0[\hat{A}, \hat{A}^\dagger \hat{A}] = \omega_0 \hat{A}, \tag{5.2}$$

nich is consistent with $\hat{A}(t) = \hat{A}(0)\exp(-i\omega_0 t)$.

Eq. (5.1) is in frequency units so it is a simple matter to convert it to energy units multiplying by the *reduced Planck constant*[1], $\hbar$. We can then use $\hbar\hat{\Omega}$ as our energy erator for the single category system. The operator for energy is generally termed e *Hamiltonian* operator, and is given the symbol, $\hat{H}$. Thus we define

$$\hat{H} := \hbar\hat{\Omega} = \hbar\omega_0\hat{N} = \hbar\omega_0\hat{A}^\dagger \hat{A}. \tag{5.3}$$

The S-type equation with the Hamiltonian is now the *Schrödinger equation*, nich is

$$i\hbar\frac{d\Phi_n}{dt} = \hat{H}\Phi_n. \tag{5.4}$$

---

[1] $\hbar = \frac{h}{2\pi} = 1.05457266 \times 10^{-34}$ Js, where h is the Planck constant. When $\hbar$ is treated as a conversion tor, its value may be obtained using a variety of experiments. The most straightforward are those olving the photoelectric effect or Compton scattering [3].

Notice that, as has been stressed in previous examples of S-type equations, Eq. (5.4) is not a definition of $\hat{H}$, i.e., $i\hbar\frac{d}{dt} :\neq \hat{H}$, rather $\hat{H}$ is defined Eq. (5.3).

The complementary H-type equation for a time-dependent operator, like $\hat{A}$ is

$$i\hbar\frac{d\hat{A}}{dt} = [\hat{A}, \hat{H}]. \tag{5.5}$$

The H-type equation, Eq. (5.5), in this case, is called the *Heisenberg equation of motion*.

Applying the Hamiltonian in Eq. (5.3) to the Fock state $|n\rangle$ gives

$$\hat{H}|n\rangle = \hbar\omega_0\hat{N}|n\rangle = \hbar\omega_0 n|n\rangle. \tag{5.6}$$

$\hat{H}$ in Eq. (5.6) serves as a perfectly good Hamiltonian for an n-body system. The total energy of an n-body system is $\hbar\omega_0 n$, which is interpreted as a collection of n particles each with energy $\hbar\omega_0$. The *bodies* in this system are often thought of as particles in physical systems. However, Eq. (5.6) is not the conventional one for an n-particle bosonic system. We will see why in the next section.

5.2 THE EMERGENCE OF CONFIGURATION SPACE

The conventional bosonic n-item Hamiltonian, is obtained as follows. The natural number amplitude operator, $\hat{A}$ and its adjoint for bosonic systems obey the commutation relation as in Eq. (3.17). Again we can represent $\hat{A}$ in terms of Hermitian operators, $\hat{U}$ and $\hat{V}$, as in Eq. (3.38). However, here we identify this pair of Hermitian operators respectively with the operators representing, the scaled co-ordinate, $\hat{X}$, of a one-dimensional *configuration space* and the complementary scaled *linear momentum*, $\hat{P}$. We will show shortly that this interpretation of $\hat{X}$ and $\hat{P}$ is justified. For the moment, then, instead of Eq. (3.38), we write

$$\hat{A} = \frac{1}{\sqrt{2}}(\hat{X} + i\hat{P}) \text{ and } \hat{A}^\dagger = \frac{1}{\sqrt{2}}(\hat{X} - i\hat{P}). \tag{5.7}$$

The canonical commutation relation, Eq. (3.39) then becomes

$$[\hat{P}, \hat{X}] = -i. \tag{5.8}$$

The Hamiltonian would then be

$$\hat{H} = \hbar\omega_0\hat{N} = \hbar\omega_0\frac{1}{2}(\hat{P}^2 + \hat{X}^2 - 1).$$

However, we can see that $\hbar\omega_0\frac{1}{2}(\hat{P}^2 + \hat{X}^2)$ represents the energy of a classical harmonic oscillator in scaled[2] operator form. So, if we redefine the Hamiltonian by adding $\frac{1}{2}\hbar\omega_0$, i.e., $\hat{H} + \frac{1}{2}\hbar\omega_0 \rightarrow \hat{H}$, then we can write

$$\hat{H} = \hbar\omega_0(\frac{1}{2}\hat{P}^2 + \frac{1}{2}\hat{X}^2) = \hbar\omega_0(\hat{N} + \frac{1}{2}), \tag{5.9}$$

---

[2]By *scaled* is meant that $\hat{X}$ and $\hat{P}$ are in dimensionless form, for the moment.

d the Hamiltonian now represents the harmonic oscillator, but also retains its in-
rpretation as the energy of a system n bosonic items. This, rather than the Hamil-
nian in Eq. (5.6) is the conventional representation of a system of n-bosonic par-
les. This identification with the harmonic oscillator has important consequences.
 means that the energy of a system with no particles in not zero. When $n = 0$, we
d that $\hat{H}|0\rangle = \frac{1}{2}\hbar\omega_0|0\rangle$, so the system has an energy of $\frac{1}{2}\hbar\omega_0$. This is sometimes
ferred to as the energy of the vacuum. It is also referred to as the *zero point* energy.

It is important to emphasize here that the zero point energy, $\frac{1}{2}\hbar\omega_0$, of a system
 n particles mentioned in Section 5.1 is contingent on their being associated with
e harmonic oscillator. Using natural number dynamics, does not force us into this
terpretation, since we begin with n items, not the harmonic oscillator. Conventional
antum mechanics begins with the oscillator and discovers n particles and so is
rced into the zero point energy interpretation.

It is the fact that an n-item bosonic system can be associated with a harmonic
cillator that allows a completely different interpretation. To see this, we begin by
rning the Hamiltonian in Eq. (5.9) into a more dimensionally appropriate form by
aling the co-ordinate and momentum operators as follows. We let

$$\hat{P} = \hat{p}/(\hbar\sqrt{m\omega_0}) \text{ and } \hat{X} = \hat{x}\sqrt{m\omega_0}, \tag{5.10}$$

here m represents the mass of the oscillator, then Hamiltonian then takes the form

$$\hat{H} = (\frac{\hat{p}^2}{2m} + \frac{1}{2}m\omega_0^2\hat{x}^2). \tag{5.11}$$

e recognize the first term on the rhs of Eq. (5.11) as the kinetic energy, where $\hat{p}$
 the linear momentum operator and the second one as the potential energy for the
rmonic oscillator, where $\hat{x}$ is the operator that represents a co-ordinate of configu-
tion space. The canonical commutation relation, Eq. (5.8) then becomes

$$[\hat{p}, \hat{x}] = -i\hbar, \tag{5.12}$$

hich is the conventional quantization condition between the operator of the x-
mponent of linear momentum, $\hat{p}$ and the co-ordinate, $\hat{x}$. We can check that is inter-
etation is correct by evaluating the time derivatives with the Heisenberg equation
 motion and get

$$\frac{d\hat{x}}{dt} = -\frac{i}{\hbar}[\hat{x}, \hat{H}] = \frac{\hat{p}}{m}, \tag{5.13}$$

d

$$\frac{d\hat{p}}{dt} = -\frac{i}{\hbar}[\hat{p}, \hat{H}] = -m\omega_0^2\hat{x}. \tag{5.14}$$

. (5.13) is the expected definition of momentum and Eq. (5.14) gives the restoring
rce on the simple harmonic oscillator. The simple harmonic nature of Eqs. (5.13)
d (5.14) can be seen by substituting Eq. (5.14) into Eq. (5.13) and vice versa to
ve

$$\frac{d^2\hat{x}}{dt^2} + \omega_0^2\hat{x} = \frac{d^2\hat{p}}{dt^2} + \omega_0^2\hat{p} = 0,$$

which shows that the quantum simple harmonic oscillator obeys the same equation of motion as its classical counterpart and that ω_0 is the oscillator frequency[3]. This is clearly a specific, time-dependent, example of the harmonic oscillator in Eqs. (3.44).

Eq. (5.12) also implies that we can represent $\hat{x}$ by a scalar variable x and then

$$\hat{p} := -i\hbar \frac{\partial}{\partial x}, \tag{5.15}$$

as is conventionally assumed in standard quantum theory, although here no assumption is involved, since the definition emerges from the commutation relation between the number amplitude operators, as explained in Chapter 3. Now, operating with $\hat{H}$ on the eigenfunction, $\Phi_n(t,x)$ yields

$$\hat{H}\Phi_n(t,x) = i\hbar \frac{\partial}{\partial t}\Phi_n(t,x) = \left(-\frac{\hbar^2}{2m}\frac{\partial^2}{\partial x^2} + \frac{1}{2}m\omega_0^2 x^2\right)\Phi_n(t,x), \tag{5.16}$$

which is immediately recognizable as the Schrödinger equation for a single particle in a quadratic potential, in analytic representation. Eq. (5.16) is clearly closely related to the differential operator representation of $\hat{N}$ in Eq. (3.58), and the eigenfunctions of Eq. (5.16) have the same form as those in Eq. (3.60). The ground state eigenfunction, for example is

$$\Phi_0(x,t) = \left(\frac{m\omega_0}{\pi\hbar}\right)^{\frac{1}{4}} \exp\left(-\left(i\omega_0 t + \frac{m\omega_0 x^2}{2\hbar}\right)\right), \tag{5.17}$$

where the time dependence of the state-function has been included explicitly.

The energy eigenvalues, E_n, come directly from Eq. (5.6) and have the form

$$E_n = \hbar\omega_0(n + \frac{1}{2}). \tag{5.18}$$

Here we interpret n, not as a number of individual particles but as the ordinal number of a mode in a ladder of energy levels. These energy levels can be thought of as the energy states of a single particle, which itself can be thought of as the oscillator. So, the n items themselves are no longer to be thought of as particles. Indeed, here they are clearly not material objects at all, but rather are harmonics of the fundamental frequency, ω_0. We have to be very careful about meaning now. We have to remember that these results have come from natural number dynamics, which are not tied to specific physical systems from the outset. So, to be clear, the bosonic character of the system is not associated with the harmonic oscillator which is now, in this new interpretation, a single particle, but rather with the number of harmonics, each with a label, n. This is rather like counting the notes in a musical scale (recall comments made in the prologue about this point). The single particle that can occupy any of the energy levels defined by Eq. (5.18) can be a boson or a fermion, since n is no longer counting particles. After all, in this simple case, the particle number is just one, which is allowed for both fermionic and bosonic particles. As we shall see later

---

[3] Strictly speaking ω_0 is and angular frequency.

Chapter 11, when relativistic phenomena emerge, the Hamiltonian in Eq. (5.16) at describes the behaviour of a particle in a quadratic potential is a non-relativistic proximation, as are the cases for more generalized potentials that are treated in the xt section.

From what we have learnt in this section, we can see that the quantum harmonic cillator sits at the junction between quantum many-body theory and quantum me-anics. There really is no substantial difference in the mathematical structure in the 'o interpretations, apart from the different conventions in the representation of the te, Φ_n. The difference is mostly in the eye of the beholder. However, because we d not start out with preconceived ideas, these two rather different interpretations n items have emerged in a natural and unified way. In the next section we look a generalization of the Hamiltonian to include more general potentials than that the quantum harmonic oscillator. This comes about by considering the operator $= f(\hat{N}) = \hat{B}^\dagger \hat{B}$ that was introduced in Section 3.8. This will also provide stronger idence for the interpretation of the parameter t as time.

3 GENERALIZING THE HAMILTONIAN

1e Hamiltonian can be generalized by utilizing a combination of Eqs. (3.141) and .142), which we can write in the form

$$i\frac{d\Phi_n}{d\tau} = \hat{F}\Phi_n = (-\frac{1}{2}\frac{\partial^2}{\partial w^2} + Y(w))\Phi_n = f(n)\Phi_n, \qquad (5.19)$$

1ere

$$\hat{F} = \hat{B}^\dagger \hat{B} = f(\hat{N}),$$

$$\hat{B} = \frac{1}{\sqrt{2}}(\frac{\partial}{\partial w} + W(w)),$$

$$2Y(w) = W^2(w) - W'(w). \qquad (5.20)$$

early, Eq. (5.19) reduces to the linear bosonic case that we treated in Section 5.1 1en $f(\hat{N}) = \hat{N}$.

Now we can follow a similar course to what was done in Section 5.1 and param-erize τ using t by putting $\tau = \omega_0 t$, $w = x$ and letting $\omega_0\hat{F} = \hat{\Omega}$. Then

$$i\frac{\partial\Phi_n}{\partial t} = \hat{\Omega}\Phi_n = \omega_0(-\frac{1}{2}\frac{\partial^2}{\partial x^2} + Y(x))\Phi_n = \omega_0 f(n)\Phi_n \qquad (5.21)$$

It is now a simple step to generalize the Hamiltonian in Eq. (5.11) to the case presented by Eq. (5.21). We can put $\hat{H} = \hbar(\hat{\Omega} + \lambda)$ and $V(x) = \hbar(Y(x) + \lambda)$ to eld

$$\hat{H} := (\frac{\hat{p}^2}{2m} + V(x)) = (-\frac{\hbar^2}{2m}\frac{\partial^2}{\partial x^2} + V(x)),$$

1ere λ is just a fixed scalar, whose purpose will be discussed shortly and we can ll have $\hat{p} := -i\hbar\frac{\partial}{\partial x}$, as in Eq. (5.11). Notice that Φ_n is still an eigenfunction of $\hat{N}$) and $\hat{H}$, so

$$\hat{H}\Phi_n = \hbar\omega(n)\Phi_n = E(n)\Phi_n, \qquad (5.22)$$

where $\omega(n) = \omega_0 f(n) + \lambda$ and $E(n)$ is the energy eigenvalue of the system. Then we can take $\hat{H}$ as the Hamiltonian of a system with a general one-dimensional *scalar potential*, $V(x)$, and so the Schrödinger equation for such a one-dimensional system is

$$i\hbar \frac{\partial}{\partial t} \Phi_n(t,x) = \hat{H}\Phi_n(t,x) = (-\frac{\hbar^2}{2m}\frac{\partial^2}{\partial x^2} + V(x))\Phi_n(t,x). \qquad (5.23)$$

Notice that λ plays an important role here, since, if λ were set to zero we would be forced to take the energy level for $n = 0$ to be zero, which is generally not the case. This situation is analogous to the zero point energy in the case of the quantum harmonic oscillator. It is also interesting to note that even in the case of the general potential, $V(x)$, $\hat{N}$ necessarily commutes with $\hat{H}$ and so is constant in time. So, the system eigenfunction keeps the same label n, in time and we can regard this state as a *stationary state*.

We can now check that x and $\hat{p}$ represent a co-ordinate and linear momentum, respectively, by using the H-type equation to evaluate their time derivatives. We get

$$\frac{dx}{dt} = -\frac{i}{\hbar}[x,\hat{H}] = \frac{\hat{p}}{m}, \qquad (5.24)$$

and

$$\frac{d\hat{p}}{dt} = -\frac{i}{\hbar}[\hat{p},\hat{H}] = -\frac{\partial V(x)}{\partial x}. \qquad (5.25)$$

Eqs. (5.24) and (5.25) can be recognized as a scaled form of Ehrenfest's theorem [126] that represents Newton's second law in operator form, which reduce to Eqs. (5.13) and (5.14) for the case of the simple harmonic oscillator. These results show clearly that $\hat{H}$ plays the role of the Hamiltonian operator of the system and that x and $\hat{p}$ are indeed the co-ordinate and linear momentum of the system. This also makes it clear that t is exactly what is expected, if we interpret it as time.

So, starting with the operator for the natural numbers, operators that mimic the behaviour of the energy, $\hat{H}$, linear momentum, $\hat{p}$ and a coordinate, $\hat{x}$, of configuration space have emerged quite naturally, and the relationships between them, agree with equations that govern the behaviour of physical systems, under the influence of potentials, at least in one dimension of configuration space. There is one spatial dimension because, so far, we have only treated single category systems that have a single number operator. We will need to examine multi-category systems to see the emergence of systems with more than one dimension of configuration space. This will be done in the ensuing chapters.

It is worth pointing out that the derivation of the Schrödinger equation, Eq. (5.23), which takes us from the natural numbers to a general potential, is a kind of back-to-front use of the factorization that leads to a method of solving the Schrödinger equation for certain types of potential that was developed by Schrödinger [112, 26]. A well-known example occurs for, (a) $W(x) = \alpha\tan(\alpha x)$ or, (b) $W(x) = \beta\cot(\beta x)$, in Eqs. (5.20), where α and β are constants. These forms are associated with the case of a particle in a one-dimensional infinite square well. To see this, let's take (a) as an example. This solution is associated with $Y(w)$ is a constant in Eq. (5.19), but within a restricted range of x. So, we solve

$$W^2(x) - W'(x) = -\alpha^2.$$

This can be integrated to give $W(x) = \alpha \tan(\alpha x)$, as required in case (a). Then

$$Y(x) = \frac{\alpha^2}{2}(\tan^2(\alpha x) - \sec^2(\alpha x)) = -\frac{\alpha^2}{2},$$

so if we make $\lambda = \frac{\alpha^2}{2}$, then $V(x) = \hbar(Y(x) + \lambda) = 0$. However, $\tan(\alpha x)$ becomes infinite at $\alpha x = \pm\frac{\pi}{2}$, which limits $\hat{B}$ and therefore $\hat{H}$ to the range of values of x between $\pm\frac{\pi}{2\alpha}$. Hence, this represents the infinite square well, with a zero of potential inside the well, as is usually the case in the standard example [95]. The corresponding ground state is then found by solving

$$\hat{B}\Phi_0 = \frac{1}{\sqrt{2}}(\frac{\partial\Phi_0}{\partial x} + \alpha \tan(\alpha x)\Phi_0) = 0.$$

This differential equation has an unnormalized solution, $\Phi_0 = \cos(\alpha x)$, which is well-known as the form of the ground state eigenfunction of the infinite square well, with nodes at $\pm\frac{\pi}{2\alpha}$. This and the second form of $W(x)$ then lead to a set of intertwined eigenfunctions of alternating even (cosine) and odd (sine) symmetry and the resulting bound state energy levels, $E(n)$, are proportional to $(n+1)^2$, with $n = 0, 1, 2, \ldots$. The relationship between n and the energy levels in this case is illustrated in Fig. 5.1[4] This case is treated in detail in ref. [61], which also treats the finite square well.

Here, the interest is not in the factorization technique *per se*, as a rather, it must be admitted, complicated way of getting eigenfunctions that can, in the case of the infinite square well, at least, be got much more simply and directly by solving the differential equation that constitutes the Schrödinger equation. Rather, the importance of the nonlinear factorization is to show that there is a legitimate reason to generalize the potential from the case where $\hat{N} = \hat{A}^\dagger\hat{A}$ that leads to a quadratic potential, to that for which $\hat{B}^\dagger\hat{B} = f(\hat{N})$, from which the more general potential, $V(x)$, emerges quite naturally. In other words, general potentials are rooted in the natural number operator too. The physical origin of the potential will be discussed later in the context of many-particle systems with interactions (see Chapters 6 and 9).

It would be interesting to know, given a particular functional relationship, $\hat{B}^\dagger\hat{B} = f(\hat{N})$, what form the potential $V(x)$ would take. Of course, given a trapping potential, it is in principle possible to find the n-dependent energy levels of the resulting bound states, using the Schrödinger equation. In practice only a small number of exact solutions is known, such as the harmonic oscillator potential and the infinite square well, but approximate methods are available in other cases.

As far as the author is aware, there is no systematic way of starting with the n-dependence of the energy levels and arriving at the form of the potential via the

---

[4]In standard, introductory texts [3, 22, 95, 118], the energy levels E_n are proportional to n^2, but with $n = 1, 2, 3, \ldots$. There is a slight semantic advantage of starting with $n = 0$, since the eigenfunctions, which, as is well-known, alternate between cosine and sine functions, are even for even n and odd for odd n, whereas in the standard form, the even functions have odd n and vice versa.

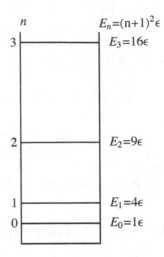

Figure 5.1 The energy levels for a particle in an infinite one-dimensional square well in scaled form. The even numbered levels are associated with $W(x) = \alpha \tan(\alpha x)$ in Eqs. (5.20) and the odd numbered levels occur when $W(x) = \beta \cot(\beta x)$.

Schrödinger equation. However, one can get an idea about this by using the virial theorem. Then, for a potential of the form $V = |x|^q$, where x is a displacement from a mean position and q is a real number, one finds that the energy levels, $E(n)$ are related approximately to the natural numbers by $E(n) \sim \varepsilon(n+\delta)^{\left(\frac{2q}{q+2}\right)}$, where ε and δ are constants, and so $\hat{H} \sim \hat{N}^{\left(\frac{2q}{q+2}\right)}$. This implies that, given a relationship of the form, $\hat{H} = \hat{N}^p$, then this is expected to come from a potential of the form $V \sim |x|^{\frac{2p}{(2-p)}}$. Details of how this result is obtained can be found in Appendix B. Remarkably, this simple relationship between $E(n)$ and V agrees with several well-known cases treated in standard quantum mechanics, including the harmonic oscillator, a particle in an infinite square well, the Coulomb potential, a free particle and quark interactions in a nucleus (see Appendix B). However, we want to stress that what is important here is not so much the relationship between a specific trapping potential and the n-dependence of the energy levels but rather that there is a systematic relationship at all.

Finally, in this section, recall that in Chapter 3, it was pointed out that there was a trivial form of Eq. (5.19) with $Y(w) = 0$ that lead to the eigenvalues of $\hat{F}$ being independent of n. Similarly here, there is a consistent trivial system in which $V(x) = 0$ which leads to

$$\hat{H} := \frac{1}{2m}\hat{p}^2. \tag{5.26}$$

The eigenvalues of Eq. (5.26) are also independent of n, as might be expected. This system, of course, corresponds to a free particle without a trapping potential. Despite being independent of n, there is still a constraint on the eigenvalues of energy since Eq. (5.26) is equivalent to

$$i\hbar\frac{\partial\Phi}{\partial t} = \frac{1}{2m}\hat{p}^2\Phi = -\frac{\hbar^2}{2m}\frac{\partial^2\Phi}{\partial x^2}. \qquad (5.27)$$

Eq. (5.27) admits a plane wave solution of the form $\exp(i(kx - \omega t))$. Then we can write the eigenvalues for the lhs of Eq. (5.27) as $\hbar\omega$, where the angular frequency ω can take on any real value. The eigenvalues for the rhs are $\frac{\hbar^2 k^2}{2m}$, where the wave number, k, can take on any real value, just like the linear momentum eigenvalue, $p = \hbar k$. However, in order to satisfy Eq. (5.27), ω is constrained to positive values such that

$$\hbar\omega = \frac{\hbar^2 k^2}{2m},$$

so, even though k itself can have any real value, positive or negative, ω must be positive and so, even in the case of a free particle, its energy must be bound from below[5]. Here, the wavelength,

$$\lambda = \frac{2\pi}{k} = \frac{h}{p},$$

is identified with the *de Broglie wavelength*.

5.4 EIGENFUNCTIONS, EXPECTATION VALUES AND x-p DUALITY

The properties of the operators, $\hat{x}$ and $\hat{p}$ and the relationship between them may be summarized by noting that they are identical in structure to $\hat{U}$ and $\hat{V}$ that were dealt with in Chapter 3. Indeed we have obtained the Schrödinger equation above by simply substituting x for u and p for $\hbar v$. Then we can simply take the u-v relationships from Chapter 3 as the x-p relationships here, apart from the factor $\hbar$, which presents no problem. So, for example, the eigenfunctions of the operator form, $\hat{x}$ and $\hat{p} = -i\frac{\partial}{\partial x}$ are repectively, $\delta(x - x_0)$ and $\exp(ikx)$, where $k = \frac{p}{\hbar}$. Then

$$\hat{x}\delta(x - x_0) = x_0\delta(x - x_0) \text{ and } -i\hbar\frac{\partial\exp(ikx)}{\partial x} = \hbar k\exp(ikx), \qquad (5.28)$$

so that x_0 and $\hbar k = p$ are the respective eigenvalues, bearing in mind the mathematical drawbacks these eigenfunctions have, with regard to their lack of normalizability, as mentioned in Chapter 3.

In general, $\hat{x}$ and $\hat{p}$ do not commute with the Hamiltonian, $\hat{H}$, and so the eigenfunctions of $\hat{x}$ and $\hat{p}$ are not in general eigenfunctions of $\hat{H}$. It is common practice to use the eigenfunctions of energy in x representation to represent the states of one-dimensional systems. This is because information about the energy of a system is

---

[5] As we shall see later, this restriction needs reassessing when we encounter relativistic behaviour.

commonly the best way to characterize it. So, a general state is then usually represented by a linear superposition of energy eigenstates. If we suppose that such a state has a normalized state function $\Psi(x)$, then the expectation value of a general operator $\hat{O}(\hat{x}, \hat{p})$ which depends on the operators $\hat{x} = x$ and $\hat{p} = -i\frac{\partial}{\partial x}$ is just

$$\langle \hat{O}(\hat{x}, \hat{p}) \rangle = \int \Psi^*(x) \hat{O}(\hat{x}, \hat{p}) \Psi(x) dx. \tag{5.29}$$

Note that just as in u-representation in Chapter 3, here the product $\Psi^*(x)\Psi(x)$ serves as probability density associated with the probability of finding the system at x. Recall that there was a u-v duality identified in Chapter 3 which showed that the system could be equivalently represented with a state function of either u or v and that these two state functions formed a Fourier pair. It is therefore obvious that we can use a state function $\Phi(p)$ to represent the system where $\Phi(p)$ is a Fourier transform of $\Psi(x)$. It is convenient to use $k = \frac{p}{\hbar}$ in this relationship so that by direct comparison with Eqs. (3.72) and (3.73), then

$$\Psi(x) = \frac{1}{\sqrt{2\pi}} \int_{-\infty}^{\infty} \Phi(k) \exp(ikx) dk \tag{5.30}$$

and

$$\Phi(k) = \frac{1}{\sqrt{2\pi}} \int_{-\infty}^{\infty} \Psi(x) \exp(-ikx) dx. \tag{5.31}$$

The consequence of the Fourier pair relationship between $\Psi(x)$ and $\Phi(p)$ is that there is a x-p duality for any one-dimensional system. This means that a system that is defined as a distribution in configuration space, x, by $\Psi(x)$ can be also represented in momentum space, p by $\Phi(k)$. As an example, suppose

$$\Psi(x) = \begin{cases} 1 & \text{if } a \leq x \leq a+\Delta \\ 0 & \text{otherwise.} \end{cases} \tag{5.32}$$

The complementary p-representation is then

$$\Phi(k) = \frac{1}{\sqrt{2\pi}} \int_{a}^{a+\Delta} \exp(-ikx) dx = \exp(ika) \frac{\exp(ik\Delta) - 1}{i\sqrt{2\pi}k}, \tag{5.33}$$

from which the p-probability density is

$$\Phi^*(k)\Phi(k) = 2 \frac{\sin^2(\frac{k\Delta}{2})}{\pi k^2}. \tag{5.34}$$

We can see that while the probability density $\Psi^*(x)\Psi(x)$ indicates that the system is confined to a region of width Δ along the x-axis, the momentum values are largely confined to the central maximum of $\Phi^*(k)\Phi(k)$ (see panel (a) in Fig. 5.2), which has a width of $\frac{4\pi}{\Delta}$ along the k-axis. So, the broader the confinement region in x, the smaller the possible range of k values.

Suppose now, that there are two regions of uniform probability in x-representation, one as in the above example and a second of equal amplitude to the first, but in the second region between $-(a+\Delta)$ and $-a$. The probability amplitude in p-representation is then given by

$$\Phi(k) = \frac{1}{\sqrt{2\pi}} \left(\int_{a}^{a+\Delta} \exp(-ikx)\mathrm{d}x + \int_{-a-\Delta}^{-a} \exp(-ikx)\mathrm{d}x \right)$$

(5.35)

$$= 4\cos(k(a+\frac{\Delta}{2})) \frac{\sin(\frac{k\Delta}{2})}{\sqrt{2\pi}k},$$

resulting in a probability density in p-representation of

$$\Phi^*(k)\Phi(k) = 8\cos^2(k(a+\frac{\Delta}{2})) \frac{\sin^2(\frac{k\Delta}{2})}{\pi k^2}.$$

(5.36)

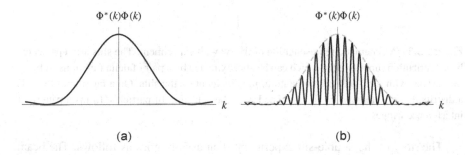

(a)   (b)

Figure 5.2   The shape of the probability density, $\Phi^*(k)\Phi(k)$ associated with the results (a) from Eq. (5.34) and (b) from Eq. (5.36) for $\Delta = 0.2a$.

The functional shapes of $\Phi^*(k)\Phi(k)$ for the two cases above are displayed in Fig. 5.2. The envelopes of the two cases (the solid curve in panel (a) and the dashed curve in panel (b)) are the same. However, when there are two separate regions of confinement in x, it can be seen that the momentum distribution becomes periodic, so that the system is most likely to have momenta with values located where the maxima in $\Phi^*(k)\Phi(k)$ occur. This behaviour has important implications for the so-called *wave-particle duality paradox* which causes such confusion in standard quantum mechanics [34]. We will examine this issue next.

5.5   THE DOUBLE-SLIT EXPERIMENT

The so-called wave-particle duality paradox has caused a lot of confusion in quantum theory, throughout its history. In many popular science books that deal with quantum ideas, it is used as an example of how weird quantum mechanics is supposed to be. The archetypal example of this paradox is the double-slit experiment. It is usually described in terms of a beam of particles of a well-defined momentum approaching a pair of slits. Detectors placed on the far side of the slits then reveal an apparent interference pattern in the form of intensity fringes indicating where the particles have impacted. This is illustrated in Fig. 5.3. Even Feynman [34], in spite of the clarity of much of his popular exposition of quantum phenomena, has used it as an example to underpin the idea that quantum mechanics is difficult to understand, at least as far as its interpretation goes.

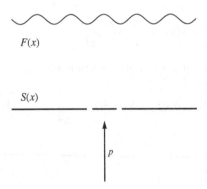

Figure 5.3 A schematic representation of the two-slit experiment. The slits are represented by a distribution function $S(x)$ which can be approximated by a pair of delta functions. A beam of particles with well-defined momentum, p, is incident on the slits. On a far side of the slits a distribution of intensity, $F(x)$, associated with the detection of particles, forms a pattern of interference fringes.

The story of the double-slit experiment then usually goes as follows. The beam of particles, and these are usually identified as photons or electrons, although any type of particle will do, as long as they have a well defined momentum, so that the particle beam can be thought of as a plane wave with a well defined wave number and phase fronts parallel to the line of the slits. Then, according to the Huygens-Fresnel principle [13, 73, 44], the plane wave fronts excite secondary wavelets with circular phase fronts at each slit. The phase fronts then impinge on a screen on the other side of the slits, where they interfere through wave superposition and form the set of interference fringes. Then, as the story continues, the intensity of the beam is slowly reduced. At first this just leads to the intensity of the bright fringes being reduced, but eventually, when only a few particles are incident on the slits, a striking change occurs in what is observed. The particles appear singly within the fringe pattern. It is just that they are most likely to appear where the bright fringes were detected when the beam was intense. If the particle detection is integrated over time, then the accumulation of particles reconstitutes the original fringe pattern.

These observations are beyond dispute. What is disputable is the usual interpretation. This is that the particles arrive at the slits in a state that is described as a wave, even if there is only one particle. This plane wave is often identified with the wave function of the particle state with a well defined momentum. Then, it is argued, even if there is only one particle it somehow interferes with itself so that it is guided to impact the detecting screen where the bright fringes are, or, even more ambiguously, the incident plane wave (function) turns into a particle on the other side of the slits, when it impacts the detecting screen. It is this version of the story that gives rise to the wave-particle paradox. The discussion becomes even more clouded when, as in some texts, the situation is compared to bullets passing through a pair of holes in a screen, when, it is argued no interference pattern is seen in the distribution of the bullets on the far side and hence they obey classical rules that differ dramatically from the predictions of quantum theory. Clearly, bullets are not particles in the sense that photons and electrons are. They are at best complex amalgamations of the order of 10^{23} particles in the quantum sense, held together by electromagnetic interactions that are themselves associated with photon exchange. We will examine the bullet analogy at the end of this section, but first we will look at the double-slit experiment in a different way from the wave-particle paradox point of view.

Suppose a one-dimensional configuration space distribution is characterized by $\Psi(x) = \delta(x-a) + \delta(x+a)$. This is obviously not a normalized distribution, in fact it is just the eigenfunction for the system being confined at $x = a$, added to the eigenfunction of a system confined at $x = -a$. These two options are equally likely when $\Psi(x) = \delta(x-a) + \delta(x+a)$. Now we can equally well represent this system in momentum space by

$$\Phi(k) = \frac{1}{\sqrt{2\pi}} \int\limits_{-\infty}^{\infty} (\delta(x-a) + \delta(x+a)) \exp(-ikx) \mathrm{d}x = \sqrt{\frac{2}{\pi}} \cos(ka), \qquad (5.37)$$

which is just the Fourier transform of $\Psi(x)$. This means that the momentum probability distribution density associated with the x-component of momentum is $\Phi^*(k)\Phi(k) = \frac{2}{\pi} \cos^2(ka)$. This probability density has peaks at

$$k = \frac{n\pi}{a}, \qquad (5.38)$$

where $n = 0, \pm 1, \pm 2, \ldots$. Thus the resulting momentum state $\Delta p = \frac{\hbar n \pi}{a}$, in the x-direction. This means that the system is equally likely to have a momentum in the x-direction of $\Delta p = 0, \pm \frac{\hbar \pi}{a}, \pm \frac{2\hbar \pi}{a}, \pm \frac{3\hbar \pi}{a}, \ldots$. This result has a profound implication for the interpretation of the so called *double-slit* experiment as represented in Fig. 5.4. We can interpret $\Psi(x) = \delta(x-a) + \delta(x+a)$ as the probability of where a particle can penetrate an otherwise impenetrable barrier along the x-axis, as illustrated in panel (a) of Fig. 5.4. Here $S(x)$ is the x-distribution of the slits and is represented by the delta functions[6] in $\Psi(x)$. The state of momentum of the incident particles is depicted in panel (a) by the plane wave. This is monochromatic and so represents a single value of momentum. On the far side of the slits the state of the system is governed

---

[6]Strictly speaking $S(x)$ is proportional to $|\Psi(x)|^2$, but this is essentially the same distribution and $\Psi(x)$.

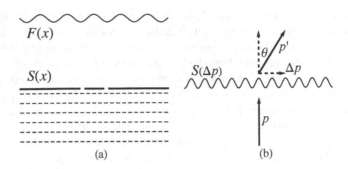

Figure 5.4 The double slit experiment, (a) in configuration space, (b) in momentum space.

by the interference fringes that result from the superposition of the secondary waves that emerge from the site of the slits.

The equivalent momentum space representation of the double-slit experiment is depicted in panel (b) of Fig. 5.4. It is important to understand that the momentum state refers to the system as a whole, that is the particle and the slits together, as a single system, and not just the incident particle. The state of momentum of the system on the side of the incident particles is represented by the arrow, p, and on the far side of the slits by p'. Here, p' is constructed from a component, p, normal to the line of the slits and a component Δp parallel to the line of the slits that is acquired from the momentum distribution associated with the slits, which is determined by Eq. (5.37). A further important point is that there is no time dependence involved, so there is no energy change in the system. This means that, $p^2 = p'^2$, and so the length of p' is the same as the length of p. Then, all that is involved is a deflection through an angle θ that depends on Δp. The sinusoid, $S(p)$ in panel (b) is proportional to $\Phi^*(k)\Phi(k)$ and hence to $\cos^2(ka)$. Since the most likely values that Δp can take will be those associated with peaks in $\cos^2(ka)$, then we expect the most likely places to find the particle on the far side of the slits is where

$$\sin\theta = \frac{\Delta p}{p'} = \frac{\Delta p}{p} = \frac{n\hbar\pi}{pa} = \frac{n\lambda}{2a}, \qquad (5.39)$$

where $\lambda = \frac{2\pi\hbar}{p}$ is the de Broglie wavelength associated with the particle momentum, p. Eq. (5.39) indicates that the deflection caused by the momentum space representation of the slits is identical to the location of intensity maxima that would be expected from a constructive interference calculation using the Huygens-Fresnel principle. In fact one can see from Eq. (5.39) an important result. The interference fringes map exactly to the momentum pattern of the slits, which shows that the interference pattern is just the Fourier transform of the configuration space pattern of the slits.

So, there is no wave-particle paradox here. We are not forced to assume that waves are incident on the screen in order to explain the interference pattern on the detector side, and then say that these waves miraculously turn into particles as the hit the detecting screen. We simply have two complementary representations of the experiment, one in which wave interference by the slit distribution in configuration space gives rise to an intensity pattern of detected particles and another in which a momentum-space distribution associated with the slits gives rises to deflections that send particles into positions at the detector which exactly match the wave interference pattern from the configuration space picture. This is no more weird than a signal being represented in the frequency domain or equivalently in the time domain. No magic or quantum weirdness is involved, just a Fourier transform! The quantum picture is clear and unambiguous. It is not quantum mechanics that is at fault here, it is the fact that we really do not know what photons and electrons are, which is a different matter entirely. Not only that, we cannot be sure that the particle that is incident on the slits is the same one that emerges. This screen with which the particles interact is full of electrons and photons[7]. All we know is that one enters and one emerges.

In practice, slits have a non-zero width, then the momentum space representation takes the form of the probability distribution in Eq. (5.36), where Δ is the width of each slit in configuration space. Then the momentum space representation of the slits $S(\Delta p)$ has the form in panel (a) of Fig 5.5, which is a sinusoid inside a sinc function envelope. Panel(b) of Fig. 5.5 shows the momentum space distribution when one slit is closed, then the interference pattern disappears, as is well known.

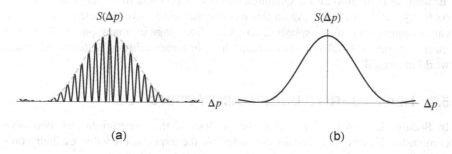

(a)   (b)

Figure 5.5 Panel (a) The momentum probability function, $S(\Delta p)$, for the two slit case with nonzero slit widths. Panel (b) illustrates what happens when only one slit is open.

The above interpretation is not new. It is discussed in detail in the book by Landé [74]. What is surprising is that it is virtually unknown in modern introductory texts on quantum mechanics. It is as if there is some desire to preserve the reputation of a weird quantum theory. The interpretation was in fact suggested by Duane [27] as early as 1923, well before Heisenberg and Schrödinger formulated their versions of quantum mechanics. Thus, we have seen the wave-particle duality paradox, which has caused so much controversy among physicists with regard to the interpretation of quantum mechanics, can be resolved by regarding a system as being represented in two mathematically equivalent ways. So, x-p duality interpreted simply as a pair

---

[7]That the material of the screen is full of electrons is obvious. That the screen is full of photons may be less obvious. However, it should be recalled that the atoms in the screen are held together by electromagnetic (Coulomb) forces that are carried by photons [135]. In addition, the incident photons cause electron accelerations that create new photons.

of equivalent mathematical representations of a system avoids the need to treat a part of a system, namely the particle, as, all on its own, being regarded as both a particle and a wave.

Finally, just a further comment on the comparison with bullets, mentioned earlier. Even if there were a fundamental particle with the mass of a bullet, say around 0.01 kg, travelling at say 1000 ms^{-1}, so that the momentum was 10 kg ms^{-1}, then the de Broglie wavelength of such a particle would be of the order of 10^{-34} m. Eq. (5.39) indicates that to get a detectable deflection one would need slits separated by no more than around 1000 wavelength, i.e., about 10^{-31} m. This is seventeen orders of magnitude smaller than the nucleus. If we were actually dealing with bullets the slits separation would have to be a few bullet diameters apart, say 0.1 m, then the deflection angles would be, by Eq. (5.39), separated by the order of 10^{-33} radians. At a screen 1 m from the slits this would mean the fringes would be separated by 10^{-33} m. This is nineteen orders of magnitude smaller than a nucleus. What is more, the slits would have to be of finite width to allow a bullet through. Suppose this were 0.01m. Eq. (5.36) then indicates that only 20 fringes would be visible under the sinc function envelope, which would spread a distance of 10^{-32} m across the detecting screen and thus be totally undetectable. This is tantamount to saying that a bullet passing though one of a pair of holes in a screen would suffer no quantum mechanical deflection at all! Quantum mechanics predicts the behaviour of bullets perfectly well in this case. Again this leads to no weird wave-particle paradox. One can imagine the headline, splashed across the front page of a newspaper: 'Quantum mechanics predicts bullet passes through hole in screen without deflection. Physics world in turmoil'.

5.6 HEISENBERG'S UNCERTAINTY PRINCIPLE

In Section 2.5 it was shown that the product of the uncertainties in two non-commuting Hermitian operators depended on the expectation value of their commutator as in Eq. (2.44). In the case of $\hat{x}$ and $\hat{p}$ then we expect

$$\Delta_x \Delta_p \geq \frac{1}{2}|\langle \Psi |[\hat{x}, \hat{p}]\Psi\rangle|. \tag{5.40}$$

So, in the present case, with Eq. (5.12), Eq. (5.40) implies

$$\Delta_x \Delta_p \geq \frac{\hbar}{2}, \tag{5.41}$$

or equivalently

$$\Delta_x \Delta_k \geq \frac{1}{2}. \tag{5.42}$$

As is well known, Eq. (5.41) is a mathematical statement of the *Heisenberg uncertainty principle*. It has profound implications for measurement with regard to quantum phenomena. It implies that, for a system in a state described by a particular wave function, a measurement of the x co-ordinate and a measurement of the corresponding linear momentum component cannot have unlimited accuracy. Indeed the product of the uncertainties in these two parameters must obey the inequality in Eq. (5.41). It is also interesting to note that the uncertainty principle in its x-k form in Eq. (5.42) can be obtained from the Fourier duality of the functions in Eqs. (5.30) and (5.31), without recourse to the commutation relation between x and $\hat{p}$. However, in the context of the development of quantum physics in the present book it should be remembered that the Heisenberg uncertainty principle follows from the commutation relation between the number operator amplitudes that were introduced in Chapter 3 and is clearly a special case of the u-v uncertainty relation discussed in Section 3.6.4.

It is instructive to evaluate the mutual x-p uncertainties for some specific cases. Let us first look at the case of the harmonic oscillator. This is very straightforward in the Fock state representation.

5.6.1 UNCERTAINTY IN FOCK SPACE

By combining Eqs. (5.7) and (5.10) the we can write $\hat{x}$ and $\hat{p}$ in terms of $\hat{A}$ and $\hat{A}^\dagger$ as

$$\hat{x} = \frac{1}{\sqrt{2m\omega_0}}(\hat{A}^\dagger + \hat{A}) \text{ and } \hat{p} = \frac{i\hbar}{\sqrt{m\omega_0}}(\hat{A}^\dagger - \hat{A}), \tag{5.43}$$

then one finds that $\langle \hat{x} \rangle = \langle n|\hat{x}|n \rangle = 0$ and $\langle \hat{p} \rangle = \langle n|\hat{p}|n \rangle = 0$. We also have

$$\langle \hat{x}^2 \rangle = \langle n|\hat{x}^2|n \rangle = \frac{1}{m\omega_0}\left(n + \frac{1}{2}\right) \tag{5.44}$$

and

$$\langle \hat{p}^2 \rangle = \langle n|\hat{p}^2|n \rangle = \frac{\hbar^2 m\omega_0}{2}\left(n + \frac{1}{2}\right). \tag{5.45}$$

Now, because $\langle \hat{x} \rangle = \langle \hat{p} \rangle = 0$, then $\Delta_x^2 = \langle \hat{x}^2 \rangle$ and $\Delta_p^2 = \langle \hat{p}^2 \rangle$ so

$$\Delta_x \Delta_p = \hbar\left(n + \frac{1}{2}\right). \tag{5.46}$$

The result in Eq. (5.46) clearly satisfies the Heisenberg uncertainty relation in Eq. (5.41) for all the values of the natural number, n, including zero.

5.6.2 UNCERTAINTY IN ANALYTIC REPRESENTATION

The probability amplitude for the ground state of the harmonic oscillator from Section 5.1 is Gaussian in form in x. We can write it as

$$\Phi_0(x) = \left(\frac{1}{\pi\sigma^2}\right)^{\frac{1}{4}} \exp\left(-\frac{x^2}{2\sigma^2}\right), \tag{5.47}$$

where $\sigma = \sqrt{\frac{\hbar}{m\omega_0}}$ is the Gaussian width. Now, because the Gaussian is symmetrical about $x = 0$, then we must have $\langle x \rangle = 0$, so the variance is just $\langle x^2 \rangle$. This is obtained from

$$\Delta_x^2 = \langle x^2 \rangle = \int_{-\infty}^{\infty} \Phi_0^*(x)x^2\Phi_0(x)dx = \frac{\sigma^2}{2}. \tag{5.48}$$

Taking the Fourier transform of $\Phi_0(x)$ in Eq. (5.47) yields

$$\Xi(k) = \left(\frac{\sigma^2}{\pi}\right)^{\frac{1}{4}} \exp\left(-\frac{k^2\sigma^2}{2}\right), \tag{5.49}$$

which is also a Gaussian but now with a width represented by σ^{-1}. Because of symmetry then $\langle k \rangle = 0$ and so

$$\Delta_k^2 = \langle k^2 \rangle = \int_{-\infty}^{\infty} \Xi(k)^* k^2 \Xi(k) dk = \frac{1}{2\sigma^2}. \tag{5.50}$$

Thus we find

$$\Delta_x \Delta_k = \frac{1}{2}, \tag{5.51}$$

or eqivalently

$$\Delta_x \Delta_p = \frac{\hbar}{2}, \tag{5.52}$$

which agrees with the result obtained using Fock state representation, in Eq. (5.46), for the case when $n = 0$. It is interesting to note that, for the harmonic oscillator in its ground state $(n = 0)$, the Heisenberg uncertainty principle gives the minimum possible product of the uncertainties in the position and momentum variables of the system.

We can compare this with the example treated in Section 5.4. Referring then to $\Psi(x)$ in Eq. (5.32), $\Psi^*(x)\Psi(x)$ is integrable and so $\Psi(x)$ can be normalized. The normalizing factor for $\Psi(x)$ is $\Delta^{-\frac{1}{2}}$. Now

$$\langle x \rangle = \frac{1}{\Delta} \int_{a}^{a+\Delta} x dx = a + \frac{\Delta}{2} \tag{5.53}$$

and

$$\langle x^2 \rangle = \frac{1}{\Delta} \int_{a}^{a+\Delta} x^2 dx = a^2 + a\Delta + \frac{\Delta^2}{3}. \tag{5.54}$$

Then

$$\Delta_x = \sqrt{\langle x^2 \rangle - \langle x \rangle^2} = \frac{\Delta}{\sqrt{12}}. \tag{5.55}$$

In order to calculate the corresponding values of $\langle k \rangle$ and $\langle k^2 \rangle$, $\Phi(k)$ in Eq. (5.33) needs to be normalized. $\Phi^*(k)\Phi(k)$ is integrable and involves a standard integral i.e.

$$\int_{-\infty}^{\infty} \Phi^*(k)\Phi(k) dk = \int_{-\infty}^{\infty} 2\frac{\sin^2(\frac{k\Delta}{2})}{\pi k^2} dk = \Delta, \tag{5.56}$$

so the normalization constant for $\Phi(k)$ is $\Delta^{-\frac{1}{2}}$. Now because $\Phi^*(k)\Phi(k)$ is symmetrical about $k = 0$, then $\langle k \rangle = 0$ so $\Delta_k = \langle k^2 \rangle$. However, we find

$$\langle k^2 \rangle = \frac{1}{\Delta} \int\limits_{-\infty}^{\infty} \Phi^*(k)k^2\Phi(k)dk = \frac{2}{\Delta\pi} \int\limits_{-\infty}^{\infty} \sin^2(\frac{k\Delta}{2})dk, \tag{5.57}$$

but this final integral is infinite, so Δ_k is strictly infinite in this case. So, we end up with the product $\Delta_x\Delta_k = \infty$, which although is strictly correct and obeys the Heisenberg uncertainty principle since $\infty > \frac{1}{2}$, it is not a very useful result. In cases like this another estimate of uncertainty may be used. We may take the half-widths of the two probability densities involved as estimates of the uncertainties. For the x-representation this is just $\Delta_x \sim \frac{\Delta}{2}$. For the p-distribution, the bulk of the probability density lies within the region defined by the first two zeros of $\frac{\sin^2(\frac{k\Delta}{2})}{k^2}$ outside of $k = 0$. These are at $\frac{k\Delta}{2} = \pm\pi$, so we can take $\Delta_k \sim \frac{2\pi}{\Delta}$. Then we have

$$\Delta_x\Delta_k \sim \pi,$$

or

$$\Delta_x\Delta_p \sim \pi\hbar,$$

which is well within the limits set by the Heisenberg uncertainty principle.

5.7 $\hat{H}$-t DUALITY?

Just as there is a Fourier duality between x and k, there is also Fourier duality between, time, t and frequency, ω, that is well-known in signal processing [77]. Using standard Fourier integral methods it is possible to show that the product of the frequency bandwidth, Δ_ω of a signal, and its temporal length, Δ_t, must satisfy the inequality

$$\Delta_\omega\Delta_t \geq \frac{1}{2}. \tag{5.58}$$

Just as in the x-k case, the minimum product occurs when the signal has a Gaussian envelope, which means it has a Gaussian spectrum.

The widths Δ_ω and Δ_t in Eq. (5.58) may be viewed respectively as uncertainties in ω and t. If we then use the quantum mechanical energy-frequency relation, $E = \hbar\omega$, then we can infer from Eq. (5.58) an energy-time uncertainty relation of the form

$$\Delta_E\Delta_t \geq \frac{\hbar}{2}. \tag{5.59}$$

Now since the x-p uncertainty relation can be inferred either from Fourier duality or from the commutation relation between x and $\hat{p}$, then it is plausible that one could define a time operator, $\hat{t}$, by a commutation relation of the form

$$[\hat{H},\hat{t}] = i\hbar \ ? \tag{5.60}$$

Just as in the case of a suggested commutation relation between the number operator and a proposed phase operator that we examined in Section 3.7, this possibility fails too and for the same reasons that $\hat{N}$-ξ failed. To see this, consider a system with a simple eigenvalue energy equation of the form

$$\hat{H}\Psi = E\Psi \tag{5.61}$$

Now, the $\hat{H}$-$\hat{t}$ commutator implies

$$[\hat{H}, f(\hat{t})] = i\hbar f'(\hat{t}), \tag{5.62}$$

which means that, for any real number, γ,

$$[\hat{H}, \exp(i\gamma\hat{t})] = -\hbar\gamma\exp(i\gamma\hat{t}). \tag{5.63}$$

With the aid of Eqs. (5.61) and (5.63), we can write

$$\hat{H}\exp(i\gamma\hat{t})\Phi = (\exp(i\gamma\hat{t})(\hat{H} - \hbar\gamma)\Phi) = (E - \hbar\gamma)\exp(i\gamma\hat{t})\Phi,$$

which would imply that there is an energy state, $\exp(i\gamma t)\Phi$, with an energy eigenvalue, $E - \hbar\gamma$, that is unbound from below, which is not permissible. So, we can rule out a time operator based on $[\hat{H}, \hat{t}] = i\hbar$. The above argument was originated by Pauli [3, 91]. We can further add that the supposed analogy between $\hat{H}$-$\hat{t}$ duality and $\hat{p}$-x duality is also incorrect, since $\hat{p}$ is the operator $-i\hbar\frac{\partial}{\partial x}$, whereas $\hat{H}$ is not equivalent to $i\hbar\frac{\partial}{\partial t}$, as was previously pointed out. We also note that it is also fruitless to construct a time operator on the basis of

$$[\frac{\partial}{\partial t}, t] = 1,$$

since, as in the similar procedure in Section 3.7 for attempting to construct a phase operator, the above relation just produces a new lowering operator

$$\hat{A}_t = \frac{1}{\sqrt{2}}(t + \frac{\partial}{\partial t}).$$

The new number operator, $\hat{N}_t = \hat{A}_t^\dagger\hat{A}_t$, that results, leads to the conclusion that t can no longer mean time.

So, far the quantum mechanics that has emerged involves only one dimension of configuration space and one component of linear momentum. To see how more dimensions emerge, we will need to consider systems with more than one category. We will examine two-category systems next.

6 Two-category systems

6.1 A FIRST LOOK AT INTERACTIONS

6.1.1 THE BOSONIC CASE

Having established the framework for a system which comprises an arbitrary number of categories in Chapter 4, it turns out to be quite instructive to examine the properties of a system with just two categories. What we learn with two we can generalize to more categories later. Studying two-category systems also serves a second purpose in that it leads to systems with two dimensions of configuration space, as we shall see in Chapter 7.

Consider then a bosonic system with a total number of items represented by an operator $\hat{N}$, divided into two categories, Cat_a and Cat_b, with corresponding number operators, $\hat{N}_a$ and $\hat{N}_b$, such that

$$\hat{N} = \hat{N}_a + \hat{N}_b, \tag{6.1}$$

where, $\hat{A}_a^{\dagger}\hat{A}_a = \hat{N}_a$ and $\hat{A}_b^{\dagger}\hat{A}_b = \hat{N}_b$. Using the multi-category result from Chapter 4, but with $S = 2$, the Hamiltonian for the whole system will take the form

$$\hat{\Omega} = \omega_a \hat{A}_a^{\dagger}\hat{A}_a + \omega_b \hat{A}_b^{\dagger}\hat{A}_b = \omega_a \hat{N}_a + \omega_b \hat{N}_b, \tag{6.2}$$

where ω_a and ω_b are constant scalar frequencies. The appropriate commutation relations are $[\hat{A}_a, \hat{A}_a^{\dagger}] = 1$ and $[\hat{A}_b, \hat{A}_b^{\dagger}] = 1$, with all other operator pair combinations in commutator brackets being zero. It is then straightforward to check that

$$[\hat{N}_a, \hat{\Omega}] = [\hat{N}_b, \hat{\Omega}] = 0. \tag{6.3}$$

So, both $\hat{N}_a$ and $\hat{N}_b$ individually commute with $\hat{\Omega}$, which means that the individual number of items in each category, as well as the total number of items will remain constant with time, as a result of the Heisenberg equation of motion, i.e.

$$i\frac{d\hat{N}_a}{dt} = [\hat{N}_a, \hat{\Omega}] = 0, \tag{6.4}$$

with a corresponding equation for $\hat{N}_b$. An operator which commutes with $\hat{\Omega}$ is said to be a *constant of the motion*. We can also use the Heisenberg equation to show that $\hat{A}_a(t) = \hat{A}_a(0)\exp(-i\omega_a t)$ and $\hat{A}_b(t) = \hat{A}_b(0)\exp(-i\omega_b t)$, which is what we would get for each category if it were a separate system with $\hat{\Omega}_a = \omega_a \hat{N}_a$ for Cat_a and $\hat{\Omega}_b = \omega_b \hat{N}_b$ for Cat_b. So, the two categories can be treated totally separately, which is what is really meant by their being independent of one another, even though we could formally treat the two categories as a single system by $\hat{\Omega} = \hat{\Omega}_a + \hat{\Omega}_b$.

The state of this system may be specified in the occupation number notation, by $|\Psi_N\rangle = |n_a, n_b\rangle$. The set of two-category states satisfies the orthogonality condition

$$\langle m_a, m_b | n_a, n_b \rangle = \delta_{n_a m_a} \delta_{n_b m_b}. \tag{6.5}$$

DOI: 10.1201/9781003377504-6

Then

$$\hat{N}_a|n_a,n_b\rangle = n_a|n_a,n_b\rangle \text{ and } \hat{N}_b|n_a,n_b\rangle = n_b|n_a,n_b\rangle,$$

$$\hat{A}_a|n_a,n_b\rangle = \sqrt{n_a}|n_a-1,n_b\rangle \text{ and } \hat{A}_a^\dagger|n_a,n_b\rangle = \sqrt{n_a+1}|n_a+1,n_b\rangle$$

and

$$\hat{A}_b|n_a,n_b\rangle = \sqrt{n_b}|n_a,n_b-1\rangle \text{ and } \hat{A}_b^\dagger|n_a,n_b\rangle = \sqrt{n_b+1}|n_a,n_b+1\rangle. \qquad (6.6)$$

As it stands this is not a particularly interesting system. Nothing much happens to it as time progresses. The two populations remain fixed in their two categories. Next we want to show that the representation of the system, in terms of $\hat{N}_a$ and $\hat{N}_b$ and their corresponding creation and annihilation operators, is not unique. We can apply a linear transformation to $\hat{A}_a$ and $\hat{A}_b$ to produce a new pair, $\hat{A}_c$ and $\hat{A}_d$, by

$$\begin{pmatrix} \hat{A}_c \\ \hat{A}_d \end{pmatrix} = \begin{pmatrix} \varepsilon & \eta \\ \lambda & \mu \end{pmatrix} \begin{pmatrix} \hat{A}_a \\ \hat{A}_b \end{pmatrix}, \qquad (6.7)$$

where ε, η, λ and μ are constant coefficients, which we will take as real numbers for the moment. Then it is straightforward to show that, $[\hat{A}_c,\hat{A}_c^\dagger] = \varepsilon^2 + \eta^2$, $[\hat{A}_d,\hat{A}_d^\dagger] = \lambda^2 + \mu^2$ and $[\hat{A}_c,\hat{A}_d^\dagger] = \varepsilon\lambda + \eta\mu$. At this stage the constants are arbitrary, but with an appropriate choice of the constants, we can make, $[\hat{A}_c,\hat{A}_c^\dagger] = 1$, $[\hat{A}_d,\hat{A}_d^\dagger] = 1$ and $[\hat{A}_c,\hat{A}_d^\dagger] = 0$, so that $\hat{A}_c$ and $\hat{A}_d$ act as annihilation operators and their respective adjoints as creation operators, obeying bosonic commutation rules. Then we find

$$\begin{pmatrix} \varepsilon & \eta \\ \lambda & \mu \end{pmatrix} = \begin{pmatrix} \cos\alpha & \sin\alpha \\ -\sin\alpha & \cos\alpha \end{pmatrix} = \hat{R}(\alpha), \qquad (6.8)$$

where the matrix (operator), $\hat{R}(\alpha)$, can be recognized as corresponding to a rotation through an angle α, which at this stage is arbitrary. It also follows that,

$$\hat{A}_c^\dagger\hat{A}_c = \hat{A}_a^\dagger\hat{A}_a \cos^2\alpha + \hat{A}_b^\dagger\hat{A}_b \sin^2\alpha + (\hat{A}_a^\dagger\hat{A}_b + \hat{A}_b^\dagger\hat{A}_a)\cos\alpha\sin\alpha \qquad (6.9)$$

and

$$\hat{A}_d^\dagger\hat{A}_d = \hat{A}_a^\dagger\hat{A}_a \sin^2\alpha + \hat{A}_b^\dagger\hat{A}_b \cos^2\alpha - (\hat{A}_a^\dagger\hat{A}_b + \hat{A}_b^\dagger\hat{A}_a)\cos\alpha\sin\alpha, \qquad (6.10)$$

where α is an arbitrary angle. Then[1] adding Eqs. (6.9) and (6.10) yields

$$\hat{N} = \hat{A}_a^\dagger\hat{A}_a + \hat{A}_b^\dagger\hat{A}_b = \hat{A}_c^\dagger\hat{A}_c + \hat{A}_d^\dagger\hat{A}_d \qquad (6.11)$$

and we can represent the system by two new categories, Cat_c and Cat_d with population numbers, $\hat{A}_c^\dagger\hat{A}_c = \hat{N}_c$ and $\hat{A}_d^\dagger\hat{A}_d = \hat{N}_d$. Thus

$$\hat{N} = \hat{N}_a + \hat{N}_b = \hat{N}_c + \hat{N}_d. \qquad (6.12)$$

---

[1]This result is not surprising since it is a form of the well known Bogoliubov transformation that is widely used in many-body quantum physics [121].

So, we have a new pair of categories whose population numbers sum to the same value as our original pair. It important to emphasize that we are looking at the same system in two different representations. One in which the system is divided between Cat_a and Cat_b and another description represented by Cat_c and Cat_d. The descriptions are not independent of one another since they are connected by the linear transformation given by the matrix in Eq. (6.8). One can think of Cat_c as containing a proportion of Cat_a *and* Cat_b; similarly with Cat_d. However, the big difference between the two representations lies with the Hamiltonian in terms of the new categories. To see this we can reverse the rotation, $\hat{R}(\alpha)$, in Eq. (6.8), i.e. we apply a rotation $\hat{R}(-\alpha)$ as follows[2]

$$\begin{pmatrix} \hat{A}_a \\ \hat{A}_b \end{pmatrix} = \begin{pmatrix} \cos\alpha & -\sin\alpha \\ \sin\alpha & \cos\alpha \end{pmatrix} \begin{pmatrix} \hat{A}_c \\ \hat{A}_d \end{pmatrix}. \tag{6.13}$$

We can substitute these relations into Eq. (6.2) and get

$$\hat{\Omega} = \omega_c \hat{A}_c^\dagger \hat{A}_c + \omega_d \hat{A}_d^\dagger \hat{A}_d + \upsilon_{cd}(\hat{A}_c^\dagger \hat{A}_d + \hat{A}_d^\dagger \hat{A}_c), \tag{6.14}$$

where,

$$\omega_c = \omega_a \cos^2\alpha + \omega_b \sin^2\alpha,$$

$$\omega_d = \omega_b \cos^2\alpha + \omega_a \sin^2\alpha,$$

$$\upsilon_{cd} = (\omega_b - \omega_a)\cos\alpha \sin\alpha. \tag{6.15}$$

The trigonometrical functions of α may actually be eliminated from the relations in Eq. (6.15). After some surprisingly tedious algebra one finds

$$\cos\alpha = \sqrt{\frac{1}{2}\left(1 + \frac{\omega_c - \omega_d}{\sqrt{(\omega_c - \omega_d)^2 + 4\upsilon_{cd}^2}}\right)} \tag{6.16}$$

and then

$$\omega_a = \frac{\omega_c + \omega_d + \sqrt{(\omega_c - \omega_d)^2 + 4\upsilon_{cd}^2}}{2},$$

$$\omega_b = \frac{\omega_c + \omega_d - \sqrt{(\omega_c - \omega_d)^2 + 4\upsilon_{cd}^2}}{2}. \tag{6.17}$$

The significance of the relations in Eq. (6.17) will become apparent shortly when we deal with the time dependence of the system.

Again, it is important to realize that the Hamiltonian in Eq. (6.14) represents the same system as that in Eq. (6.2); it is just written in a different basis. The first two terms on the rhs of Eq. (6.14) contain number operators multiplied by frequencies, like those in Eq. (6.2). It is the third term on the rhs of Eq. (6.14) that makes the real difference between the two representations. Notice that the form in Eq. (6.14) is still Hermitian, as it should be, since $(\hat{A}_c^\dagger \hat{A}_d + \hat{A}_d^\dagger \hat{A}_c)^\dagger = (\hat{A}_d^\dagger \hat{A}_c + \hat{A}_c^\dagger \hat{A}_d)$.

---

[2]Notice that $\hat{R}(-\alpha) = \hat{R}(\alpha)^{-1}$.

Now the crucial new thing to notice about the Hamiltonian written in terms of $\hat{A}_c$ and $\hat{A}_d$ is that $\hat{N}_c$ and $\hat{N}_d$ do not commute with $\hat{\Omega}$. Indeed we find for example

$$[\hat{N}_c, \hat{\Omega}] = \upsilon_{cd}(\hat{A}_c^\dagger \hat{A}_d - \hat{A}_d^\dagger \hat{A}_c),$$

and so $\hat{N}_c$ is not a constant of the motion. It turns out that in general both $\hat{N}_c$ and $\hat{N}_d$ are time dependent. We shall explore this aspect of the system later in this chapter.

6.1.2   THE FERMIONIC CASE

The linear transformation in Eq. (6.8) that preserves bosonic commutation rules also preserves fermionic rules. If $\hat{C}_a$ and $\hat{C}_b$ are a pair of independent fermionic creation operators obeying the usual fermionic anti-commutation rules that are transformed to a new pair, $\hat{C}_c$ and $\hat{C}_d$ by

$$\begin{pmatrix} \hat{C}_c \\ \hat{C}_d \end{pmatrix} = \begin{pmatrix} \cos\alpha & \sin\alpha \\ -\sin\alpha & \cos\alpha \end{pmatrix} \begin{pmatrix} \hat{C}_a \\ \hat{C}_b \end{pmatrix}, \tag{6.18}$$

then

$$\hat{C}_c^\dagger \hat{C}_c = \cos^2\alpha \hat{C}_a^\dagger \hat{C}_a + \cos\alpha\sin\alpha(\hat{C}_a^\dagger \hat{C}_b + \hat{C}_b^\dagger \hat{C}_a) + \sin^2\alpha \hat{C}_b^\dagger \hat{C}_b \tag{6.19}$$

and

$$\hat{C}_c \hat{C}_c^\dagger = \cos^2\alpha \hat{C}_a \hat{C}_a^\dagger + \cos\alpha\sin\alpha(\hat{C}_a \hat{C}_b^\dagger + \hat{C}_b \hat{C}_a^\dagger) + \sin^2\alpha \hat{C}_b \hat{C}_b^\dagger. \tag{6.20}$$

Adding Eqs. (6.19) and (6.20) gives

$$\begin{aligned} \{\hat{C}_c, \hat{C}_c^\dagger\} &= \cos^2\alpha\{\hat{C}_a, \hat{C}_a^\dagger\} \\ &\quad + \cos\alpha\sin\alpha(\{\hat{C}_a, \hat{C}_b^\dagger\} + \{\hat{C}_b, \hat{C}_a^\dagger\}) + \sin^2\alpha\{\hat{C}_b, \hat{C}_b^\dagger\} \\ &= \cos^2\alpha + \sin^2\alpha = 1. \end{aligned} \tag{6.21}$$

Similarly one can show that $\{\hat{C}_d, \hat{C}_d^\dagger\} = 1$ and that all of the possible pair combinations in anti-commutator brackets involving $\hat{C}_c$, $\hat{C}_d$ and their respective adjoints are zero, confirming that $\hat{C}_c$ and $\hat{C}_d$ are fermionic operators.

Then one finds that for a system of fermions with just two categories with a total of items represented by $\hat{N}$, then

$$\hat{N} = \hat{C}_a^\dagger \hat{C}_a + \hat{C}_b^\dagger \hat{C}_b = \hat{C}_c^\dagger \hat{C}_c + \hat{C}_d^\dagger \hat{C}_d, \tag{6.22}$$

just as in the bosonic case. Also, we can transform the Hamiltonian too and get

$$\begin{aligned} \hat{\Omega} &= \omega_a \hat{C}_a^\dagger \hat{C}_a + \omega_b \hat{C}_b^\dagger \hat{C}_b \\ &= \omega_c \hat{C}_c^\dagger \hat{C}_c + \omega_d \hat{C}_d^\dagger \hat{C}_d + \upsilon_{cd}(\hat{C}_c^\dagger \hat{C}_d + \hat{C}_d^\dagger \hat{C}_c), \end{aligned} \tag{6.23}$$

where ω_c, ω_d and υ_{cd} are exactly the same as found in Eq. (6.15). The only differences between the bosonic and fermionic cases here are that, in the fermionic case, n_1 and n_2 can only take on the values 1 or 0, so that their sum can have the values 0, 1 or 2.

6.2 TIME DEPENDENCE

Here we treat in some detail the two-category bosonic case introduced in Subsection 6.1.1. As we noted earlier, neither $\hat{N}_c$ nor $\hat{N}_d$ are constants of the motion. This can be seen by noting that

$$i\frac{d\hat{N}_c}{dt} = [\hat{N}_c, \hat{\Omega}] = v_{cd}(\hat{A}_c^\dagger \hat{A}_d - \hat{A}_d^\dagger \hat{A}_c) \tag{6.24}$$

and

$$i\frac{d\hat{N}_d}{dt} = [\hat{N}_d, \hat{\Omega}] = v_{cd}(\hat{A}_d^\dagger \hat{A}_c - \hat{A}_c^\dagger \hat{A}_d). \tag{6.25}$$

Adding Eqs. (6.24) and (6.25) gives

$$i\frac{d(\hat{N}_c + \hat{N}_d)}{dt} = [\hat{N}_c + \hat{N}_d, \hat{\Omega}] = 0 \tag{6.26}$$

confirming that, even though $\hat{N}_c$ and $\hat{N}_d$ are not separately constants of the motion, their sum, which is of course equal to $\hat{N}$, is a constant of the motion, since it commutes with $\hat{\Omega}$.

Since the system with a total number of items represented by $\hat{N}$ can now be represented using the number operators $\hat{N}_c$ and $\hat{N}_d$ we would like to use a basis to represent it which is an eigenstate of $\hat{N}_c$ and $\hat{N}_d$. We could then consider a representation in occupation number form as, $\Phi_N = |n_c, n_d\rangle$. However, there is a technical problem here. We are using an H-type representation, with time-dependent operators and time-independent states. If $\hat{N}_c$ and $\hat{N}_d$ are time dependent, then, is some sense that we will need to explore, the expectation values of the numbers of items in the two categories will be expected to be time-dependent. To deal with this issue, we must choose a state Φ_N that is independent of time. This suggests we choose the values of n_c and n_d at a fixed time. It is most convenient to choose $t = 0$, so that [7], $\Phi_N = |n_c(0), n_d(0)\rangle$, where $n_a(0)$ represents the initial state of the items in Cat$_a$. The orthogonality condition is then

$$\langle m_c(0), m_d(0)|n_c(0), n_d(0)\rangle = \delta_{n_c(0)m_c(0)}\delta_{n_d(0)m_d(0)}. \tag{6.27}$$

Then we have

$$\hat{N}_c(0)|n_c(0), n_d(0)\rangle = n_c(0)|n_c(0), n_d(0)\rangle,$$

$$\hat{N}_d(0)|n_c(0), n_d(0)\rangle = n_d(0)|n_c(0), n_d(0)\rangle,$$

$$\hat{A}_c(0)|n_c(0), n_d(0)\rangle = \sqrt{n_c(0)}|n_c(0) - 1, n_d(0)\rangle,$$

$$\hat{A}_c^\dagger(0)|n_c(0), n_d(0)\rangle = \sqrt{n_c(0) + 1}|n_c(0) + 1, n_d(0)\rangle,$$

$$\hat{A}_d(0)|n_c(0), n_d(0)\rangle = \sqrt{n_d(0)}|n_c(0), n_d(0) - 1\rangle$$

and

$$\hat{A}_d(0)^\dagger|n_c(0), n_d(0)\rangle = \sqrt{n_d(0) + 1}|n_c(0), n_d(0) + 1\rangle. \tag{6.28}$$

In general, $|n_c(0), n_d(0)\rangle$ is not an eigenstate of $\hat{N}_c(t)$ and we cannot write a simple relation like those in Eq. (6.28) for $\hat{N}_c(t)|n_c(0), n_d(0)\rangle$. We shall investigate shortly how to calculate the expectation values of the numbers of items in the two categories.

In order to understand what $\hat{\Omega}$ in Eq. (6.14) means we need to examine the effect of the third term on the rhs on the state $\Psi_N = |n_c(0), n_{d(0)}\rangle$. First

$$\hat{A}_c^\dagger(0)\hat{A}_d(0)|n_c(0), n_d(0)\rangle = \sqrt{(n_c(0)+1)n_d(0)}|n_c(0)+1, n_d(0)-1\rangle, \qquad (6.29)$$

from which we can see that $\hat{A}_c^\dagger(0)\hat{A}_d(0)$ has the effect of removing an item from Cat_d and putting one into Cat_c. Now although we are dealing with what are traditionally called creation and annihilation operators, there is no reason to interpret the effect of these operators as creation and annihilation. We can just as well see the combination in Eq. (6.17) as the transfer of an item from one category to another. Then $\hat{A}_d^\dagger(0)\hat{A}_c(0)$ reverses the process, i.e.

$$\hat{A}_d^\dagger(0)\hat{A}_c(0)|n_c(0), n_d(0)\rangle = \sqrt{n_c(0)(n_d(0)+1)}|n_c(0)-1, n_d(0)+1\rangle. \qquad (6.30)$$

However, this does not in general result in zero effect, since $n_c(0)$ and $n_d(0)$ are in general different. This gives rise to an effective interaction between the two categories which has a profound effect on their behaviour.

6.2.1 EIGENFREQUENCIES FOR TWO-CATEGORY INTERACTIONS

It is often the case in quantum physics that one constructs the Hamiltonian for a system and then investigates its properties by finding the energy eigenvalues or equivalently the eigenfrequencies of the system. There are several ways of doing this. Suppose we were presented with a Hamiltonian in the form of an interacting pair of bosonic categories as in Eq. (6.14). The most direct way is to assume the operators $\hat{A}_c$ and $\hat{A}_b$ vary as $\exp(-i\omega t)$ and deduce the values the eigenfrequency, ω can take. We can use the Heisenberg equation of motion to find the eigenvalue equations for $\hat{A}_c$ and $\hat{A}_b$, which are

$$i\frac{d\hat{A}_c}{dt} = [\hat{A}_c, \hat{\Omega}] \implies \omega\hat{A}_c = \omega_c\hat{A}_c + \upsilon_{cd}\hat{A}_d, \qquad (6.31)$$

where we have used the Hamiltonian from Eq. (6.14). Similarly, one finds for $\hat{A}_d$.

$$\omega\hat{A}_d = \upsilon_{cd}\hat{A}_c + \omega_d\hat{A}_d, \qquad (6.32)$$

which leads to

$$\begin{pmatrix} \omega - \omega_c & -\upsilon_{cd} \\ -\upsilon_{cd} & \omega - \omega_d \end{pmatrix} \begin{pmatrix} \hat{A}_c \\ \hat{A}_d \end{pmatrix} = 0. \qquad (6.33)$$

This eigenvalue equation is satisfied if the determinant of the 2×2 matrix is zero, i.e.,

$$\begin{vmatrix} \omega - \omega_c & -\upsilon_{cd} \\ -\upsilon_{cd} & \omega - \omega_d \end{vmatrix} = (\omega - \omega_c)(\omega - \omega_d) - \upsilon_{cd}^2 = 0, \qquad (6.34)$$

which leads to

$$\omega = \frac{\omega_c + \omega_d \pm \sqrt{(\omega_c - \omega_d)^2 + 4v_{cd}^2}}{2}. \tag{6.35}$$

The behaviour of the coupled system can be evaluated by noting that the eigenfrequencies in Eq. (6.35) are precisely the frequencies associated with $\hat{A}_a$ and $\hat{A}_b$ in Eq. (6.17). However, the operators in the Hamiltonian that we are investigating involves, $\hat{A}_c$ and $\hat{A}_d$, operating on states, $|n_c, n_d\rangle$, but these are not the eigenstates of the system. This is clearly due to the interaction term which does not leave $|n_c, n_d\rangle$ unchanged. Indeed, the eigenstates are actually, $|n_a, n_b\rangle$ and the eigenvalues of $\hat{\Omega}$ are $\omega_a n_a + \omega_b n_b$. We can regard, $\hat{A}_a(t) = \hat{A}_a(0) \exp(-i\omega_a t)$ and $\hat{A}_b(t) = \hat{A}_b(0) \exp(-i\omega_b t)$ as normal modes of the system so that $\hat{A}_c(t)$ and $\hat{A}_d(t)$ will be linear combinations of the two normal modes. By applying the rotation, $R(\alpha)$, we can write this in explicit time dependent form as

$$\begin{pmatrix} \hat{A}_c(t) \\ \hat{A}_b(t) \end{pmatrix} = \begin{pmatrix} \cos\alpha & \sin\alpha \\ -\sin\alpha & \cos\alpha \end{pmatrix} \begin{pmatrix} \hat{A}_a(0) \exp(-i\omega_a t) \\ \hat{A}_b(0) \exp(-i\omega_b t) \end{pmatrix}. \tag{6.36}$$

It is worth pointing out here that one gets exactly the same eigenfrequencies if one uses the fermionic anti-commutation relations for the interacting categories case with the Hamiltonian in Eq. (6.23). In both the bosonic and fermionic cases, this behaviour is analogous to that of a pair of coupled oscillators in classical physics where the annihilation operators are behaving like the amplitudes of the oscillators. From that point of view, the creation and annihilation operators can really be considered as amplitudes of the number oscillator. When we have a single isolated category the sizes of these amplitudes do not change with time, but as we will see, when we have a pair of interacting categories in which items move from one category to another, then we have coupled oscillators the sizes of whose amplitudes do change with time.

In the next section we show how to use the relations between the two sets of annihilation operators to calculate the time varying number of items associated with Cat_c and Cat_d.

6.2.2 CALCULATING THE TIME DEPENDENT POPULATION NUMBERS

Having obtained expressions for the time dependent creation and annihilation operators the next step is to work out the time dependent number of items in the individual categories. This is essentially an initial value problem. We have assumed that initially (at $t = 0$) there are $n_c(0)$ items in Cat_c, and $n_d(0)$ items in Cat_d. Now recall that in the representation we are using that the operators are time dependent. In fact they are H-type operators and in this case the eigenstates of the system are independent of time. Now we know that

$$\hat{N}_c(0) = \hat{A}_c^\dagger(0)\hat{A}_c(0),$$

so, using the relation in Eqs. (6.28), we know

$$\hat{A}_c^\dagger(0)\hat{A}_c(0)|n_c(0), n_d(0)\rangle = n_c(0)|n_c(0), n_d(0)\rangle.$$

The next step is to try to evaluate

$$\hat{A}_c^\dagger(t)\hat{A}_c(t)|n_c(0),n_d(0)\rangle = \hat{N}_c(t)|n_c(0),n_d(0)\rangle.$$

Notice that at the moment we have $\hat{A}_c(t)$ only in terms of the time dependent operators for the normal modes of the system, Eq. (6.36). We need to get everything in terms of $\hat{A}_c(0)$ and $\hat{A}_d(0)$, since these give known results when they operate on $|n_c(0),n_d(0)\rangle$, using Eqs. (6.28). The next step is to get $\hat{A}_a(0)$ and $\hat{A}_b(0)$ in terms of $\hat{A}_c(0)$ and $\hat{A}_d(0)$ by applying the reverse rotation R$(-\alpha)$ as we did earlier in Eq. (6.13), then

$$\begin{pmatrix} \hat{A}_a(0) \\ \hat{A}_b(0) \end{pmatrix} = \begin{pmatrix} \cos\alpha & -\sin\alpha \\ \sin\alpha & \cos\alpha \end{pmatrix} \begin{pmatrix} \hat{A}_c(0) \\ \hat{A}_d(0) \end{pmatrix}. \tag{6.37}$$

The result of Eq. (6.37) may be substituted into Eq. (6.36) to give

$$\begin{pmatrix} \hat{A}_c(t) \\ \hat{A}_d(t) \end{pmatrix} = \begin{pmatrix} \cos\alpha & \sin\alpha \\ -\sin\alpha & \cos\alpha \end{pmatrix} \begin{pmatrix} (\hat{A}_c(0)\cos\alpha - \hat{A}_d(0)\sin\alpha)\exp(-i\omega_a t) \\ (\hat{A}_c(0)\sin\alpha + \hat{A}_d(0)\cos\alpha)\exp(-i\omega_b t) \end{pmatrix}. \tag{6.38}$$

So, we get

$$\begin{aligned} \hat{A}_c(t) = &(\hat{A}_c(0)\cos^2\alpha - \hat{A}_d(0)\sin\alpha\cos\alpha)\exp(-i\omega_a t) + \\ &(\hat{A}_c(0)\sin^2\alpha + \hat{A}_d(0)\sin\alpha\cos\alpha)\exp(-i\omega_b t), \end{aligned} \tag{6.39}$$

with a corresponding equation for $\hat{A}_d(t)$. Now, Eq. (6.39) is already quite complicated and next we need to use it to evaluate $\hat{A}_c^\dagger(t)\hat{A}_c(t)|n_c(0),n_d(0)\rangle$. We can simplify things by noting that from, Eq. (6.39), $\hat{A}_c^\dagger(t)\hat{A}_c(t)$ will contain terms like $\hat{A}_c^\dagger(0)\hat{A}_c(0)$ and $\hat{A}_d^\dagger(0)\hat{A}_d(0)$ for which $|n_c(0),n_d(0)\rangle$ is an eigenstate. However, the expression for $\hat{A}_c^\dagger(t)\hat{A}_c(t)$ will also contain terms like $\hat{A}_c^\dagger(0)\hat{A}_d(0)$, for which $|n_c(0),n_d(0)\rangle$ is not an eigenstate. This means that $|n_c(0),n_d(0)\rangle$ is not an eigenstate of $\hat{A}_c^\dagger(t)\hat{A}_c(t)$. However, we can still obtain useful information about the system if we look for time dependent expectation values of the form [7]

$$\begin{aligned} \tilde{n}_c(t) &= \langle n_c(0),n_d(0)|\hat{A}_c^\dagger(t)\hat{A}_c(t)|n_c(0),n_d(0)\rangle \\ &= \langle n_c(0),n_d(0)|\hat{N}_c(t)|n_c(0),n_d(0)\rangle. \end{aligned} \tag{6.40}$$

Noting that

$$\langle n_c(0),n_d(0)|\hat{A}_c^\dagger(0)\hat{A}_d(0)|n_c(0),n_d(0)\rangle = 0$$

and

$$\langle n_c(0),n_d(0)|\hat{A}_d^\dagger(0)\hat{A}_c(0)|n_c(0),n_d(0)\rangle = 0,$$

and after substituting the expression for $\cos\alpha$ from Eq. (6.16), we find

$$\tilde{n}_c(t) = n_c(0) + 2(n_d(0) - n_c(0))\left(\frac{v_{cd}}{\omega_n}\right)^2(1 - \cos(\omega_n t)), \tag{6.41}$$

where $\omega_n = \omega_a - \omega_b = \sqrt{(\omega_c - \omega_d)^2 + 4v_{cd}^2}$.

Notice that $\tilde{n}_c(0) = n_c(0)$, as it should and that $\tilde{n}_c(t)$ oscillates with a frequency of ω_n that is equal to the difference in the frequencies of the normal modes. Thus the number of items oscillates like the *beats* of the two normal modes. Fortunately, we do not need to go through the whole process again to find $\tilde{n}_d(t) = \langle n_c(0), n_d(0) | \hat{A}_d^\dagger(t) \hat{A}_d(t) | n_c(0), n_d(0) \rangle$ for the time variation of the number of items in Cat$_d$, since we already know that the total number of items is conserved, so $\tilde{n}_d(t) = n_c(0) + n_d(0) - \tilde{n}_c(t)$, from which we find

$$\tilde{n}_d(t) = n_d(0) + 2(n_c(0) - n_d(0)) \left(\frac{v_{cd}}{\omega_n} \right)^2 (1 - \cos(\omega_n t)). \qquad (6.42)$$

From Eq. (6.42) we can conclude that $\tilde{n}_d(t)$ also oscillates with frequency ω_n and in anti-phase with $\tilde{n}_c(t)$ so that the total number of items remains constant. Notice that two factors drive the fluctuations of the time dependent expectation values of the number of items in the two categories. One is the coefficient v_{cd} of the interaction term in the Hamiltonian. When this coefficient is zero then $\tilde{n}_c(t)$ and $\tilde{n}_d(t)$ do not fluctuate, as might be expected. However, if $n_c(0) = n_d(0)$ then the populations do not fluctuate either. We can understand this result since it implies the transfer of items from Cat$_c$ to Cat$_d$ exactly cancels the reverse process, leading to stable populations, on average. So, the fluctuation level is driven by the initial difference in the populations of the two categories, as well as the strength of their interaction.

A typical example of the interaction between two categories of items, calculated from Eqs. (6.41) and (6.42) is illustrated in Fig. 6.1. Both sets of expectation values of population numbers oscillate sinusoidally in anti-phase. This behaviour ensures that the total number of items in the two categories combined is an invariant. This situation leads to some interesting consequences for the higher order fluctuations in the context of *quantum noise* that we examine next.

6.2.3 QUANTUM NOISE

It was pointed out above that products of operators with the same subscript like $\hat{A}_c^\dagger(0) \hat{A}_c(0)$ and $\hat{A}_d^\dagger(0) \hat{A}_d(0)$ contribute to $\tilde{n}_a(t)$, but that operator products like $\hat{A}_c^\dagger(0) \hat{A}_d(0)$ with different subscripts contribute nothing. However, if we want to evaluate the expectation values of higher powers of $\hat{N}_c(t)$ such as $\hat{N}_c^2(t)$, then not only $\hat{A}_c^\dagger(0) \hat{A}_c(0)$ and $\hat{A}_d^\dagger(0) \hat{A}_d(0)$ contribute but now $\hat{A}_c^\dagger(0) \hat{A}_d(0)$ and $\hat{A}_d^\dagger(0) \hat{A}_c(0)$ contribute also, because $\hat{N}_c^2(t)$ contains products like $\hat{A}_d^\dagger(0) \hat{A}_c(0) \hat{A}_c^\dagger(0) \hat{A}_d(0) = \hat{N}_d(0)(\hat{N}_c(0) + 1)$, which do contribute. As a result we find that the square of the expectation value of $\hat{N}_c(t)$ is not equal to the expectation value of $\hat{N}_c^2(t)$. The difference between these two quantities is a measure of what is termed *quantum noise*. Let us define

$$\Delta \tilde{n}_c^2(t) = \widetilde{n_c^2(t)} - \tilde{n}_c^2(t), \qquad (6.43)$$

where

$$\begin{aligned}
\widetilde{n_c^2(t)} &= \langle n_c(0), n_d(0) | (\hat{A}_c^\dagger(t) \hat{A}_c(t))^2 | n_c(0), n_d(0) \rangle \\
&= \langle n_c(0), n_d(0) | \hat{N}_c^2(t) | n_c(0), n_d(0) \rangle,
\end{aligned} \qquad (6.44)$$

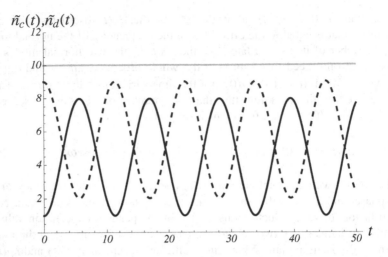

Figure 6.1 An example of the variation with time of the population values from Eq. (6.41) (dark curve), and Eq. (6.42) (dark dashes), with $n_c(0) = 1$, $n_d(0) = 9$, $\omega_c = 0.3$, $\omega_d = 0.1$, and $v_{cd} = 0.26$. The horizontal grey line indicates the total number of items.

and $\tilde{n}_c^2(t)$ is found by just squaring $\tilde{n}_c(t)$ from Eq. (6.40). Then $\sqrt{\Delta \tilde{n}_c^2(t)}$ is the quantum noise. Substituting for $\hat{A}_c(t)$ from Eq. (6.39) one finds

$$\Delta \tilde{n}_c^2(t) = 2(2n_c(0)n_d(0) + n_c(0) + n_d(0)) \times$$
$$\left(\frac{v_{cd}}{\omega_n}\right)^2 \left(\left(\frac{\Delta_{cd}}{\omega_n}\right)^2 (1 - \cos(\omega_n t) + 2\left(\frac{v_{cd}}{\omega_n}\right)^2 \sin^2(\omega_n t)\right), \quad (6.45)$$

where $\Delta_{cd} = \omega_d - \omega_c$.

Since $\Delta \tilde{n}_c^2(t)$ is in general time varying, it is useful to define a time average,

$$\overline{\Delta \tilde{n}_c^2(t)} = 2(2n_c(0)n_d(0) + n_c(0) + n_d(0)) \left(\frac{v_{cd}}{\omega_n}\right)^2 \left(\frac{\Delta_{cd}^2 + v_{cd}^2}{\omega_n^2}\right). \quad (6.46)$$

Then $\sqrt{\overline{\Delta \tilde{n}_c^2(t)}}$ is the root-mean-square (rms) quantum noise of the Cat_c items. Notice that $\sqrt{\overline{\Delta \tilde{n}_c^2(t)}} = 0$ when $v_{cd} = 0$, so that the quantum noise is entirely due to the interaction between the two populations of items.

It is also important to know the value of the quantum noise that is associated with the Cat_d population. This actually is easy to evaluate by using the conservation rule $\hat{N}_c(t) + \hat{N}_d(t) = \hat{N}_c(0) + \hat{N}_d(0) = \hat{N}$. Then

$$\tilde{n}_d(t) = n - \tilde{n}_c(t), \quad (6.47)$$

where $n = n_c(0) + n_d(0)$. Now

$$\langle n_c(0), n_d(0)|\hat{N}_d^2(t)|n_c(0), n_d(0)\rangle = \langle n_c(0), n_d(0)|(\hat{N} - \hat{N}_c(t))^2|n_c(0), n_d(0)\rangle, \quad (6.48)$$

and so

$$\widetilde{n_d^2(t)} = n^2 + \widetilde{n_c^2(t)} - 2n\tilde{n}_c(t). \tag{6.49}$$

Now, squaring Eq. (6.47) gives

$$\tilde{n}_d^2(t) = n^2 + \tilde{n}_c^2(t) - 2n\tilde{n}_c(t). \tag{6.50}$$

Combining Eqs. (6.49) and (6.50) yields

$$\widetilde{n_d^2(t)} - \tilde{n}_d^2(t) = \widetilde{n_c^2(t)} - \tilde{n}_c^2(t), \tag{6.51}$$

from which we conclude that the quantum noise levels of the Cat_c and Cat_d populations of items are identical.

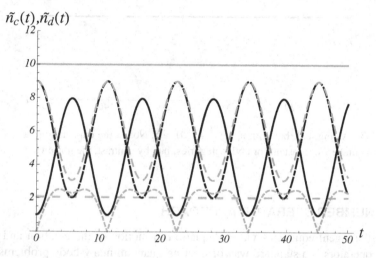

Figure 6.2 The dark curves are identical with those in Fig. 6.1. The grey dashed curves are various representations of quantum noise. The horizontal dashed line is the mean noise evaluated from Eq. (6.46), the short dashed grey curve is $\sqrt{\Delta \tilde{n}_c^2(t)}$, evaluated from Eq. (6.45), and the long dashed curve is the value of $\sqrt{\widetilde{n_c^2(t)}}$, evaluated from Eq. (6.44).

The behaviour of the quantum noise for the system that was illustrated in Fig. 6.1 is illustrated in Fig. 6.2. It is clear that in this case the values of $\tilde{n}_c(t)$ and $\sqrt{\widetilde{n_c^2(t)}}$ are very similar, as they are for Cat_d. This means that the quantum noise does not affect our information about the fluctuation levels of the system, that much, in this case.

It was pointed out above that two factors determine the amplitudes of the fluctuations in the expectation values $n_c(t)$ and $n_d(t)$. These are the strength of the interaction, υ_{cd}, and also the initial difference, $n_c(0) - n_d(0)$, in the populations of the two categories. However, the quantum noise levels of the two populations, while they do go to zero when the interaction strength υ_{cd} goes to zero, they do not disappear when $n_c(0) = n_d(0)$. This means that the values of $\widetilde{n_c^2(t)}$ and $\widetilde{n_d^2(t)}$ still fluctuate when

$n_c(0) = n_d(0)$, as long as $v_{cd} \neq 0$. This situation is illustrated in Fig. 6.3. This shows that even though the expectation values of the two categories do not vary with time, there is a non-zero level of quantum noise, so there are still fluctuations associated with the system.

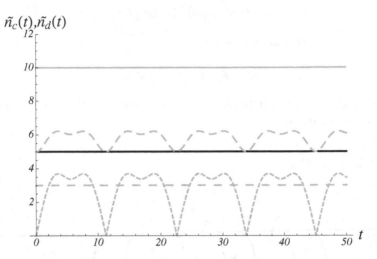

Figure 6.3 As Fig. 6.2, but with $n_c(0) = n_d(0) = 5$. Notice the expectation values of the population numbers is constant for both categories, but, by contrast, the noise values are non-zero.

6.2.4 NUMBER OPERATOR APPROACH

Finding the eigenfrequencies via the equation of motion of the creation and annihilation operators is a standard way of treating quantum many-body problems. The example in the previous section can be regarded as a toy model where we only have two categories of items. The method is also used in more complicated situations, though approximations are then sometimes necessary. However, the relatively simple system we have here can be treated more directly by applying the Heisenberg equation of motion to the number operators involved. This approach is a somewhat more direct way of getting the result in Eq. (6.41), which proves useful in more complicated situations that we will meet later, so the method is worth exploring in the simple case now under consideration. We begin with the time derivative of $\hat{N}_c$ with the Heisenberg equation of motion and the Hamiltonian from Eq. (6.14), then

$$i\frac{d\hat{N}_c}{dt} = [\hat{N}_c, \hat{\Omega}] = v_{cd}(\hat{A}_c^\dagger \hat{A}_d - \hat{A}_d^\dagger \hat{A}_c). \tag{6.52}$$

Next we find the second derivative of $\hat{N}_c$ by applying the Heisenberg equation of motion to the first derivative in Eq. (6.52) to get

$$-\frac{d^2\hat{N}_c}{dt^2} = \left[i\frac{d\hat{N}_c}{dt}, \Omega\right] = v_{cd}[(\hat{A}_c^\dagger \hat{A}_d - \hat{A}_d^\dagger \hat{A}_c), \Omega], \tag{6.53}$$

which leads to

$$\frac{d^2 \hat{N}_c}{dt^2} = (\omega_c - \omega_d) v_{cd} (\hat{A}_c^\dagger \hat{A}_d + \hat{A}_d^\dagger \hat{A}_c) - 2v_{cd}^2 (\hat{N}_c - \hat{N}_d)$$

$$= (\omega_c - \omega_d)(\Omega - \omega_c \hat{N}_c - \omega_d \hat{N}_d) - 2v_{cd}^2 (\hat{N}_c - \hat{N}_d),$$

(6.54)

where we have used the Hamiltonian, Eq. (6.14), to eliminate $v_{cd}(\hat{A}_c^\dagger \hat{A}_d + \hat{A}_d^\dagger \hat{A}_c)$ from Eq. (6.54). The next step is to evaluate the expectation value of each term it Eq. (6.54). Recalling that

$$\frac{d}{dt}\langle n_c(0), n_d(0)|\hat{N}_c|n_c(0), n_d(0)\rangle = \langle n_c(0), n_d(0)|\frac{d\hat{N}_c}{dt}|n_c(0), n_d(0)\rangle$$

(6.55)

and hence that,

$$\frac{d^2}{dt^2}\langle n_c(0), n_d(0)|\hat{N}_c|n_c(0), n_d(0)\rangle = \langle n_c(0), n_d(0)|\frac{d^2\hat{N}_c}{dt^2}|n_c(0), n_d(0)\rangle,$$

(6.56)

yields

$$\frac{d^2 \tilde{n}_c(t)}{dt^2} + ((\omega_d - \omega_c)^2 + 4v_{cd}^2)\tilde{n}_c(t) = (\omega_d - \omega_c)^2 n_c(0) + 2v_{cd}^2(n_c(0) + n_d(0)).$$

(6.57)

Eq. (6.57) is a second order inhomogeneous differential equation, with boundary conditions

$$\tilde{n}_c(0) = n_c(0)$$

and

$$\frac{d\tilde{n}_c(t)}{dt}\bigg|_{t=0} = -iv_{cd}\langle n_c(0), n_d(0)|\hat{A}_c^\dagger(0)\hat{A}_d(0) - \hat{A}_d^\dagger(0)\hat{A}_c(0)|n_c(0), n_d(0)\rangle = 0.$$

The general solution of Eq. (6.57) is

$$\tilde{n}_c(t) = a_0 + a_1 \cos(\omega t) + a_2 \sin(\omega t),$$

(6.58)

where ω is the frequency of oscillation and a_0, a_1 and a_2 are arbitrary constants to be determined by the initial conditions. After substituting the general solution into Eq. (6.57) and applying the initial values one finds that $\omega = \omega_n$ and we get Eq. (6.41), i.e., unsurprisingly, exactly the same solution as we obtained using the eigenfrequency approach, but with markedly less work. This number operator approach proves useful in a number of applications that we will treat later.

6.3 COMPLEX ROTATION TRANSFORMATION

We can relax the assumption that we used in Eq. (6.7) that the components of the matrix that produced the linear transformation are all real and assume they are complex. We only need to look at the bosonic case, so let

$$\begin{pmatrix} \hat{A}_e \\ \hat{A}_f \end{pmatrix} = \begin{pmatrix} \varepsilon & \eta \\ \lambda & \mu \end{pmatrix} \begin{pmatrix} \hat{A}_a \\ \hat{A}_b \end{pmatrix},$$

(6.59)

where now ε, ν, λ and μ are complex scalars. Then we find, $[\hat{A}_e, \hat{A}_e^\dagger] = \varepsilon^* \varepsilon + \eta^* \eta$, $[\hat{A}_f, \hat{A}_f^\dagger] = \lambda^* \lambda + \mu^* \mu$ and $[\hat{A}_e, \hat{A}_f^\dagger] = \varepsilon^* \lambda + \eta^* \mu$. At this stage the constants are arbitrary, but with an appropriate choice of the constants, we can again make, $[\hat{A}_e, \hat{A}_e^\dagger] = 1$, $[\hat{A}_f, \hat{A}_f^\dagger] = 1$ and $[\hat{A}_e, \hat{A}_f^\dagger] = 0$, so that $\hat{A}_e$ and $\hat{A}_f$ are bosonic in character. There are several possibilities that satisfy these conditions. One such is

$$\begin{pmatrix} \varepsilon & \eta \\ \lambda & \mu \end{pmatrix} = \begin{pmatrix} \cos\beta & i\sin\beta \\ i\sin\beta & \cos\beta \end{pmatrix} = \hat{C}(\beta), \tag{6.60}$$

where again β is an arbitrary angle and $\hat{C}(\beta)$ represents a type of rotation matrix, but with complex coefficients in contrast to the real rotation matrix $\hat{R}(\alpha)$, that we met earlier. Then one finds

$$\hat{A}_e^\dagger \hat{A}_e = \hat{A}_a^\dagger \hat{A}_a \cos^2\beta + \hat{A}_b^\dagger \hat{A}_b \sin^2\beta + i(\hat{A}_a^\dagger \hat{A}_b - \hat{A}_b^\dagger \hat{A}_a)\cos\beta \sin\beta \tag{6.61}$$

and

$$\hat{A}_f^\dagger \hat{A}_f = \hat{A}_a^\dagger \hat{A}_a \sin^2\beta + \hat{A}_b^\dagger \hat{A}_b \cos^2\beta - i(\hat{A}_a^\dagger \hat{A}_b - \hat{A}_b^\dagger \hat{A}_a)\cos\beta \sin\beta. \tag{6.62}$$

So, we can see that adding Eqs. (6.61) and (6.62), then

$$\hat{A}_e^\dagger \hat{A}_e + \hat{A}_f^\dagger \hat{A}_f = \hat{A}_a^\dagger \hat{A}_a + \hat{A}_b^\dagger \hat{A}_b$$

so the total number of items is invariant, as before. However, now

$$\begin{aligned} \hat{\Omega} &= \omega_a \hat{A}_a^\dagger \hat{A}_a + \omega_b \hat{A}_b^\dagger \hat{A}_b \\ &= \omega_e \hat{A}_e^\dagger \hat{A}_e + \omega_f \hat{A}_f^\dagger \hat{A}_f + i\upsilon_{ef}(\hat{A}_e^\dagger \hat{A}_f - \hat{A}_f^\dagger \hat{A}_e), \end{aligned} \tag{6.63}$$

where, ω_e, ω_f and υ_{ef} are equal to

$$\begin{aligned} \omega_e &= \omega_a \cos^2\beta + \omega_b \sin^2\beta, \\ \omega_f &= \omega_b \cos^2\beta + \omega_a \sin^2\beta, \\ \upsilon_{ef} &= (\omega_b - \omega_a)\cos\beta \sin\beta. \end{aligned} \tag{6.64}$$

and then

$$\cos\beta = \sqrt{\frac{1}{2}\left(1 + \frac{\omega_e - \omega_f}{\sqrt{(\omega_e - \omega_f)^2 + 4\upsilon_{ef}^2}}\right)} \tag{6.65}$$

and

$$\begin{aligned} \omega_a &= \frac{\omega_e + \omega_f + \sqrt{(\omega_e - \omega_f)^2 + 4\upsilon_{ef}^2}}{2}, \\ \omega_b &= \frac{\omega_e + \omega_f - \sqrt{(\omega_e - \omega_f)^2 + 4\upsilon_{ef}^2}}{2}. \end{aligned} \tag{6.66}$$

These clearly have the same form as found in Eqs. (6.16) and (6.17). Also notice that Hamiltonian in Eq. (6.63) is still Hermitian, since $(i(\hat{A}_e^\dagger \hat{A}_f - \hat{A}_f^\dagger \hat{A}_e))^\dagger = -i(\hat{A}_f^\dagger \hat{A}_e - \hat{A}_e^\dagger \hat{A}_f)$. So, Eqs. (6.14) and (6.63) are essentially equivalent to one another, but it is sometimes useful to use one rather than the other in certain situations, as we shall see later.

6.4 COUPLED CREATION AND ANNIHILATION OPERATORS

Another entirely real linear transformation that has some useful applications is one in which a creation operator is coupled with an annihilation operator. Unlike the previous linear transformations that we have looked at so far in this chapter, the bosonic and fermionic cases have very different characters, so we will need to look at them both. We will look at the bosonic case first.

6.4.1 BOSONIC CASE

Consider

$$\begin{pmatrix} \hat{A}_g \\ \hat{A}_h^\dagger \end{pmatrix} = \begin{pmatrix} \varepsilon & \eta \\ \lambda & \mu \end{pmatrix} \begin{pmatrix} \hat{A}_a \\ \hat{A}_b^\dagger \end{pmatrix}, \tag{6.67}$$

where $\hat{A}_a$ and $\hat{A}_b$ are bosonic annihilation operators, that along with their corresponding creation operators, obey the usual commutation rules. Then we find, $[\hat{A}_g, \hat{A}_g^\dagger] = \varepsilon^2 - \eta^2$, $[\hat{A}_h, \hat{A}_h^\dagger] = \mu^2 - \lambda^2$ and $[\hat{A}_g, \hat{A}_h] = \varepsilon\lambda - \eta\mu$. Ensuring that the bosonic character of $\hat{A}_g$ and $\hat{A}_h$ is preserved leads to

$$\begin{pmatrix} \varepsilon & \eta \\ \lambda & \mu \end{pmatrix} = \begin{pmatrix} \cosh\gamma & \sinh\gamma \\ \sinh\gamma & \cosh\gamma \end{pmatrix} = \hat{P}(\gamma), \tag{6.68}$$

where γ is an arbitrary parameter and this time $\hat{P}(\gamma)$ is not a rotation, but a matrix with hyperbolic functions as components that is referred to as a *pseudo-rotation*. Now we find

$$\hat{A}_g^\dagger \hat{A}_g = \hat{A}_a^\dagger \hat{A}_a \cosh^2\gamma + \hat{A}_b \hat{A}_b^\dagger \sinh^2\gamma + (\hat{A}_a^\dagger \hat{A}_b^\dagger + \hat{A}_b \hat{A}_a)\cosh\gamma\sinh\gamma \tag{6.69}$$

and

$$\hat{A}_h^\dagger \hat{A}_h = \hat{A}_a \hat{A}_a^\dagger \sinh^2\gamma + \hat{A}_b^\dagger \hat{A}_b \cosh^2\gamma + (\hat{A}_a^\dagger \hat{A}_b^\dagger + \hat{A}_b \hat{A}_a)\cosh\gamma\sinh\gamma. \tag{6.70}$$

Here, the sum of the populations in Cat_g and Cat_h is no longer equal to the sum of those in Cat_a and Cat_b. Rather, the differences are conserved, since

$$\hat{A}_g^\dagger \hat{A}_g - \hat{A}_h^\dagger \hat{A}_h = \hat{A}_a^\dagger \hat{A}_a(\cosh^2\gamma - \sinh^2\gamma) + \hat{A}_b^\dagger \hat{A}_b(\sinh^2\gamma - \cosh^2\gamma)$$
$$= \hat{A}_a^\dagger \hat{A}_a - \hat{A}_b^\dagger \hat{A}_b. \tag{6.71}$$

So, in this case

$$\hat{N}_g - \hat{N}_h = \hat{N}_a - \hat{N}_b. \tag{6.72}$$

Further, if we note that

$$\begin{pmatrix} \hat{A}_a \\ \hat{A}_b^\dagger \end{pmatrix} = \begin{pmatrix} \cosh\gamma & -\sinh\gamma \\ -\sinh\gamma & \cosh\gamma \end{pmatrix} \begin{pmatrix} \hat{A}_g \\ \hat{A}_h^\dagger \end{pmatrix}, \tag{6.73}$$

where the matrix is, $\hat{P}(-\gamma)$ and is the inverse of $\hat{P}(\gamma)$ in Eq. (6.68)[3], then the Hamiltonian in Eq. (6.2) now transforms into

$$\hat{\Omega} = \omega_g \hat{A}_g^\dagger \hat{A}_g + \omega_h \hat{A}_h^\dagger \hat{A}_h + \upsilon_{gh}(\hat{A}_g^\dagger \hat{A}_h^\dagger + \hat{A}_h \hat{A}_g) + \mu_{gh}, \tag{6.74}$$

---

[3]Recall that $\sinh(-\gamma) = -\sinh\gamma$ and $\cosh(-\gamma) = \cosh\gamma$.

where,

$$\omega_g = \omega_a \cosh^2 \gamma + \omega_b \sinh^2 \gamma,$$

$$\omega_h = \omega_b \cosh^2 \gamma + \omega_a \sinh^2 \gamma,$$

$$v_{gh} = -(\omega_a + \omega_b) \cosh \gamma \sinh \gamma,$$

$$\mu_{gh} = (\omega_a + \omega_b) \sinh^2 \gamma, \tag{6.75}$$

and

$$\cosh \gamma = \sqrt{\frac{1}{2}\left(1 + \frac{\omega_g + \omega_h}{\sqrt{(\omega_g + \omega_h)^2 - 4v_{gh}^2}}\right)}. \tag{6.76}$$

There are some important differences in the structure of the Hamiltonian in Eq. (6.74) and the transformed ones we encountered earlier. This is largely due to the fact that the transform involves a mixture of creation and annihilation operators which leads to the interaction term that pairs the creation operators of Cat_g and Cat_h and then pairs the corresponding annihilation operators in the interaction term of the Hamiltonian. The transform is no longer actually unitary, although the determinant of the transform matrix is still unity, just as in the earlier two examples. The additional constant μ_{gh} in the Hamiltonian is unimportant in practice since it commutes with all operators and hence does not contribute to the Heisenberg equation of motion and hence cannot affect the dynamics of the system. To all intents and purposes it can be left out altogether or simply set equal to zero. The eigenvalues associated with the normal modes of the system need to be evaluated carefully, because the annihilation of Cat_g is coupled to the creation operator of Cat_h. This can be seen by noting that

$$i\frac{d\hat{A}_g}{dt} = [\hat{A}_g, \hat{\Omega}] = \omega_g \hat{A}_g + v_{gh}\hat{A}_h^\dagger. \tag{6.77}$$

Now in order to close the system of equations we need

$$i\frac{d\hat{A}_h^\dagger}{dt} = [\hat{A}_h^\dagger, \hat{\Omega}] = -\omega_h \hat{A}_h^\dagger - v_{gh}\hat{A}_g. \tag{6.78}$$

To find the eigenfrequencies of the system we now need to take both $\hat{A}_g$ and $\hat{A}_h^\dagger$ proportional to $\exp(-i\omega t)$! So, the eigenvalue equation is

$$\begin{pmatrix} \omega - \omega_g & -v_{gh} \\ v_{gh} & \omega + \omega_h \end{pmatrix} \begin{pmatrix} \hat{A}_g \\ \hat{A}_h^\dagger \end{pmatrix} = 0. \tag{6.79}$$

This eigenvalue equation is satisfied if the determinant of the 2×2 matrix is zero, i.e.,

$$\begin{vmatrix} \omega - \omega_g & -v_{gh} \\ v_{gh} & \omega + \omega_h \end{vmatrix} = (\omega - \omega_g)(\omega + \omega_h) + v_{gh}^2 = 0, \tag{6.80}$$

which leads to

$$\omega = \frac{\omega_g - \omega_h \pm \sqrt{(\omega_g + \omega_h)^2 - 4v_{gh}^2}}{2}. \tag{6.81}$$

Although the eigenfrequencies in Eq. (6.81) look similar to those in Eq. (6.35), their behaviour is quite different. The frequencies in Eq. (6.35) are entirely real so the normal modes are stable oscillations. The modes in Eq. (6.81) are only stable if $2|v_{gh}| < (\omega_g + \omega_h)$ is satisfied. Otherwise the eigenfrequencies become complex and then exponential growth or damping is possible. Thus, if the interaction term is strong enough the system may be unstable. This result will be explored in more detail when we look at some important applications in a later chapter.

6.4.2 FERMIONIC CASE

We begin with two category fermionic system with a Hamiltonian of the form

$$\hat{\Omega} = \omega_a \hat{C}_a^\dagger \hat{C}_a + \omega_b \hat{C}_b^\dagger \hat{C}_b, \tag{6.82}$$

where $\hat{C}_a$ and $\hat{C}_b$ are a pair of fermionic annihilation operators with their corresponding creation operators, obeying the usual anti-commutation rules. The corresponding fermionic number operators, $\hat{N}_a = \hat{C}_a^\dagger \hat{C}_a$ and $\hat{N}_b = \hat{C}_b^\dagger \hat{C}_b$. Now consider the linear transformation

$$\begin{pmatrix} \hat{C}_g \\ \hat{C}_h^\dagger \end{pmatrix} = \begin{pmatrix} \varepsilon & \eta \\ \lambda & \mu \end{pmatrix} \begin{pmatrix} \hat{C}_a \\ \hat{C}_b^\dagger \end{pmatrix}, \tag{6.83}$$

Then, $\{\hat{C}_g, \hat{C}_g^\dagger\} = \varepsilon^2 + \eta^2$, $\{\hat{C}_h, \hat{C}_h^\dagger\} = \mu^2 + \lambda^2$ and $\{\hat{C}_g, \hat{C}_h\} = \varepsilon\lambda + \eta\mu$. Ensuring that the fermionic character of $\hat{C}_g$ and $\hat{C}_h$ is preserved leads to

$$\begin{pmatrix} \varepsilon & \eta \\ \lambda & \mu \end{pmatrix} = \begin{pmatrix} \cos\gamma & \sin\gamma \\ -\sin\gamma & \cos\gamma \end{pmatrix} = \hat{R}(\gamma), \tag{6.84}$$

where γ is now an arbitrary angle and this time $\hat{R}(\gamma)$ is again a rotation. So, we can immediately see that the coupling of fermionic creation and annihilation operators has an entirely different character to the bosonic case in Section 6.4.1. Note that in the fermionic case the transform is unitary, unlike the bosonic case when a creation operator is coupled to an annihilation operator.

Now we find

$$\hat{N}_g = \hat{N}_a \cos^2\gamma + (1 - \hat{N}_b)\sin^2\gamma - (\hat{C}_a^\dagger \hat{C}_b^\dagger + \hat{C}_b \hat{C}_a)\cos\gamma\sin\gamma \tag{6.85}$$

and

$$\hat{N}_h = (1 - \hat{N}_a)\sin^2\gamma + \hat{N}_b \cos^2\gamma - (\hat{C}_a^\dagger \hat{C}_b^\dagger + \hat{C}_b \hat{C}_a)\cos\gamma\sin\gamma. \tag{6.86}$$

where $\hat{N}_g = \hat{C}_g^\dagger \hat{C}_g$ and $\hat{N}_h = \hat{C}_h^\dagger \hat{C}_h$. As in the bosonic case in Section 6.4.1, the sum of the populations in Cat$_g$ and Cat$_h$ is no longer equal to the sum of those in Cat$_a$ and Cat$_b$, and again it is the differences that are conserved, since

$$\begin{aligned} \hat{N}_g - \hat{N}_h &= \hat{N}_a(\cos^2\gamma + \sin^2\gamma) - \hat{N}_b(\sin^2\gamma + \cos^2\gamma) \\ &= \hat{N}_a - \hat{N}_b. \end{aligned} \tag{6.87}$$

Further, if we note that

$$\begin{pmatrix} \hat{C}_a \\ \hat{C}_b^\dagger \end{pmatrix} = \hat{R}(-\gamma) \begin{pmatrix} \hat{C}_g \\ \hat{C}_h^\dagger \end{pmatrix} = \begin{pmatrix} \cos\gamma & -\sin\gamma \\ \sin\gamma & \cos\gamma \end{pmatrix} \begin{pmatrix} \hat{C}_g \\ \hat{C}_h^\dagger \end{pmatrix}, \qquad (6.88)$$

where the matrix is, $\hat{R}(-\gamma)$ and is the inverse of $\hat{R}(\gamma)$ in Eq. (6.84), then the Hamiltonian in Eq. (6.82) now transforms into

$$\hat{\Omega} = \omega_g \hat{C}_g^\dagger \hat{C}_g + \omega_h \hat{C}_h^\dagger \hat{C}_h + \upsilon_{gh}(\hat{C}_h^\dagger \hat{C}_g^\dagger + \hat{C}_g \hat{C}_h) + \mu_{gh}, \qquad (6.89)$$

where,

$$\omega_g = \omega_a \cos^2\gamma - \omega_b \sin^2\gamma,$$

$$\omega_h = \omega_b \cos^2\gamma - \omega_a \sin^2\gamma,$$

$$\upsilon_{gh} = (\omega_a + \omega_b)\cos\gamma\sin\gamma,$$

$$\mu_{gh} = (\omega_a + \omega_b)\sin^2\gamma, \qquad (6.90)$$

and

$$\cos\gamma = \sqrt{\frac{1}{2}\left(1 + \frac{\omega_g + \omega_h}{\sqrt{(\omega_g + \omega_h)^2 + 4\upsilon_{gh}^2}}\right)}. \qquad (6.91)$$

It is now instructive to evaluate the eigenvalues of the coupled fermionic system. We can again neglect the constant μ_{gh} in the Hamiltonian as it has no effect on the Heisenberg equation of motion. So, proceeding with the equation for $\hat{C}_g$ we find

$$i\frac{d\hat{C}_g}{dt} = [\hat{C}_g, \hat{\Omega}] = \omega_g \hat{C}_g - \upsilon_{gh} \hat{C}_h^\dagger. \qquad (6.92)$$

So, in order to close the system of equations we need

$$i\frac{d\hat{C}_h^\dagger}{dt} = [\hat{C}_h^\dagger, \hat{\Omega}] = -\omega_h \hat{C}_h^\dagger - \upsilon_{gh} \hat{C}_g. \qquad (6.93)$$

To find the eigenfrequencies of the system we again need to take both $\hat{C}_g$ and $\hat{C}_h^\dagger$ proportional to $\exp(-i\omega t)$. So, the eigenvalue equation is

$$\begin{pmatrix} \omega - \omega_g & \upsilon_{gh} \\ \upsilon_{gh} & \omega + \omega_h \end{pmatrix} \begin{pmatrix} \hat{C}_g \\ \hat{C}_h^\dagger \end{pmatrix} = 0. \qquad (6.94)$$

This eigenvalue equation is satisfied if the determinant of the 2×2 matrix is zero, i.e.,

$$\begin{vmatrix} \omega - \omega_g & \upsilon_{gh} \\ \upsilon_{gh} & \omega + \omega_h \end{vmatrix} = (\omega - \omega_g)(\omega + \omega_h) - \upsilon_{gh}^2 = 0, \qquad (6.95)$$

which leads to

$$\omega = \frac{\omega_g - \omega_h \pm \sqrt{(\omega_g + \omega_h)^2 + 4\upsilon_{gh}^2}}{2}. \qquad (6.96)$$

The crucial difference between the eigenfrequencies in Eq. (6.96) and those for the bosonic case in Eq. (6.81), is the sign of the v_{gh}^2 term inside the square root. In the fermionic case it is positive, which means that the frequencies can only be real and hence there is no instability, unlike what we found in the bosonic case. It turns out that the fermonic case with coupled creation and annihilation operators plays a key role in the explanation of both superconductivity and fermionic mass, as we shall see later.

Finally, we can just check the time dependence of the numbers of items in the two categories by evaluating $[\hat{N}_g, \hat{\Omega}]$ and $[\hat{N}_g, \hat{\Omega}]$ and find

$$[\hat{N}_g, \hat{\Omega}] = [\hat{N}_h, \hat{\Omega}] = v_{gh}(\hat{C}_g^\dagger \hat{C}_h^\dagger - \hat{C}_h \hat{C}_g), \tag{6.97}$$

so the number operators of the individual categories vary with time as does the total number $\hat{N} = \hat{N}_g + \hat{N}_h$, since Eq. (6.97) implies that $[\hat{N}, \hat{\Omega}] \neq 0$. However, Eq. (6.96) does indicate that

$$\frac{d(\hat{N}_g - \hat{N}_h)}{dt} = 0 \tag{6.98}$$

which is consistent with Eq. (6.87).

7 Degenerate Two-category Systems

In this chapter we look at what happens when we apply the important symmetry condition of *degeneracy* to a two-category system. This involves imposing the condition that the frequencies in the Hamiltonian of the two categories are identical, i.e., we let the frequencies in the Hamiltonians in Eqs. (6.2) and (6.23) be equal, so $\omega_a = \omega_b = \omega_0$. We are first going to examine bosonic systems, but in the last section of the chapter we will look at an important fermionic case.

7.1 DEGENERATE BOSONIC TWO-CATEGORY SYSTEMS

There are two distinct situations we want to look at for bosonic case. The first involves the rotation transformation, $\hat{R}(\alpha)$. Substituting degeneracy condition into Eqs. (6.15), we immediately get $\omega_c = \omega_d = \omega_0$. Also, importantly, $v_{cd} = 0$, so the interaction term in the Hamiltonian disappears! This leads to

$$\begin{aligned}
\hat{\Omega} &= \omega_0(\hat{A}_a^\dagger \hat{A}_a + \hat{A}_b^\dagger \hat{A}_b) = \omega_0(\hat{N}_a + \hat{N}_b) \\
&= \omega_0(\hat{A}_c^\dagger \hat{A}_c + \hat{A}_d^\dagger \hat{A}_d) = \omega_0(\hat{N}_c + \hat{N}_d).
\end{aligned} \tag{7.1}$$

From Eq. (7.1), we can immediately see that $\hat{N}_c$ and $\hat{N}_d$ both individually commute with $\hat{\Omega}$, just like the original number operators, $\hat{N}_a$ and $\hat{N}_b$ do. This means that, in the degenerate case, unlike the non-degenerate case we treated in Chapter 6, the individual population numbers for the new categories, Cat_c and Cat_d, are constants, just like the original categories, Cat_a and Cat_b. This means that the occupation numbers in the eigenstates, $|n_c, n_d\rangle$ are constants, so we can use them as an alternative representation of the system to the original Fock states represented by $|n_a, n_b\rangle$. This greatly simplifies things compared to the difficulties of dealing with time dependent occupation numbers that were found in the non-degenerate cases in Chapter 6.

Now notice also that in the degenerate case

$$[\hat{A}_c^\dagger \hat{A}_d, \hat{\Omega}] = \omega_0[\hat{A}_c^\dagger \hat{A}_d, \hat{N}_c + \hat{N}_d] = \omega_0(\hat{A}_c^\dagger \hat{A}_d - \hat{A}_c^\dagger \hat{A}_d) = 0 \tag{7.2}$$

and

$$[\hat{A}_d^\dagger \hat{A}_c, \hat{\Omega}] = \omega_0[\hat{A}_d^\dagger \hat{A}_c, \hat{N}_c + \hat{N}_d] = \omega_0(\hat{A}_d^\dagger \hat{A}_c - \hat{A}_d^\dagger \hat{A}_c) = 0, \tag{7.3}$$

so $\hat{A}_c^\dagger \hat{A}_d$ and $\hat{A}_d^\dagger \hat{A}_c$ are both individually invariant in the degenerate case. It is also important to note that neither is invariant in the non-degenerate case. Similarly, we can see that

$$[\hat{A}_a^\dagger \hat{A}_b, \hat{\Omega}] = \omega_0[\hat{A}_a^\dagger \hat{A}_b, \hat{N}_a + \hat{N}_b] = \omega_0(\hat{A}_a^\dagger \hat{A}_b - \hat{A}_a^\dagger \hat{A}_b) = 0 \tag{7.4}$$

DOI: 10.1201/9781003377504-7

and

$$[\hat{A}_b^\dagger\hat{A}_a,\hat{\Omega}] = \omega_0[\hat{A}_b^\dagger\hat{A}_a,\hat{N}_a+\hat{N}_b] = \omega_0(\hat{A}_b^\dagger\hat{A}_a - \hat{A}_b^\dagger\hat{A}_a) = 0, \tag{7.5}$$

So, like $\hat{A}_c^\dagger\hat{A}_d$ and $\hat{A}_d^\dagger\hat{A}_c$, $\hat{A}_a^\dagger\hat{A}_b$ and $\hat{A}_b^\dagger\hat{A}_a$ are also invariant, but none of them is Hermitian. However, we can construct pairs of Hermitian operators that are invariants. These are

$$\hat{K}_{ab} = \hat{A}_a^\dagger\hat{A}_b + \hat{A}_b^\dagger\hat{A}_a \text{ and } \hat{L}_{ab} = i(\hat{A}_b^\dagger\hat{A}_a - \hat{A}_a^\dagger\hat{A}_b), \tag{7.6}$$

and

$$\hat{K}_{cd} = \hat{A}_c^\dagger\hat{A}_d + \hat{A}_d^\dagger\hat{A}_c \text{ and } \hat{L}_{cd} = i(\hat{A}_d^\dagger\hat{A}_c - \hat{A}_c^\dagger\hat{A}_d). \tag{7.7}$$

The importance and meaning of these pairs of Hermitian operators in physical dynamics will become apparent later. The degeneracy condition leaves the value of the rotation angle, α, unaffected, so we can choose it as we wish. The properties of the operators in Eqs. (7.4) to (7.7) are best illustrated by making a specific choice of α, in $\hat{R}(\alpha)$. After applying $\hat{R}(\alpha)$,

$$\begin{pmatrix}\hat{A}_c \\ \hat{A}_d\end{pmatrix} = \hat{R}(\alpha)\begin{pmatrix}\hat{A}_a \\ \hat{A}_b\end{pmatrix},$$

we find

$$\hat{A}_c^\dagger\hat{A}_d = \sin\alpha\cos\alpha(\hat{N}_b - \hat{N}_a) + \cos^2\alpha\hat{A}_a^\dagger\hat{A}_b - \sin^2\alpha\hat{A}_b^\dagger\hat{A}_a, \tag{7.8}$$

and

$$\hat{A}_d^\dagger\hat{A}_c = \sin\alpha\cos\alpha(\hat{N}_b - \hat{N}_a) + \cos^2\alpha\hat{A}_b^\dagger\hat{A}_a - \sin^2\alpha\hat{A}_a^\dagger\hat{A}_b. \tag{7.9}$$

Then

$$\hat{K}_{cd} = \hat{A}_c^\dagger\hat{A}_d + \hat{A}_d^\dagger\hat{A}_c = \sin 2\alpha(\hat{N}_b - \hat{N}_a) + \cos 2\alpha(\hat{A}_b^\dagger\hat{A}_a + \hat{A}_a^\dagger\hat{A}_b), \tag{7.10}$$

and

$$\hat{L}_{cd} = i(\hat{A}_c^\dagger\hat{A}_d - \hat{A}_d^\dagger\hat{A}_c) = i(\hat{A}_a^\dagger\hat{A}_b - \hat{A}_b^\dagger\hat{A}_a) = \hat{L}_{ab}. \tag{7.11}$$

Notice that, unlike Eq. (7.10), Eq. (7.11) is entirely independent of α and it does not give us any clue to the properties of $\hat{L}_{cd}$, nor of $\hat{L}_{ab}$. We will have to find an alternative route if we are to find out more about $\hat{L}_{ab}$. However, Eq. (7.10) is dependent on α and tells us a lot about the eigenvalues and eigenstates of $\hat{K}_{cd}$. We can see this if we choose $\alpha = \frac{\pi}{4}$ and then Eq. (7.10) becomes

$$\hat{K}_{cd} = \hat{N}_b - \hat{N}_a, \tag{7.12}$$

which shows that $\hat{K}_{cd}$ has integer eigenvalues, since it is the difference between two natural numbers. It is also clear from Eq. (7.12) that the eigenstate of $\hat{K}_{cd}$ is $|n_a,n_b\rangle$, and not $|n_c,n_d\rangle$.

If we construct $\hat{N}_c = \hat{A}_c^\dagger\hat{A}_c$ and $\hat{N}_d = \hat{A}_d^\dagger\hat{A}_d$ with $\alpha = \frac{\pi}{4}$, then we find

$$\hat{N}_c - \hat{N}_d = \hat{A}_c^\dagger\hat{A}_c - \hat{A}_d^\dagger\hat{A}_d = \hat{A}_a^\dagger\hat{A}_b + \hat{A}_b^\dagger\hat{A}_a = \hat{K}_{ab}, \tag{7.13}$$

so that the eigenvalues of $\hat{K}_{ab}$ must also be integers, and its eigenstates are $|n_c, n_d\rangle$. So, their is reciprocity between $\hat{K}_{ab}$ and $\hat{K}_{cd}$.

Now we will return to $\hat{L}_{ab}$. We can find eigenvalues and eigenvectors for $\hat{L}_{ab}$, if we employ $\hat{C}(\beta)$, rather than $\hat{R}(\alpha)$. This is the second of the cases referred to earlier. Putting $\omega_a = \omega_b = \omega_0$ into Eqs. (6.64) gives $\omega_e = \omega_f = \omega_0$ and $v_{ef} = 0$, so

$$\begin{aligned}
\hat{\Omega} &= \omega_0(\hat{A}_a^{\dagger}\hat{A}_a + \hat{A}_b^{\dagger}\hat{A}_b) = \omega_0(\hat{N}_a + \hat{N}_b) \\
&= \omega_0(\hat{A}_e^{\dagger}\hat{A}_e + \hat{A}_f^{\dagger}\hat{A}_f) = \omega_0(\hat{N}_e + \hat{N}_f).
\end{aligned} \tag{7.14}$$

Notice that, just like $\hat{N}_c$ and $\hat{N}_d$, here $\hat{N}_e$ and $\hat{N}_e$ both individually commute with $\hat{\Omega}$ and so are invariants. This means that we can use $|n_e, n_f\rangle$ as another Fock state representation of the system.

Applying $\hat{C}(\beta)$, so that

$$\begin{pmatrix} \hat{A}_e \\ \hat{A}_f \end{pmatrix} = \hat{C}(\beta) \begin{pmatrix} \hat{A}_a \\ \hat{A}_b \end{pmatrix},$$

then we find

$$\hat{A}_e^{\dagger}\hat{A}_f = i\sin\beta\cos\beta(\hat{N}_a - \hat{N}_b) + \cos^2\beta\hat{A}_a^{\dagger}\hat{A}_b + \sin^2\beta\hat{A}_b^{\dagger}\hat{A}_a, \tag{7.15}$$

and

$$\hat{A}_f^{\dagger}\hat{A}_e = i\sin\beta\cos\beta(\hat{N}_b - \hat{N}_a) + \cos^2\beta\hat{A}_b^{\dagger}\hat{A}_a - \sin^2\beta\hat{A}_a^{\dagger}\hat{A}_b. \tag{7.16}$$

Then we get

$$\hat{L}_{ef} = i(\hat{A}_e^{\dagger}\hat{A}_f - \hat{A}_f^{\dagger}\hat{A}_e) = \sin 2\beta(\hat{N}_b - \hat{N}_a) + i\cos 2\beta(\hat{A}_a^{\dagger}\hat{A}_b + \hat{A}_b^{\dagger}\hat{A}_a), \tag{7.17}$$

and

$$\hat{K}_{ef} = \hat{A}_e^{\dagger}\hat{A}_f + \hat{A}_f^{\dagger}\hat{A}_e = \hat{A}_a^{\dagger}\hat{A}_b + \hat{A}_b^{\dagger}\hat{A}_a = \hat{K}_{ab}. \tag{7.18}$$

Notice that Eq. (7.18) is independent of β and does not yield any information about the eigenstates and eigenvalues of either $\hat{K}_{ef}$ or $\hat{K}_{ab}$. However, putting $\beta = \frac{\pi}{4}$ in Eq. (7.17) gives

$$\hat{L}_{ef} = \hat{N}_b - \hat{N}_a. \tag{7.19}$$

This shows that $\hat{L}_{ef}$ has integer eigenvalues and that its eigenstates are $|n_a, n_b\rangle$. Also, when $\beta = \frac{\pi}{4}$, we find that

$$\hat{N}_e - \hat{N}_f = \hat{A}_e^{\dagger}\hat{A}_e - \hat{A}_f^{\dagger}\hat{A}_f = i(\hat{A}_a^{\dagger}\hat{A}_b - \hat{A}_b^{\dagger}\hat{A}_a) = \hat{L}_{ab}, \tag{7.20}$$

and so, $\hat{L}_{ab}$ has integer eigenvalues and its eigenstates are $|n_e, n_f\rangle$.

The remarkable symmetry properties of degenerate systems that have been pointed out in this section, have important dynamical consequences, which we will examine next.

7.2 SYMMETRY PROPERTIES OF THE TWO-CATEGORY BOSONIC SYSTEM

Note that there are three distinct eigenvectors in the degenerate two-category examples we have looked at above, namely, $|n_a, n_b\rangle$, $|n_c, n_d\rangle$, and $|n_e, n_f\rangle$. All three are eigenstates of $\hat{N}$ and $\hat{\Omega}$. It is useful to define the key operators that appear in these examples. Let the pair of subscripts (i, j) stand for any of (a, b), (c, d) or (e, f). Then let

$$\hat{N}_{ij} = \hat{N}_i + \hat{N}_j = \hat{A}_i^\dagger \hat{A}_i + \hat{A}_j^\dagger \hat{A}_j, \tag{7.21}$$

$$\hat{K}_{ij} = \hat{A}_i^\dagger \hat{A}_j + \hat{A}_j^\dagger \hat{A}_i, \tag{7.22}$$

$$\hat{L}_{ij} = i(\hat{A}_j^\dagger \hat{A}_i - \hat{A}_i^\dagger \hat{A}_j) \tag{7.23}$$

and

$$\hat{Z}_{ij} = \hat{N}_j - \hat{N}_i = \hat{A}_j^\dagger \hat{A}_j - \hat{A}_i^\dagger \hat{A}_i, \tag{7.24}$$

where $\hat{N}_i = \hat{A}_i^\dagger \hat{A}_i$. Notice that $\hat{K}_{ij}$ and $\hat{L}_{ij}$ are the two invariant Hermitian operators that we identified at the beginning of this chapter in Section 7.1.1. We can interpret $\hat{Z}_{ij}$ in Eq. (7.24) as an *integer operator*. In addition we define a Hamiltonian, $\hat{\Omega}_{ij} = \omega_0 \hat{N}_{ij}$.

So far we have shown that

$$\hat{\Omega} = \hat{\Omega}_{ab} = \hat{\Omega}_{cd} = \hat{\Omega}_{ef} \tag{7.25}$$

$$\hat{N} = \hat{N}_{ab} = \hat{N}_{cd} = \hat{N}_{ef} \tag{7.26}$$

and

$$\hat{K}_{ab} = \hat{Z}_{cd}, \hat{K}_{cd} = \hat{Z}_{ab}, \hat{L}_{ab} = \hat{Z}_{ef} \text{ and } \hat{L}_{ef} = \hat{Z}_{ab} \tag{7.27}$$

It is important to note that an operator with subscripts (i, j) is explicitly defined on the Fock space defined by state vectors with the same subscripts, i.e., $|n_i, n_j\rangle$, but that this does not necessarily mean that the eigenvector in question is an eigenvector of that operator. For example, $\hat{L}_{ab}$ is explicitly defined on $|n_a, n_b\rangle$, but it is not an eigenvector of $\hat{L}_{ab}$. However, all three Fock spaces are eigenvectors of $\hat{\Omega}$ and $\hat{N}$ and they have the same form and the same eigenvalues for all three states. On the other hand $\hat{K}_{ab}$ and $\hat{L}_{ab}$ are explicitly defined on $|n_a, n_b\rangle$, but their respective eigenvectors are, $|n_c, n_d\rangle$ and $|n_e, n_f\rangle$.

We have shown that $\hat{K}_{ab}$ and $\hat{L}_{ab}$ both commute with $\hat{N}_{ab}$ and $\hat{\Omega}_{ab}$ and hence are constants of the motion, but, as we have seen, $|n_a, n_b\rangle$, whilst it is an eigenvector of $\hat{N}_{ab}$ and $\hat{\Omega}_{ab}$ is not an eigenvector of $\hat{K}_{ab}$ and $\hat{L}_{ab}$. This may seem puzzling, since we have learned that operators that commute must have common eigenstates. The point is that $\hat{K}_{ab}$ and $\hat{\Omega}$ do have a common eigenvectors, i.e., $|n_c, n_d\rangle$, but only when $\hat{K}_{ab}$ is transformed to $\hat{Z}_{cd}$ by the linear transformation, $\hat{R}\left(\frac{\pi}{4}\right)$, and only then are the eigenvalues of $n_d - n_c$ revealed. Similarly, $\hat{L}_{ab}$ has a common eigenvector of $|n_e, n_f\rangle$ with $\hat{\Omega}$ when it is transformed to $\hat{Z}_{ef}$ by the linear transformation, $\hat{C}\left(\frac{\pi}{4}\right)$, which shows that $\hat{L}_{ab}$ has integer eigenvalues, $n_f - n_e$. The integer operator that does have an eigenvector of $|n_a, n_b\rangle$ is in fact $\hat{Z}_{ab} = \hat{N}_b - \hat{N}_a$, whose eigenvalues are then

$n_b - n_a$. So, we have three different integer operators, $\hat{K}_{ab}$, $\hat{L}_{ab}$ and $\hat{Z}_{ab}$, with three different corresponding eigenvectors, i.e., $|n_c, n_d\rangle$, $|n_e, n_f\rangle$ and $|n_a, n_b\rangle$, respectively. The implication is then of course, that none of these three commutes with any of the other two. However, there is an important relationship between them. We first define

$$\frac{\hat{L}_{ab}}{2} = \hat{J}_1, \quad \frac{\hat{K}_{ab}}{2} = \hat{J}_2 \quad \text{and} \quad \frac{\hat{Z}_{ab}}{2} = \hat{J}_3. \tag{7.28}$$

Then we find

$$[\hat{J}_i, \hat{J}_j] = i\hat{J}_k \tag{7.29}$$

for $i = 1$, $j = 2$, $k = 3$ and further two cyclic combinations of the subscripts. This means that the three operators, $\hat{K}_{ab}/2$, $\hat{L}_{ab}/2$ and $\hat{Z}_{ab}/2$ form a closed finite group. In fact it is a Lie group just like the one we found in Section 4.4 for the fermionic Hermitian operators (see Appendix D). Such a result shows that there is an important internal symmetry to the system. This is a consequence of the degeneracy of the system. So, if we take the eigenfrequency of the system as simply[1] $\omega_0 n$, this can come about with a variety of different pairs of (n_a, n_b) as long as $n_a + n_b = n$. There are in fact $n + 1$ such combinations. Then $n + 1$ is the degeneracy of the system.

We can explore the symmetry properties of the Lie group further by defining a vector $\hat{\mathbf{J}} = (\hat{J}_1, \hat{J}_2, \hat{J}_3)$ and a corresponding square modulus, $\hat{\mathbf{J}}^2 = \hat{J}_1^2 + \hat{J}_2^2 + \hat{J}_3^2$. Then we can establish the following

$$\begin{aligned}
\hat{J}_1^2 &= -\frac{1}{4}(\hat{A}_b^\dagger \hat{A}_a - \hat{A}_a^\dagger \hat{A}_b)^2 \\
&= -\frac{1}{4}((\hat{A}_b^\dagger \hat{A}_a)^2 + (\hat{A}_a^\dagger \hat{A}_b)^2 - \hat{A}_b^\dagger \hat{A}_a \hat{A}_a^\dagger \hat{A}_b - \hat{A}_a^\dagger \hat{A}_b \hat{A}_b^\dagger \hat{A}_a \\
&= -\frac{1}{4}((\hat{A}_b^\dagger \hat{A}_a)^2 + (\hat{A}_a^\dagger \hat{A}_b)^2) + \frac{1}{4}((\hat{N}_b(\hat{N}_a + 1) + \hat{N}_a(\hat{N}_b + 1)).
\end{aligned} \tag{7.30}$$

Similarly

$$\hat{J}_2^2 = \frac{1}{4}((\hat{A}_b^\dagger \hat{A}_a)^2 + (\hat{A}_a^\dagger \hat{A}_b)^2) + \frac{1}{4}((\hat{N}_b(\hat{N}_a + 1) + \hat{N}_a(\hat{N}_b + 1)). \tag{7.31}$$

and also

$$\hat{J}_3^2 = \frac{1}{4}(\hat{N}_b - \hat{N}_a)^2. \tag{7.32}$$

Then one finds

$$\begin{aligned}
\hat{\mathbf{J}}^2 = \hat{J}_1^2 + \hat{J}_2^2 + \hat{J}_3^2 &= \left(\frac{\hat{N}_a + \hat{N}_b}{2}\right)\left(\frac{\hat{N}_a + \hat{N}_b}{2} + 1\right) \\
&= \left(\frac{\hat{N}}{2}\right)\left(\frac{\hat{N}}{2} + 1\right).
\end{aligned} \tag{7.33}$$

Since the rhs of Eq. (7.33) only depends on $\hat{N}$, then any of the three state vectors are its eigenstates and all give the same eigenvalue of $\frac{n}{2}\left(\frac{n}{2} + 1\right)$. Thus although the

---

[1] Here, the eigenvalue, n, is not being associated with a Fock state, $|n\rangle$.

individual components of $\hat{\mathbf{J}}$ have no common eigenvectors, $\hat{\mathbf{J}}^2$ has the same eigen-value with all three eigenvectors. We also note that

$$[\hat{\mathbf{J}}^2, \hat{J}_i] = 0 \tag{7.34}$$

for any i, since for example, with $i = 1$

$$\begin{aligned}
[\hat{\mathbf{J}}^2, \hat{J}_1] &= [\hat{J}_1^2 + \hat{J}_2^2 + \hat{J}_3^2, \hat{J}_1] \\
&= [\hat{J}_2^2 + \hat{J}_3^2, \hat{J}_1] \\
&= \hat{J}_2[\hat{J}_2, \hat{J}_1] + [\hat{J}_2, \hat{J}_1]\hat{J}_2 + \hat{J}_3[\hat{J}_3, \hat{J}_1] + [\hat{J}_3, \hat{J}_1]\hat{J}_3 \\
&= -i\hat{J}_2\hat{J}_3 - i\hat{J}_3\hat{J}_2 + i\hat{J}_3\hat{J}_2 + i\hat{J}_2\hat{J}_3 = 0.
\end{aligned} \tag{7.35}$$

Clearly, we get the same result for $i = 2$ and $i = 3$.

We began this section by characterizing the degenerate two category system by a pair of number operators, $\hat{N}_a$ and $\hat{N}_b$. Now recalling that $\hat{N} = \hat{N}_a + \hat{N}_b$ and $\hat{Z}_{ab} = \hat{N}_b - \hat{N}_a$ we could instead use the pair of operators $\hat{N}$ and $\hat{Z}_{ab}$ which contain the same information as $\hat{N}_a$ and $\hat{N}_b$. Further, we can replace $\hat{N}$ and $\hat{Z}_{ab}$ by $\hat{\mathbf{J}}^2$ and $\hat{J}_3$ since $\hat{\mathbf{J}}^2$ depends only on $\hat{N}$ and $\hat{Z}_{ab}$ depends only on $\hat{J}_3$. So now, putting $\frac{(n_b - n_a)}{2} = m$ and $\frac{(n_a + n_b)}{2} = j$, then instead of the eigenstate $|n_a, n_b\rangle$ we can use $|j, m\rangle$. Then

$$\hat{\mathbf{J}}^2|j, m\rangle = j(j+1)|j, m\rangle \quad \text{and} \quad \hat{J}_3|j, m\rangle = m|j, m\rangle. \tag{7.36}$$

There is a further interesting property of the operators in Eq. (7.28) which can be seen by defining $\hat{R} = \hat{J}_1 - i\hat{J}_2$. Then one finds

$$[\hat{J}_3, \hat{R}] = -\hat{R}. \tag{7.37}$$

Furthermore

$$\hat{J}_3\hat{R}|j, m\rangle = \hat{R}\hat{J}_3|j, m\rangle - \hat{R}|j, m\rangle = (m-1)\hat{R}|j, m\rangle, \tag{7.38}$$

which shows that $\hat{R}|j, m\rangle = \alpha(j, m)|j, m-1\rangle$, where $\alpha(j, m)$ is a scalar which can depend on j and m. So, $\hat{R}$ acts as a lowering operator for $\hat{J}_3$. Similarly we find $\hat{R}^\dagger|j, m\rangle = \beta(j, m)|j, m+1\rangle$, where $\beta(j, m)$ is a scalar that can depend on j and m. So, $\hat{R}^\dagger$ acts as a raising operator for $\hat{J}_3$. This shows that the eigenstates of $\hat{J}_3$ are separated by ± 1. So, since $m = \frac{(n_b - n_a)}{2}$, the system can be in a state which m is either integer or half integer. However, the raising and lowering operator properties mean that, if the system is in a state with m as an integer, then it can be changed to any other integer, but if it is in a state with m as half-integer, then it can only be raised or lowered to another half-integer state. The significance of this result will not become apparent until we have looked at relativistic phenomena in a later chapter.

It can also be seen that

$$[\hat{R}, \hat{R}^\dagger] = [\hat{J}_1 - i\hat{J}_2, \hat{J}_1 + i\hat{J}_2] = -2\hat{J}_3 \tag{7.39}$$

and

$$\hat{R}^\dagger\hat{R} = \hat{J}_1^2 + \hat{J}_2^2 + \hat{J}_3 = \hat{\mathbf{J}}^2 - \hat{J}_3^2 + \hat{J}_3. \tag{7.40}$$

Then we find

$$\hat{R}^\dagger \hat{R}|j,m\rangle = (j(j+1) - m(m-1))|j,m\rangle, \qquad (7.41)$$

and so

$$\langle j,m|\hat{R}^\dagger \hat{R}|j,m\rangle = \| \hat{R}|j,m\rangle \|^2 = \alpha^2(j,m)$$
$$= (j(j+1) - m(m-1)), \qquad (7.42)$$

from which we get $\alpha(j,m) = \sqrt{j(j+1) - m(m-1)}$. Furthermore

$$\hat{R}^\dagger \hat{R}|j,m\rangle = \beta(j,m-1)\alpha(j,m)|j,m\rangle, \qquad (7.43)$$

which makes $\beta(j,m) = \sqrt{j(j+1) - m(m+1)}$.

For the pair (n_1, n_2) and the pair (j,m) to be equivalent representations of the two-category system, they should contain exactly the same information about the system. This means we should count the same number of distinct pairs in each case. In other words, the two representations should have the same degeneracy. We have already noted that the degeneracy of the (n_1, n_2) representation, where $n_1 + n_2 = n$ is $n+1$. In the case of the (j,m) representation, we know that $j = \frac{n}{2}$ so there are the same number of values of n as j. So, the degeneracy depends on how many values of m there are for a given value of j. Now we have found that the m values are separated by ± 1 and can have values between $\pm j$. If n is even then j is a natural number and m has integer values. For example if $n = 4$, then $j = 2$ and so m can have values of $0, \pm 1$ and ± 2, so the degeneracy is $5 = 4+1$, which is $n+1$, as required. It is easy to see that when n is even, then the degeneracy of the (j,m) representation is $n+1$. Similarly, if n is odd, then j takes on half integer values, as does m. Notice this rules out $m = 0$ when n is odd, because the integer step between $m = \pm\frac{1}{2}$ must be preserved. For example if $n = 3$, then $j = \frac{1}{2}$ and there are four possible values of m, i.e., $m = \pm\frac{1}{2}, \pm\frac{3}{2}$. So, in this case too, the degeneracy is $n+1$. This is clearly the general case, so we can see that the degeneracy of the two representations is identical and that there is a one-to-one correspondence between them. Also, the fact that m can take on half-integer values when n is odd is not a matter of concern, since it should be recalled that m is the eigenvalue of $\frac{\hat{L}_{ab}}{2}$, so $\hat{L}_{ab}$ always has integer eigenvalues as expected from Eq. (7.27). However, these integer values are now separated by ± 2, so that, for example, when $n = 4$, the allowed eigenvalues of $\hat{L}_{ab}$ are $0, \pm 2$, and ± 4 and when $n = 3$, they are ± 1 and ± 3.

It should also be noted that the symmetry properties of the system may actually be derived from the Lie group properties in Eq. (7.29) alone, without the specific representations Eq. (7.21)-(7.24). We should also stress that for the two-category system, the eigenvalue, $j(j+1)$ of the operator, $\hat{J}^2$ has no particular dynamical significance. It simply is an alternative way of representing $n = 2j$, the total number of items, just as $m = \frac{n_2 - n_1}{2}$ is an alternative way of representing the difference in the numbers in each category. We will see that these operators do take on important dynamical roles in the case of degenerate three-category systems that will be treated in Chapter 8.

7.3 THE EMERGENCE OF TWO-DIMENSIONAL CONFIGURATION SPACE

Next we express the non-Hermitian operators, $\hat{A}_a$, and $\hat{A}_b$ in a form involving a pair each of Hermitian operators, as was done in Chapter 3. Let $\hat{A}_a = (\hat{X} + i\hat{P}_x)/\sqrt{2}$ and $\hat{A}_b = (\hat{Y} + i\hat{P}_y)/\sqrt{2}$, then

$$\hat{N} = \hat{N}_a + \hat{N}_b = \hat{A}_a^\dagger \hat{A}_a + \hat{A}_b^\dagger \hat{A}_b$$
$$= \frac{1}{2}(\hat{P}_x^2 + \hat{X}^2 + \hat{P}_y^2 + \hat{Y}^2 + 2), \tag{7.44}$$

$$\hat{K}_{ab} = \hat{A}_a^\dagger \hat{A}_b + \hat{A}_b^\dagger \hat{A}_a$$
$$= \frac{1}{2}(\hat{P}_x \hat{P}_y + \hat{X}\hat{Y}), \tag{7.45}$$

$$\hat{L}_{ab} = i(\hat{A}_b^\dagger \hat{A}_a - \hat{A}_a^\dagger \hat{A}_b)$$
$$= \hat{X}\hat{P}_y - \hat{Y}\hat{P}_x, \tag{7.46}$$

and

$$\hat{Z}_{ab} = \hat{N}_b - \hat{N}_a = \hat{A}_b^\dagger \hat{A}_b - \hat{A}_a^\dagger \hat{A}_a$$
$$= \frac{1}{2}(\hat{P}_y^2 + \hat{Y}^2) - \frac{1}{2}(\hat{P}_x^2 + \hat{X}^2). \tag{7.47}$$

In addition we have

$$\hat{\Omega} = \omega_0 \hat{N} = \frac{\omega_0}{2}(\hat{P}_x^2 + \hat{X}^2 + \hat{P}_y^2 + \hat{Y}^2 + 2), \tag{7.48}$$

and it is then straightforward to show, that

$$\frac{1}{4}(\hat{K}_{ab}^2 + \hat{L}_{ab}^2 + \hat{Z}_{ab}^2) = \frac{\hat{\Omega}}{2}\left(\frac{\hat{\Omega}}{2} + 1\right), \tag{7.49}$$

which is equivalent to Eq. (7.33).

Two dimensions of configuration space have now emerged in the form of the operators, $\hat{X}$ and $\hat{Y}$, together with their corresponding components of linear momentum, $\hat{P}_x$ and $\hat{P}_y$. As was done in Chapter 5, the configuration space components can be represented by corresponding scalar coordinates, x and y. Because $\hat{X}$ and $\hat{Y}$ commute with one another, the co-ordinates are mathematically orthogonal.

Finally, we note that the operator $\hat{L}_{ab}$ in Eq. (7.46) has the form of the component of a vector product of the two-dimensional configuration space vector, $(\hat{X}, \hat{Y})$, and the two-dimensional momentum space vector, $(\hat{P}_x, \hat{P}_y)$, which suggests that it can be interpreted as a component of angular momentum, L_z, in a new direction that is orthogonal to the plane containing the configuation space coordinates, (x, y). This indicates that our picture is incomplete and we need a third coordinate z, so that x, y and z could form a set of three mutually orthogonal, and thus independent coordinates. This suggests we explore a 3-category system. We will postpone this step until Chapter 8. We will investigate the significance of $\hat{L}_z$ a little further in the next section.

7.4 THE EMERGENCE OF ANGULAR MOMENTUM

Let us go back to the non-degenerate 2-category system for a moment, with a Hamiltonian of the form

$$\hat{\Omega} = \frac{1}{2}(\omega_x(\hat{P}_x^2 + \hat{X}^2 + 1) + \omega_y(\hat{P}_y^2 + \hat{Y}^2 + 1)). \tag{7.50}$$

Now let us evaluate the time derivative of $\hat{L}_z$. We find

$$i\frac{\partial \hat{L}_z}{\partial t} = [\hat{L}_z, \hat{\Omega}] = i(\omega_x - \omega_y)(\hat{P}_x\hat{P}_y + \hat{X}\hat{Y}). \tag{7.51}$$

So, unless $\omega_x = \omega_y$, then $\hat{L}_z$ does not commute with $\hat{\Omega}$ and hence is not a constant of the motion, i.e., not a conserved quantity in time. However, in the degenerate case, $\omega_x = \omega_y = \omega_0$ and then $[\hat{L}_z, \hat{\Omega}] = 0$. So, in the degenerate two-dimensional system, $\hat{L}_z$ is conserved and can be considered an important element of its dynamics. There is another important significance to the commutation of $\hat{L}_z$ and the Hamiltonian. The degenerate Hamiltonian is invariant to rotations in the x-y plane. This, it turns out, is the meaning of $\hat{L}_z$. If an operator commutes with $\hat{L}_z$, then it guarantees invariance to rotation about about the z-axis. This can be seen more directly if we switch to a polar co-ordinate system. Let us first define a degenerate two-dimensional Hamiltonian, $\hat{H}_2$, in dimensionless form, by[2]

$$\hat{H}_2 = \frac{\hat{\Omega}}{\omega_0} + 1 = \hat{N} + 1 = \frac{1}{2}(\hat{P}_x^2 + \hat{X}^2 + \hat{P}_y^2 + \hat{Y}^2), \tag{7.52}$$

where $\hat{N}$ is defined by Eq. (7.48). In terms of the co-ordinates[3] (x, y), then

$$\hat{H}_2\Psi = \frac{1}{2}(-\frac{\partial^2}{\partial x^2} + x^2 - \frac{\partial^2}{\partial y^2} + y^2)\Psi \tag{7.53}$$

and

$$\hat{L}_z\Psi = -i(x\frac{\partial}{\partial y} - y\frac{\partial}{\partial x})\Psi. \tag{7.54}$$

The polar co-ordinates that we require are a radial co-ordinate, ρ, and an azimuthal angle, ϕ, in the x-y plane, such that $x = \rho\cos\phi, y = \rho\sin\phi$. Then

$$\hat{H}_2\Psi = \frac{1}{2}(-\frac{\partial^2}{\partial \rho^2} - \frac{1}{\rho}\frac{\partial}{\partial \rho} - \frac{1}{\rho^2}\frac{\partial^2}{\partial \phi^2} + \rho^2)\Psi \tag{7.55}$$

and

$$\hat{L}_z\Psi = -i\frac{\partial \Psi}{\partial \phi}. \tag{7.56}$$

---

[2]Adding the $1 = \frac{1}{2} + \frac{1}{2}$ to the rhs of Eq. (7.52), puts it on the same footing as the single harmonic oscillator case in Section 5.2. Each half represents the (dimensionless) zero point energy of each of the two oscillators. The rhs of Eq. (7.52) then correctly represents the energy of the pair of oscillators in a form that corresponds to classical harmonic oscillators.

[3]Strictly speaking, these co-ordinates should be regarded, here and in what follows as dimensionless.

It is obvious that $\hat{H}_2$ is independent of ϕ and that $\hat{L}_z$ is independent of both ρ and ϕ, and so we can infer that

$$[\hat{L}_z, \hat{H}_2] = 0. \tag{7.57}$$

Eq. (7.57) not only tells us that $\hat{L}_z$ is in independent of time, and is therefore a constant of the motion, but also that $\hat{H}_2$ is independent of ϕ and so has rotational symmetry about the z-axis. It is also worth noting that it is the degeneracy of the system that leads to the conservation rules that we have found in this two-dimensional system.

It is clear that $\hat{L}_z$ represents a component of angular momentum about a z-axis and that it is associated with a rotation about this axis. However, although we have found an interesting symmetry in the two-dimensional space, there is something unsatisfactory about it. We have two dimensions of configuration space, x and y, together with two corresponding components of linear momentum that are associated with differentiation with respect to x and y. However, we only have a single axis of rotation of this space, which we have identified with a z-axis, that is necessarily orthogonal to x and y. So, we really have a third dimension, but we do not see rotation around the x and y directions, nor is there a component of linear momentum associated with z. Something is clearly missing. To see what is required, we will need to examine a degenerate three-category system. This will be done in the next chapter. However, before we take this issue further we will examine the degenerate two-category case for fermions as this also has some interesting and important consequences.

7.5   A DEGENERATE FERMIONIC CASE

If we impose the degeneracy condition, $\omega_a = \omega_b = \omega_0$, on the fermionic system with coupled creation and annihilation operator in Section 6.4.2, we get $\omega_g = \omega_h = \omega_0 \cos(2\gamma)$ and $\upsilon_{gh} = \omega_0 \sin(2\gamma)$. The Hamiltonian, Eq. (6.89) then becomes[4]

$$\hat{\Omega} = U(\hat{C}_g^\dagger \hat{C}_g + \hat{C}_h^\dagger \hat{C}_h) + V(\hat{C}_h^\dagger \hat{C}_g^\dagger + \hat{C}_g \hat{C}_h), \tag{7.58}$$

where $U = \omega_0 \cos(2\gamma)$ and $V = \omega_0 \sin(2\gamma)$. Unlike degenerate bosonic examples examples in the previous sections of this chapter, here interaction term does not, in general, disappear in the degenerate case. So, here the number of items in the individual categories is not a constant.

Applying the Heisenberg equation of motion to $\hat{C}_g$ gives

$$i\frac{d\hat{C}_g}{dt} = [\hat{C}_g, \hat{\Omega}] = U\hat{C}_g - V\hat{C}_h^\dagger. \tag{7.59}$$

In order to close the system of equations we need the equation of motion for $\hat{C}_h^\dagger$.

$$i\frac{d\hat{C}_h^\dagger}{dt} = [\hat{C}_h^\dagger, \hat{\Omega}] = -U\hat{C}_h^\dagger - V\hat{C}_g. \tag{7.60}$$

---

[4]We have neglected the additive constant μ_{gh}, since it has no effect on the equations of motion.

To find the eigenfrequencies of the system we again need to take both $\hat{C}_g$ and $\hat{C}_h^\dagger$ proportional to $\exp(-i\omega t)$. So, the eigenvalue equation is

$$\begin{pmatrix} \omega - U & V \\ V & \omega + U \end{pmatrix} \begin{pmatrix} \hat{C}_g \\ \hat{C}_h^\dagger \end{pmatrix} = 0. \tag{7.61}$$

This eigenvalue equation is satisfied if the determinant of the 2×2 matrix is zero, i.e.,

$$\begin{vmatrix} \omega - U & V \\ V & \omega + U \end{vmatrix} = (\omega - U)(\omega + U) - V^2 = 0, \tag{7.62}$$

which leads to

$$\omega^2 = U^2 + V^2. \tag{7.63}$$

We can write Eq. (7.61) in the form

$$\hat{\Omega} \begin{pmatrix} \hat{C}_g \\ \hat{C}_h^\dagger \end{pmatrix} = \begin{pmatrix} U & V \\ V & -U \end{pmatrix} \begin{pmatrix} \hat{C}_g \\ \hat{C}_h^\dagger \end{pmatrix}, \tag{7.64}$$

where

$$\hat{\Omega} = \begin{pmatrix} \omega & 0 \\ 0 & \omega \end{pmatrix}. \tag{7.65}$$

Then we note that

$$\begin{pmatrix} U & V \\ V & -U \end{pmatrix} = U \begin{pmatrix} 1 & 0 \\ 0 & -1 \end{pmatrix} + V \begin{pmatrix} 0 & 1 \\ 1 & 0 \end{pmatrix}. \tag{7.66}$$

We can recognize the two 2×2 matrices on the rhs of Eq. (7.66) as two of the Pauli matrices, i.e.

$$\hat{\Omega}\phi = (U\hat{\sigma}_3 + V\hat{\sigma}_1)\phi, \tag{7.67}$$

where

$$\phi = \begin{pmatrix} \hat{C}_g \\ \hat{C}_h^\dagger \end{pmatrix}. \tag{7.68}$$

Operating again with $\hat{\Omega}$ on Eq. (7.67) yields

$$\hat{\Omega}^2 \phi = (U\hat{\sigma}_3 + V\hat{\sigma}_1)(U\hat{\sigma}_3 + V\hat{\sigma}_1)\phi = (U^2 + V^2)\phi, \tag{7.69}$$

in agreement with Eq. (7.63). The form of Eq. (7.67) is identical to that of Dirac equation for relativistic fermions in reduced dimensions [26, 67]. The fact that it has appeared in the context of a pair of coupled fermionic creation and annihilation operators has an important significance. We can explore this a little further by noting that, in the degenerate case here, just as in the non-degenerate fermionic case that was treated in Section 6.4.2, the individual category number operators, do not commute with $\hat{\Omega}$ and so vary with time. Also, just as in the non-degenerate case, the total number of items, $\hat{N} = \hat{N}_g + \hat{N}_h$ is not conserved, but we do have

$$\frac{d(\hat{N}_g - \hat{N}_h)}{dt} = 0,$$

so again the difference, $\hat{N}_g - \hat{N}_h$ is a conserved quantity. This rather anomalous behaviour does have an interesting interpretation. If we suppose that each item of Cat_g carries a *charge*, q and that each item of Cat_h carries a charge of $-q$, then the total charge $\hat{Q} = q(\hat{N}_g - \hat{N}_h)$ is conserved. Given that Eq. (7.67) suggests that we are dealing with relativistic fermions and that Eq. (7.63) implies a pair of eigenvalue for ω of

$$\omega = \pm \sqrt{U^2 + V^2}, \tag{7.70}$$

then we could suppose we are dealing with the excitation of a pair of fermions with equal and opposite charge. We will explore this picture further in Chapter 9 in the context of the theory of *superconductivity* and in Chapter 11 in the context of an explanation for the origins of fermionic mass, where we deal with a mechanism that can explain the coupling between the fermionic creation and annihilation operators in terms of the exchange of bosons between a fermion and its anti-particle. We will also meet this picture again in Chapter 12, in a dynamic context, where an explanation of the phenomenon of *zitterbewegung* is found.

7.6 COMMENT

In this chapter we have uncovered some important new features of two-category systems, in both the bosonic and fermionic cases. However, it is important to note that these have come via simple unitary transformations of the natural number amplitude operators (i.e., creation and annihilation operators) that define the number operators in the original equation, Eq. (6.1), for the total population, together with the degenerate Hamiltonian, Eq. (7.1), in which the two categories are completely independent. Thus, the structure that leads to new quantities like angular momentum in the bosonic case and fermions that are seen to be related to the Dirac equation, is already embedded in the simple system of two independent categories and hence in the simplest description of the two category system as two independent sets of items. The new phenomenology is simply revealed by the unitary transformation, which as we have seen, takes the form of a simple rotation in the space defined by the pair of number amplitude operators. This further emphasizes just how fundamental the number operators and their constituent amplitude operators are to the behaviour of physical systems. These results ultimately go back directly to the universal quantum equation, Eq. (1.1). The next few chapters will reveal even more of its fundamental importance to the quantum nature of things.

8 Degenerate three-category systems

8.1 OPERATORS IN DEGENERATE THREE-CATEGORY SYSTEMS

In the previous chapter we identified a component of angular momentum that was orthogonal to the two-dimensional configuration space that had emerged from a degenerate two-category bosonic system. This implied that we needed a third dimension of configuration space to complete our configuration space picture. This suggests we explore a bosonic three-category system. So, we now let $\hat{N}$ be

$$\hat{N} = \hat{N}_1 + \hat{N}_2 + \hat{N}_3, \tag{8.1}$$

where $\hat{N}_i = \hat{A}_i^\dagger \hat{A}_i$, with $i = 1, 2$ or 3. The system is represented by the eigenstate of $\hat{N}$ which is $|n_1, n_2, n_3\rangle$ in occupation number representation. It is interesting to note that we can define a three-dimensional vector operator $\hat{\mathbf{A}} = (\hat{A}_1, \hat{A}_2, \hat{A}_3)$ with its Hermitian conjugate, $\hat{\mathbf{A}}^\dagger = (\hat{A}_1^\dagger, \hat{A}_2^\dagger, \hat{A}_3^\dagger)$. Then we have

$$\begin{aligned}\hat{N} = \hat{\mathbf{A}}^\dagger . \hat{\mathbf{A}} &= \hat{A}_1^\dagger \hat{A}_1 + \hat{A}_2^\dagger \hat{A}_2 + \hat{A}_3^\dagger \hat{A}_3 \\ &= \hat{N}_1 + \hat{N}_2 + \hat{N}_3.\end{aligned} \tag{8.2}$$

There is no huge gain in using $\hat{\mathbf{A}}$ in this context, apart from a certain degree of formal neatness, but it does prove useful later in the context of calculating the degeneracy of the three-category system.

Having learned some of the symmetry properties of the two-category system, that were based on the constants of the motion of that system, we can investigate symmetries of the three-category case by investigating the effect on the state, $|n_1, n_2, n_3\rangle$, of the operators

$$\hat{K}_i = \hat{A}_j^\dagger \hat{A}_k + \hat{A}_k^\dagger \hat{A}_j, \tag{8.3}$$

$$\hat{L}_i = i(\hat{A}_k^\dagger \hat{A}_j - \hat{A}_j^\dagger \hat{A}_k), \tag{8.4}$$

and

$$\hat{Z}_i = \hat{A}_k^\dagger \hat{A}_k - \hat{A}_j^\dagger \hat{A}_j = \hat{N}_k - \hat{N}_j, \tag{8.5}$$

where $i = 1, j = 2$ and $k = 3$ or cyclic permutation of these. The three operators in Eqs. (8.3) to (8.5) are just those in Eqs. (7.22) to (7.24), except that here they each have three components.

The Hamiltonian of the degenerate three-category system is

$$\hat{\Omega} = \omega_0 \hat{N} = \omega_0 (\hat{N}_1 + \hat{N}_2 + \hat{N}_3). \tag{8.6}$$

DOI: 10.1201/9781003377504-8

It is a straightforward matter to check that all of the quantities in Eqs. (8.3)-(8.5) commute with $\hat{\Omega}$ and so are indeed constants of the motion. Furthermore, we can construct triplets of the operators above such as $\frac{1}{2}\hat{L}_i$, $\frac{1}{2}\hat{K}_i$ and $\frac{1}{2}\hat{Z}_i$ that form a three-element Lie subgroup with properties like those in the two-category case. However, each group of three elements would not span the whole system. Consequently we will look for other combinations that do. Now we have three components of each type in Eqs. (8.3)-(8.5), so we can check if any of them form three element groups. Let's take the $\hat{L}_i$s. Then

$$\begin{aligned}
[\hat{L}_1, \hat{L}_2] &= -[\hat{A}_3^\dagger \hat{A}_2 - \hat{A}_2^\dagger \hat{A}_3, \hat{A}_1^\dagger \hat{A}_3 - \hat{A}_3^\dagger \hat{A}_1] \\
&= -[\hat{A}_3^\dagger \hat{A}_2, \hat{A}_1^\dagger \hat{A}_3] - [\hat{A}_2^\dagger \hat{A}_3, \hat{A}_3^\dagger \hat{A}_1] \\
&= \hat{A}_1^\dagger \hat{A}_2 - \hat{A}_2^\dagger \hat{A}_1 \\
&= i\hat{L}_3.
\end{aligned}$$
(8.7)

Similarly one finds $[\hat{L}_2, \hat{L}_3] = i\hat{L}_1$ and $[\hat{L}_3, \hat{L}_1] = i\hat{L}_2$, so the $\hat{L}_i$s do form a closed Lie subgroup of 3 elements. Before we explore the consequences of this symmetry between the three $\hat{L}_i$s, we will check the $\hat{K}_i$s. Then, for example, we get

$$\begin{aligned}
[\hat{K}_1, \hat{K}_2] &= [\hat{A}_2^\dagger \hat{A}_3 + \hat{A}_3^\dagger \hat{A}_2, \hat{A}_3^\dagger \hat{A}_1 + \hat{A}_1^\dagger \hat{A}_3] \\
&= [\hat{A}_2^\dagger \hat{A}_3, \hat{A}_3^\dagger \hat{A}_1] + [\hat{A}_3^\dagger \hat{A}_2, \hat{A}_1^\dagger \hat{A}_3] \\
&= \hat{A}_2^\dagger \hat{A}_1 - \hat{A}_1^\dagger \hat{A}_2 \\
&= -i\hat{L}_3,
\end{aligned}$$
(8.8)

with similar outcomes for the other $\hat{K}_i$ components. So, the three $\hat{K}_i$s do not form a closed Lie group. Rather they are part of a larger Lie group involving all of the nine components, the three $\hat{K}_i$s, three $\hat{L}_i$s and the three $\hat{Z}_i$s. Two further examples are

$$[\hat{L}_3, \hat{K}_3] = 2i\hat{Z}_3,$$
(8.9)

and

$$[\hat{Z}_1, \hat{Z}_2] = 0.$$
(8.10)

We can see from Eq. (8.10) that the $\hat{Z}_i$s also do not form a Lie subgroup. However, the fact that the three $\hat{L}_i$s form a closed subgroup gives them an extra symmetry that turns out to be of special significance, as we shall see. We already know what to expect from the symmetry of the closed three component Lie group, an example of which was developed for the two-category case (also see Appendix D). This suggests that we define a vector operator, $\hat{\mathbf{L}} = (\hat{L}_1, \hat{L}_2, \hat{L}_3)$ and then an operator $\hat{\mathbf{L}}^2$ such that

$$\hat{\mathbf{L}}^2 = \hat{L}_1^2 + \hat{L}_2^2 + \hat{L}_3^2$$
(8.11)

Using the result in Eq. (8.7) and its cyclic permutation of components, it is straightforward to show that

$$[\hat{\mathbf{L}}^2, \hat{L}_3] = 0$$
(8.12)

and hence that $\hat{\mathbf{L}}^2$ and $\hat{L}_3$ share a common eigenstate. We also know that $\hat{\mathbf{L}}^2$ and $\hat{L}_3$ commute with $\hat{N}$, so we can write a three component eigenstate, $|n,l,m\rangle$, such that

$$\hat{N}|n,l,m\rangle = n|n,l,m\rangle,$$

$$\hat{\mathbf{L}}^2|n,l,m\rangle = l(l+1)|n,l,m\rangle,$$

$$\hat{L}_3|n,l,m\rangle = m|n,l,m\rangle. \tag{8.13}$$

The first of Eqs. (8.13) is obvious, and the last two come from the properties of the Lie group characterized by the Lie products between the three $\hat{L}_i$s which are derived in Appendix D. There one finds that l can take on values of natural numbers or half-natural numbers, while m can take on integer values when l is a natural number and half-integer values when l has a value equal to half a natural number. The m values are separated by ± 1. The other condition that applies to l and m is that $|m| \leq l$.

Now in the present case we have established that the eigenvalues of the $\hat{L}_i$ operators are integers. So, we can rule out the half integer eigenvalues that the Lie group allows when applied to the 3-category system that we are dealing with. This means we must also rule out the half-natural number values of l. The half-integer m values and half-natural number l values do have a use, but to see this we will have to wait until later when we treat the Dirac equation. For the present case then we remain with the possibility that $l = 0, 1, 2, \ldots, l_{max}$ where l_{max} is the, as yet to be determined, maximum value that l can have. Then we would have, $m = 0, \pm 1, \pm 2, \ldots, \pm l_{max}$. It therefore remains to determine the value of l_{max} and which of the natural number values of l and hence the integer values of m are allowed, of the possibilities available. The way this is done is to insist, as we did in the two-category case, that the degeneracies of the two representations of the three-category system must be equal for a given value of n. This just means that we can choose n in the same number of ways in both of the representations, i.e., either $|n_1, n_2, n_3\rangle$ or $|n,l,m\rangle$. Now with the representation based on the eigenstates, $|n_1, n_2, n_3\rangle$, where $n = n_1 + n_2 + n_3$, the degeneracy, $D_n = \frac{1}{2}(n+1)(n+2)$. To see this we note that once n_1, is chosen then we essentially have a two-category choice with a total of $n - n_1$ items, so there are $n - n_1 + 1$ possibilities. Putting $n - n_1 = k$ and then

$$D_n = \sum_{k=0}^{n}(k+1) = \sum_{k=0}^{n}k + \sum_{k=0}^{n}1$$

$$= \frac{1}{2}n(n+1) + (n+1) \tag{8.14}$$

$$= \frac{1}{2}(n+1)(n+2).$$

To determine the degeneracy of the $|n,l,m\rangle$ eigenstates, it is necessary to find which states of those listed above are allowed. We can show quite simply that it cannot be all of them up to a maximum value of l. We can note that for a given value of l, the condition $|m| \leq l$ with m as any integer within this range, means that the degeneracy is $2l + 1$. So, the degeneracy of the three-category system would be

$$
\begin{aligned}
D_{l_{max}} &= \sum_{l=0}^{l_{max}} (2l + 1) = 2 \sum_{l=0}^{l_{max}} l + \sum_{l=0}^{l_{max}} 1 \\
&= l_{max}(l_{max} + 1) + (l_{max} + 1) \\
&= (l_{max} + 1)^2.
\end{aligned}
\tag{8.15}
$$

We can see that whatever natural number value we give to l_{max} it can never equal D_n in Eq. (8.14). To see what l values are allowed and thus obtain the correct degeneracy needs a little more effort. What we need is a raising operator for $\hat{\mathbf{L}}^2$.

We already know the raising and lowering operators that will work for $\hat{L}_3$, from the two-category case and more generally from Appendix D. The lowering operator can be defined by $\hat{R}_3 = \hat{L}_1 - i\hat{L}_2$, then the raising operator is $\hat{R}_3^\dagger = \hat{L}_1 + i\hat{L}_2$ and it is easy to check that $[\hat{L}_3, \hat{R}_3^\dagger] = \hat{R}_3^\dagger$, which leads to $\hat{R}_3^\dagger |n, l, m\rangle \to |n, l, m+1\rangle$. Similarly, $[\hat{L}_3, \hat{R}_3] = -\hat{R}_3$, which leads to $\hat{R}_3 |n, l, m\rangle \to |n, l, m-1\rangle$. Notice that both $\hat{R}_3$ and $\hat{R}_3^\dagger$ commute with $\hat{N}$ and $\hat{\mathbf{L}}^2$, so do not affect n or l.

We next define a new pair of Hermitian conjugate raising and lowering operators by

$$
\hat{Q}_3 = \hat{A}_1 - i\hat{A}_2 \text{ and } \hat{Q}_3^\dagger = \hat{A}_1^\dagger + i\hat{A}_2^\dagger.
\tag{8.16}
$$

Then we find

$$
[\hat{N}, \hat{Q}_3^\dagger] = \hat{Q}_3^\dagger \text{ and } [\hat{L}_3, \hat{Q}_3^\dagger] = \hat{Q}_3^\dagger,
\tag{8.17}
$$

which means that $\hat{Q}_3^\dagger$ raises the eigenvalues of both $\hat{N}$ and $\hat{L}_3$ by 1. The effect of $\hat{Q}_3^\dagger$ on the eigenvalues of $\hat{\mathbf{L}}^2$ is a little more subtle. On evaluating the corresponding commutator brackets we find

$$
[\hat{\mathbf{L}}^2, \hat{Q}_3^\dagger] = 2\hat{Q}_3^\dagger \hat{L}_3 - 2\hat{A}_3^\dagger \hat{R}_3^\dagger + 2\hat{Q}_3^\dagger.
\tag{8.18}
$$

Now we make use of the fact that $\hat{R}_3^\dagger$ raises the eigenvalue m in $|n, l, m\rangle$ by 1 while leaving n and l unchanged, but also when m reaches a value equal to l it can be raised no further and so we must have $\hat{R}_3^\dagger |n, l, l\rangle = 0$. Then we find

$$
\begin{aligned}
\hat{\mathbf{L}}^2 \hat{Q}_3^\dagger |n, l, l\rangle &= (\hat{Q}_3^\dagger \hat{\mathbf{L}}^2 + 2\hat{Q}_3^\dagger \hat{L}_3 - 2\hat{A}_3^\dagger \hat{R}_3^\dagger + 2\hat{Q}_3^\dagger) |n, l, l\rangle \\
&= (l(l+1) + 2l + 2)) \hat{Q}_3^\dagger |n, l, l\rangle \\
&= (l+1)(l+2) \hat{Q}_3^\dagger |n, l, l\rangle,
\end{aligned}
\tag{8.19}
$$

where we have used $\hat{L}_3 |n, l, l\rangle = l |n, l, l\rangle$. Eq. (8.19) shows us that $\hat{Q}_3^\dagger$ raises the eigenvalue of $\hat{\mathbf{L}}^2$ by increasing l by 1 when $m = l$. So, overall we see that

$$
\hat{Q}_3 |n, l, l\rangle \to |n+1, l+1, l+1\rangle.
\tag{8.20}
$$

The final ingredient we need is constructed from the vector operator $\hat{\mathbf{A}} = (\hat{A}_1, \hat{A}_2, \hat{A}_3)$, introduced earlier. Then we have

$$
\hat{\mathbf{A}}^{\dagger 2} = \hat{A}_1^{\dagger 2} + \hat{A}_2^{\dagger 2} + \hat{A}_3^{\dagger 2}.
\tag{8.21}
$$

By direct substitution we can then show that

$$[\hat{N}, \hat{\mathbf{A}}^{\dagger 2}] = 2\hat{\mathbf{A}}^{\dagger 2} \tag{8.22}$$

and

$$[\hat{L}_i, \hat{\mathbf{A}}^{\dagger 2}] = 0, \tag{8.23}$$

for any i, which also implies that $[\hat{\mathbf{L}}^2, \hat{\mathbf{A}}^{\dagger 2}] = 0$. These results show that $\hat{\mathbf{A}}^{\dagger 2}$ raises the eigenvalue n by 2, but does not affect l or m. So, $\hat{\mathbf{A}}^{\dagger 2}|n,l,m\rangle \rightarrow |n+2,l,m\rangle$.

We are now in a position to solve our degeneracy problem and so find l_{max}. We proceed as follows. We can construct the state $|n,l,m\rangle$ by operating on the ground state $|0,0,0\rangle$ with multiples of $\hat{\mathbf{A}}^{\dagger 2}$, $\hat{Q}_3^{\dagger}$ and $\hat{R}_3$, thus [22]

$$\hat{R}_3^{(l-m)}\hat{Q}_3^{\dagger l}\hat{\mathbf{A}}^{\dagger(n-l)}|0,0,0\rangle \rightarrow |n,l,m\rangle. \tag{8.24}$$

The key point about Eq. (8.24) is that $n - l$ must be even so that we operate a whole number, $\frac{n-l}{2}$, of multiples of $\hat{\mathbf{A}}^{\dagger 2}$. That means when n is even, then so is l, and when n is odd then so is l. Now we can evaluate the sum in Eq. (8.15) properly. When n is even, l can take on values of $0, 2, 4, \ldots, l_{max}$ and of course l_{max} must be even. So, writing $l = 2q$, where q is $0, 1, 2, \ldots, q_{max}$, where $l_{max} = 2q_{max}$, then the sum in Eq. (8.15) is instead

$$\begin{aligned}D_{q_{max}} = \sum_{q=0}^{q_{max}}(4q+1) &= 4\sum_{q=0}^{q_{max}}q + \sum_{q=0}^{q_{max}}1 \\ &= 2q_{max}(q_{max}+1) + (q_{max}+1) \\ &= (2q_{max}+1)(q_{max}+1).\end{aligned} \tag{8.25}$$

So,

$$D_{l_{max}} = \frac{1}{2}(l_{max}+1)(l_{max}+2). \tag{8.26}$$

By comparing Eqs. (8.26) and (8.14) we can see that $D_n = D_{l_{max}}$ when $l_{max} = n$. This both ensures that the degeneracy of the $|n,l,m\rangle$ and the $|n_1,n_2,n_3\rangle$ are identical and allows us to determine l_{max}. It is straightforward to check that we get the same result when n is odd. Then l must be odd too and can take on values of $1, 3, 5, \ldots, n$. The degeneracy in this case is again $\frac{1}{2}(n+1)(n+2)$ as required. These results ensure that the representations based on the eigenstates, $|n,l,m\rangle$ and $|n_1,n_2,n_3\rangle$, are completely equivalent.

8.2 THREE-DIMENSIONAL CONFIGURATION SPACE AND ANGULAR MOMENTUM

8.2.1 CARTESIAN FORM

We now let $\hat{A}_1 = (\hat{X} + i\hat{P}_x)/\sqrt{2}$, $\hat{A}_2 = (\hat{Y} + i\hat{P}_y)/\sqrt{2}$, and $\hat{A}_3 = (\hat{Z} + i\hat{P}_z)/\sqrt{2}$. Then we have

$$\hat{\Omega} = \omega_0 \hat{N} = \frac{\omega_0}{2}(\hat{P}_x^2 + \hat{X}^2 + \hat{P}_y^2 + \hat{Y}^2 + \hat{P}_z^2 + \hat{Z}^2 + 3). \tag{8.27}$$

We define a dimensionless three-category Hamiltonian, $\hat{H}_3$, along the same lines as $\hat{H}_2$ in the two-category case, as[1]

$$\hat{H}_3 = \frac{\hat{\Omega}}{\omega_0} + \frac{3}{2} = \hat{N} + \frac{3}{2} = \frac{1}{2}(\hat{P}_x^2 + \hat{X}^2 + \hat{P}_y^2 + \hat{Y}^2 + \hat{P}_z^2 + \hat{Z}^2), \qquad (8.28)$$

where the number operator, $\hat{N}$, is that in Eq. (8.1). From, Eq. (8.28), we can immediately see that the energy eigenvalues are $n = n_1 + n_2 + n_3$.

The Schrödinger equation then takes the analytic form[2]

$$\hat{H}_3 \Psi = \frac{1}{2}(-\frac{\partial^2}{\partial x^2} + x^2 - \frac{\partial^2}{\partial y^2} + y^2 - \frac{\partial^2}{\partial z^2} + z^2)\Psi \qquad (8.29)$$

and the components of angular momentum become

$$\hat{L}_x = -i(y\frac{\partial}{\partial z} - z\frac{\partial}{\partial y}),$$

$$\hat{L}_y = -i(z\frac{\partial}{\partial x} - x\frac{\partial}{\partial z}),$$

$$\hat{L}_z = -i(x\frac{\partial}{\partial y} - y\frac{\partial}{\partial x}). \qquad (8.30)$$

These components satisfy the commutation relations in Eq. (8.7) and its cyclic permutation of subscripts.

Defining, respectively, the vector linear momentum operator, $\hat{\mathbf{P}}$ and the vector angular momentum, $\hat{\mathbf{L}}$ by

$$\hat{\mathbf{P}} = (-i\frac{\partial}{\partial x}, \frac{\partial}{\partial y}, \frac{\partial}{\partial z}) = -i\nabla \qquad (8.31)$$

and

$$\hat{\mathbf{L}} = -i\hat{\mathbf{r}} \wedge \nabla, \qquad (8.32)$$

then

$$\hat{H}_3 = \frac{1}{2}(\hat{\mathbf{P}}^2 + \mathbf{r}^2) = \frac{1}{2}(-\nabla^2 + \mathbf{r}^2), \qquad (8.33)$$

where $\mathbf{r} = (x, y, z)$.

We have now found a system in which there are three independent configuration space co-ordinates, (x, y, z), with three corresponding linear momentum components, Eq. (8.31), that allow linear motion throughout that space, and three independent components of angular momentum, Eq. (8.32), that ensure that there are three independent axes of rotation in that space. This we will take as a complete description of a self-consistent space that can support linear and rotational dynamical activity. Later in the chapter we will test the possibility of a higher dimensional, but still self-consistent space that can support linear and rotational dynamics, using the symmetry techniques similar to those for the three-dimensional case.

---

[1] The addition of the $\frac{3}{2}$ corresponds to the three zero point energies of the three harmonic oscillators that makes the rhs of Eq. (8.28) consistent with the sum of their energies.

[2] Recall that the co-ordinates are still dimensionless here.

8.2.2 SPHERICAL POLAR FORM

We can now transform the system into spherical polar coordinates. Let $\rho = r\sin\theta$ and so $x = \rho\cos\phi\sin\theta$, $y = \rho\sin\phi\sin\theta$, and $z = r\cos\theta$. Then

$$\nabla^2 = \frac{1}{r^2}\frac{\partial}{\partial r}(r^2\frac{\partial}{\partial r}) + \frac{1}{r^2\sin\theta}\frac{\partial}{\partial\theta}(\sin\theta\frac{\partial}{\partial\theta}) + \frac{1}{r^2\sin^2\theta}\frac{\partial^2}{\partial\phi^2}, \tag{8.34}$$

and

$$\hat{L}^2 = -\left(\frac{1}{\sin\theta}\frac{\partial}{\partial\theta}(\sin\theta\frac{\partial}{\partial\theta}) + \frac{1}{\sin^2\theta}\frac{\partial^2}{\partial\phi^2}\right), \tag{8.35}$$

so we can write Eq. (8.34) as

$$\nabla^2 = \frac{1}{r^2}\frac{\partial}{\partial r}(r^2\frac{\partial}{\partial r}) - \frac{\hat{L}^2}{r^2}. \tag{8.36}$$

The dimensionless Hamiltonian in Eq. (8.33) then becomes

$$\hat{H}_3\Psi = \frac{1}{2}\left(-\frac{1}{r^2}\frac{\partial}{\partial r}(r^2\frac{\partial}{\partial r}) + \frac{\hat{L}^2}{r^2} + r^2\right)\Psi = E\Psi. \tag{8.37}$$

We can also write $\hat{L}_z$ in polar form and get

$$\hat{L}_z = i\frac{\partial}{\partial\phi}. \tag{8.38}$$

In this form it is easy to check that

$$[\hat{L}^2,\hat{L}_z] = 0 \tag{8.39}$$

in agreement with the results in Section 8.1. We also find that

$$[\hat{L}_z,\hat{H}_3] = 0, \tag{8.40}$$

and

$$[\hat{L}^2,\hat{H}_3] = 0, \tag{8.41}$$

which tells us that $\hat{L}_z$ and $\hat{L}^2$ are constants of the motion, but also that $\hat{H}_3$ is spherically symmetric, i.e., isotropic. When $\hat{H}_3$ is expressed in its cartesian form in Eq. (8.29), it is straightforward to check that it commutes with all three components of angular momentum, so that $\hat{L} = (\hat{L}_x,\hat{L}_y,\hat{L}_z)$ is a constant of the motion.

8.2.3 SEPARATION OF VARIABLES

Let $\Psi(r,\phi,\theta) = R(r)Y(\phi,\theta)$, then, after a little rearranging we get

$$\left(-\frac{1}{2}\frac{1}{R}\frac{\partial}{\partial r}(r^2\frac{\partial R}{\partial r}) + \frac{r^4}{2} - r^2E\right) + \left(\frac{1}{2Y}\hat{L}^2Y\right) = 0. \tag{8.42}$$

This equation can only be correct if the two bracketed terms are each equal to a constant, which add to zero. Letting these constants be $-\lambda$ and λ, respectively, leads to

$$-\frac{1}{2}\frac{1}{r^2}\frac{\partial}{\partial r}(r^2\frac{\partial R}{\partial r}) + (\frac{1}{2}r^2 + \frac{\lambda}{r^2})R = ER \qquad (8.43)$$

and

$$\frac{1}{2}\hat{L}^2 Y = \lambda Y. \qquad (8.44)$$

Now we know that the eigenvalues of $\hat{L}^2$ are equal to $l(l+1)$ so $2\lambda = l(l+1)$. Then Eq. (8.43) becomes

$$-\frac{1}{2}\frac{1}{r^2}\frac{\partial}{\partial r}(r^2\frac{\partial R}{\partial r}) + \frac{1}{2}(r^2 + \frac{l(l+1)}{r^2})R = ER. \qquad (8.45)$$

If we make the substitution, $\Upsilon = rR$, then Eq. (8.45) becomes

$$\hat{H}_r \Upsilon = -\frac{1}{2}\frac{\partial^2 \Upsilon}{\partial r^2} + \frac{1}{2}(r^2 + \frac{l(l+1)}{r^2})\Upsilon = E\Upsilon. \qquad (8.46)$$

$\hat{H}_r$ can actually be factorized by defining

$$\hat{B}_r = \frac{1}{\sqrt{2}}(\frac{\partial}{\partial r} - \frac{l+1}{r} + r). \qquad (8.47)$$

Then with

$$\hat{B}_r^\dagger = \frac{1}{\sqrt{2}}(-\frac{\partial}{\partial r} - \frac{l+1}{r} + r), \qquad (8.48)$$

one finds

$$\hat{B}_r^\dagger \hat{B}_r = -\frac{1}{2}\frac{\partial^2}{\partial r^2} + \frac{1}{2}(\frac{l(l+1)}{r^2} + r^2) - (l + \frac{3}{2}). \qquad (8.49)$$

Thus

$$\hat{H}_r = \hat{B}_r^\dagger \hat{B}_r + (l + \frac{3}{2}). \qquad (8.50)$$

Clearly, Eq. (8.50) is still satisfied when $\hat{B}_r \to \hat{B}_r \exp(-i\xi)$. Now if we consider the radial equation as a one-dimensional problem, then it can be treated as a single-category system so that we can use the single category number operator, $\hat{N}$, in the form defined by Eq. (3.14). Then we can take

$$i\frac{d}{d\xi}\hat{B}_r = [\hat{B}_r, \hat{N}] = \hat{B}_r, \qquad (8.51)$$

which implies that $\hat{B}_r$ is a lowering operator for $\hat{N}$ and $\hat{B}_r^\dagger \hat{B}_r = f(\hat{N})$. This implies that the eigenvalues of $\hat{B}_r^\dagger \hat{B}_r$ have the form $f(n)$, though, unlike in the true one-dimensional case, this does not rule out dependence on other system parameters such as l. Indeed, standard analytical methods for solving the eigenvalue equation, Eq. (8.46) [19] yield, $E = 2k + l + \frac{3}{2}$, where k is natural number. However, one finds that E only depends on the sum $2k + l$ which is another natural number, n, say, albeit with degenerate dependence on k and l. So then, $E = n + \frac{3}{2}$, which is just what we expect. Now, from Eqs. (8.50) and (8.28), we can deduce that the eigenvalues of $\hat{B}_r^\dagger \hat{B}_r$ must then be equal to $n - l$. This is always nonnegative, as it needs to be, since we showed earlier that for the three-dimensional degenerate oscillator, $n \geq l$. The ground state clearly corresponds to $n = l = 0$, in which case $E = \frac{3}{2}$, as expected. Also notice that $n - l = 2k$, showing that n and l must differ by an even integer, a result deduced earlier.

Arguments similar to those used in the one-dimensional case lead to the ground state, Υ_0 satsfying $\hat{B}_r \Upsilon_0 = 0$. Thus

$$\frac{1}{\sqrt{2}}\left(\frac{\partial}{\partial r} - \frac{l+1}{r} + r\right)\Upsilon_0 = 0, \tag{8.52}$$

which is easily integrated to give $\Upsilon_0 = \chi_0 r^{l+1} \exp(-\frac{r^2}{2})$. In the ground state, we know $l = 0$, Then, $R = \frac{\Upsilon_0}{r} = \exp(-\frac{r^2}{2})$. This is precisely what we should expect since the ground state for $n = 0$, using cartesian coordinates is $\exp(-\frac{1}{2}(x^2 + y^2 + z^2)) = \exp(-\frac{r^2}{2})$.

8.3 CENTRAL POTENTIALS

We can extend our examination of spherically symmetric systems by generalizing $\hat{B}_r$ to

$$\hat{B}_r = \frac{1}{\sqrt{2}}\left(\frac{\partial}{\partial r} - \frac{l+1}{r} + W(r)\right), \tag{8.53}$$

as we did in the one-dimensional case in Chapter 5. Then we find

$$\hat{B}_r^\dagger \hat{B}_r = -\frac{1}{2}\frac{\partial^2}{\partial r^2} + \frac{1}{2}\left(\frac{l(l+1)}{r^2} - \frac{2W(r)(l+1)}{r} + W(r)^2 - W'(r)\right), \tag{8.54}$$

where $W'(r) = \frac{\partial W(r)}{\partial r}$. Notice that with $W(r) = r$, we recover the three-dimensional spherically symmetric oscillator. We can seek other simple cases. The simplest is when $W(r)$ is a constant. If we choose $W(r) = \alpha(l+1)^{-1}$, where α is an adjustable constant, then [119]

$$\hat{B}_r = \frac{1}{\sqrt{2}}\left(\frac{\partial}{\partial r} - \frac{l+1}{r} + \frac{\alpha}{l+1}\right), \tag{8.55}$$

then we obtain

$$\hat{B}_r^\dagger \hat{B}_r = -\frac{1}{2}\frac{\partial^2}{\partial r^2} + \frac{1}{2}\frac{l(l+1)}{r^2} - \frac{\alpha}{r} + \frac{\alpha^2}{2(l+1)^2}. \tag{8.56}$$

Now we can recognize that $\hat{B}_r^\dagger \hat{B}_r$, with an appropriate choice for α, is essentially the radial part of the Hamiltonian, $\hat{H}_C$, for a (dimensionless)[3] Coulomb potential,

---

[3] Again recall that the co-ordinates are still dimensionless here.

apart from the final constant term [22, 95, 126][4]. Thus, the radial energy eigenvalue equation for a Coulomb potential may be written

$$\hat{H}_C \Upsilon_C = (\hat{B}_r^\dagger \hat{B}_r - \frac{\alpha^2}{2(l+1)^2})\Upsilon_C = E\Upsilon_C, \tag{8.57}$$

where Υ_C is the radial wavefunction. Standard methods yield[5]

$$E = -\frac{\alpha^2}{2(n+1)^2},$$

where $n = 0, 1, 2, \ldots$, which implies that

$$\hat{B}_r^\dagger \hat{B}_r \Upsilon_C = (\frac{\alpha^2}{2(l+1)^2} - \frac{\alpha^2}{2(n+1)^2})\Upsilon_C. \tag{8.58}$$

Once again, the eigenvalues of $\hat{B}_r^\dagger \hat{B}_r$ are functions of n, as well as of l, which is consistent with $\hat{B}_r^\dagger \hat{B}_r = f(\hat{N})$. The eigenvalues of $\hat{B}_r^\dagger \hat{B}_r$ must be non-negative. This condition, together with Eq. (8.57) implies $n \geq l$. This is indeed the correct relationship between the principal quantum number, n, and the angular momentum quantum number, l, for a Coulomb potential.

The ground state eigenfunction, Υ_{C_0}, can be obtained for the Coulomb potential by solving $\hat{B}_r \Upsilon_{C_0} = 0$, thus

$$\frac{1}{\sqrt{2}}(\frac{\partial}{\partial r} - \frac{l+1}{r} + \frac{\alpha}{l+1})\Upsilon_{C_0} = 0. \tag{8.59}$$

On integration this yields

$$\Upsilon_{C_0} = r^{l+1} \exp(-\frac{\alpha r}{l+1}).$$

In the ground state, $l = 0$, and so $R(r) = r^{-1}\Upsilon_{C_0} = \exp(-\alpha r)$, which is the correct form for the Coulomb potential.

In principle we can define a general central potential function of the form

$$V(r) = \frac{1}{2}(-\frac{2W(r)(l+1)}{r} + W(r)^2 - W'(r)) + \gamma,$$

where γ is an adjustable constant so that the radial part of the Hamiltonian, $\hat{H}_r$ is

$$\hat{H}_r = \hat{B}_r^\dagger \hat{B}_r + \gamma = -\frac{1}{2}\frac{\partial^2}{\partial r^2} + \frac{1}{2}(\frac{l(l+1)}{r^2}) + V(r). \tag{8.60}$$

---

[4]Notice that this approach does not rely on any need to invoke a planetary model of the atom to justify the inclusion of a Coulomb potential.

[5]The conventional result is $E = -\frac{\alpha^2}{2n^2}$ and the principal quantum number, n is taken to have values of 1, 2, 3, ... However, there is nothing wrong with taking $E = -\frac{\alpha^2}{2(n+1)^2}$ and allowing the principal quantum number to take values of 0, 1, 2, 3, ... This keeps the result in line with previous use of the eigenvalues of $\hat{N}$.

The full Hamitonian in this central potential would then simply be

$$\hat{H} = \frac{1}{2}\hat{P}^2 + V(r) = -\frac{1}{2}\nabla^2 + V(r). \tag{8.61}$$

This Hamiltonian in Eq. (8.61) is spherically symmetric and consequently commutes with both $\hat{L}^2$ and $\hat{L}_z$, which are thus constants of the motion. It is easy to check that $\hat{H}$ in Eq. (8.61) commutes with all three components of angular momentum. This means that $\hat{L} = (\hat{L}_x, \hat{L}_y, \hat{L}_z)$ is also a constant of the motion. Notice that the angular momentum components are also constants of the motion for the special case when $V(r) = 0$, i.e., in the absence of any potential.

8.4   SCALAR AND VECTOR POTENTIALS

Up to now, we have used a representation of $\hat{P}$, which, in the one-dimensional case, takes the form $\hat{P} \rightarrow -i\frac{d}{dx}$ and $\hat{X} \rightarrow x$, since

$$[i\hat{P}, \hat{X}] \rightarrow \left[\frac{d}{dx}, x\right] = 1. \tag{8.62}$$

However, it is not difficult to see that if we let $\hat{P}$ be represented by $\hat{P} \rightarrow -i\frac{d}{dx} + Q(x)$ where $Q(x)$ is a real function of x and is thus Hermitian, then $\hat{P}$ remains Hermitian with this modification, and still satisfies

$$[\frac{d}{dx} + iQ(x), x] = 1.$$

With this new representation of $\hat{P}$, the Hamiltonian eigenvalue equation now takes the form

$$\hat{H}\Psi = (\frac{1}{2}\hat{P}^2 + V(x))\Psi = (\frac{1}{2}(-i\frac{d}{dx} + Q(x))^2 + V(x))\Psi = \lambda\Psi, \tag{8.63}$$

where λ is an eigenvalue. If we let $\Psi = \Phi\exp(i\theta(x))$, where $\theta(x)$ is an arbitrary real phase that is a function of x, then we find

$$-i\frac{d\Phi\exp(i\theta(x))}{dx} + Q(x)\Phi\exp(i\theta(x)) = (-i\frac{d\Phi}{dx} + \frac{d\theta(x)}{dx}\Phi + Q(x)\Phi)\exp(i\theta(x)). \tag{8.64}$$

If we now choose $\theta(x)$ such that $\frac{d\theta(x)}{dx} + Q(x) = 0$, then

$$-i\frac{d\Phi\exp(i\theta(x))}{dx} + Q(x)\Phi\exp(i\theta(x)) = (-i\frac{d\Phi}{dx})\exp(i\theta(x)). \tag{8.65}$$

Now we can apply $(-i\frac{d}{dx} + Q(x))$ to $(-i\frac{d\Phi}{dx})\exp(i\theta(x))$ and obtain

$$(-i\frac{d}{dx} + Q(x))(-i\frac{d\Phi}{dx}\exp(i\theta(x)) = (-\frac{d^2\Phi}{dx^2})\exp(i\theta(x)). \tag{8.66}$$

The eigenvalue equation is now

$$\left(-\frac{1}{2}\frac{d^2\Phi}{dx^2}+V(x)\Phi\right)\exp(i\theta(x)) = \lambda\Phi\exp(i\theta(x)), \qquad (8.67)$$

so, after dividing throughout by $\exp(i\theta(x))$, we obtain

$$\left(-\frac{1}{2}\frac{d^2}{dx^2}+V(x)\right)\Phi = \lambda\Phi. \qquad (8.68)$$

This is precisely the one-dimensional eigenvalue equation we would have got had we not bothered to add the $Q(x)$ term to the momentum operator in the first place. In other words, in the one-dimensional case the $Q(x)$ has no effect on the eigenvalue spectrum of the system. It merely introduces an extra phase into the wave function of the form $\theta(x) = -\int Q(x)\,dx$. This phase shift of course has no effect on the probability densities, since $\Psi^*\Psi = \Phi^*\Phi$.

Also the addition of the $Q(x)$ term has no effect on the H-type equations of motion since

$$\frac{dx}{dt} = -i[x,\hat{H}] = \hat{P} \qquad (8.69)$$

and

$$\frac{d\hat{P}}{dt} = -i[\hat{P},\hat{H}] = -\frac{\partial V(x)}{\partial x}. \qquad (8.70)$$

If we now turn to the three-dimensional case the situation is quite different. With three dimensions we now have three spatial co-ordinates and three components of linear momentum. So, we can define the components of linear momentum as

$$\hat{P}_x \rightarrow -i\frac{\partial}{\partial x} + Q_x(x,y,z),$$

$$\hat{P}_y \rightarrow -i\frac{\partial}{\partial y} + Q_y(x,y,z),$$

and

$$\hat{P}_z \rightarrow -i\frac{\partial}{\partial z} + Q_z(x,y,z),$$

where $Q_x(x,y,z), Q_y(x,y,z)$ and $Q_z(x,y,z)$ are arbitrary real scalar functions. Clearly, we still have $[i\hat{P}_x,x] = [i\hat{P}_y,y] = [i\hat{P}_z,z] = 1$ and so the canonical quantum commutator brackets are still satisfied. However, now the different components of linear momentum no longer commute and so are no longer independent of one another. For example

$$[\hat{P}_x,\hat{P}_y] = \left[-i\frac{\partial}{\partial x}+Q_x,-i\frac{\partial}{\partial y}+Q_y\right] = -i\left(\frac{\partial Q_y}{\partial x}-\frac{dQ_x}{dy}\right) = iB_z, \qquad (8.71)$$

where B_z is the z-component of $-\nabla \wedge \mathbf{Q}$, with $\mathbf{Q} = (Q_x,Q_y,Q_z)$. Taking the three-dimensional Hamiltonian as

$$\hat{H} = \frac{1}{2}\hat{\mathbf{P}}^2 + V(x,y,z), \qquad (8.72)$$

where $\hat{\mathbf{P}} = (\hat{P}_x, \hat{P}_y, \hat{P}_z)$, one then finds

$$\hat{\mathbf{v}} = \frac{d\mathbf{x}}{dt} = -i[\mathbf{x}, \hat{H}] = \hat{\mathbf{P}}, \tag{8.73}$$

where $\mathbf{x} = (x, y, z)$ and $\hat{\mathbf{v}}$ is the velocity vector operator. Then, from

$$\frac{d\hat{P}_x}{dt} = -i[\hat{P}_x, \hat{H}] \tag{8.74}$$

we get

$$\begin{aligned}
\frac{d\hat{P}_x}{dt} &= -i[\hat{P}_x, \frac{1}{2}(\hat{P}_y^2 + \hat{P}_z^2 + V(x,y,z))] \\
&= -i\frac{1}{2}(\hat{P}_y[\hat{P}_x, \hat{P}_y] + [\hat{P}_x, \hat{P}_y]\hat{P}_y + \hat{P}_z[\hat{P}_x, \hat{P}_z] + [\hat{P}_x, \hat{P}_z]\hat{P}_z) - \frac{\partial V(x,y,z)}{\partial x} \\
&= \frac{1}{2}(\hat{P}_y B_z - \hat{P}_z B_y) - \frac{1}{2}(\hat{B}_y P_z - \hat{B}_z P_y) - \frac{\partial V(x,y,z)}{\partial x},
\end{aligned} \tag{8.75}$$

with corresponding equations for $\hat{P}_y$ and $\hat{P}_z$. So, in three-dimensional vector form we get

$$\frac{d\hat{\mathbf{P}}}{dt} = \frac{d\hat{\mathbf{v}}}{dt} = \hat{\mathbf{E}} + \frac{1}{2}(\hat{\mathbf{v}} \wedge \hat{\mathbf{B}} - \hat{\mathbf{B}} \wedge \hat{\mathbf{v}}), \tag{8.76}$$

where $\hat{\mathbf{E}} = -\nabla V$ and $\hat{\mathbf{B}} = -\nabla \wedge \hat{\mathbf{Q}}$. We now recognize the rhs of Eq. (8.76) as the Lorentz force [60] on a a unit positive electric charge in an electromagnetic field, if we interpret $\hat{\mathbf{E}}$ as the electric field and $\hat{\mathbf{B}}$ as the magnetic field. This is actually identical to the classical result if we use scalar components rather than an operator for the velocity vector and note that then

$$\mathbf{v} \wedge \mathbf{B} = -\mathbf{B} \wedge \mathbf{v}$$

so

$$\mathbf{v} \wedge \mathbf{B} = \frac{1}{2}(\mathbf{v} \wedge \mathbf{B} - \mathbf{B} \wedge \mathbf{v}),$$

which has exactly the same form as the quantum result. It is interesting to note that in the quantum version, $-\hat{\mathbf{B}} \wedge \hat{\mathbf{v}} \neq \hat{\mathbf{v}} \wedge \hat{\mathbf{B}}$, and so

$$\frac{1}{2}(\hat{\mathbf{v}} \wedge \hat{\mathbf{B}} - \hat{\mathbf{B}} \wedge \hat{\mathbf{v}}) \neq \hat{\mathbf{v}} \wedge \hat{\mathbf{B}}$$

because $\hat{\mathbf{B}}$ and $\hat{\mathbf{v}}$ do not commute. In fact, if we return to Eq. (8.75) and note that $\hat{P}_y B_z = B_z \hat{P}_y + [\hat{P}_y, B_z]$, then the last line of Eq. (8.75) becomes

$$\begin{aligned}
\frac{d\hat{P}_x}{dt} &= \frac{1}{2}(\hat{P}_y B_z - \hat{P}_z B_y) - \frac{1}{2}(\hat{B}_y P_z - \hat{B}_z P_y) - \frac{\partial V(x,y,z)}{\partial x} \\
&= \hat{P}_y B_z - \hat{P}_z B_y - \frac{1}{2}([\hat{P}_y, B_z] - [\hat{P}_z, B_y]) - \frac{\partial V(x,y,z)}{\partial x} \\
&= \hat{P}_y B_z - \hat{P}_z B_y + i\frac{1}{2}(\frac{\partial B_z}{\partial y} - \frac{\partial B_y}{\partial z}) - \frac{\partial V(x,y,z)}{\partial x},
\end{aligned} \tag{8.77}$$

with corresponding equations for $\hat{P}_y$ and $\hat{P}_z$. So, instead of Eq. (8.76) we get [42]

$$\frac{d\hat{\mathbf{P}}}{dt} = \frac{d\hat{\mathbf{v}}}{dt} = \hat{\mathbf{E}} + \hat{\mathbf{v}} \wedge \hat{\mathbf{B}} + i\frac{1}{2}\nabla \wedge \hat{\mathbf{B}}. \tag{8.78}$$

It would be misleading to interpret Eq. (8.78) as indicating a quantum Lorentz force equal to $\hat{\mathbf{v}} \wedge \hat{\mathbf{B}}$ plus a purely quantum force, $i\frac{1}{2}\nabla \wedge \hat{\mathbf{B}}$. The reason is that $i\frac{1}{2}\nabla \wedge \hat{\mathbf{B}}$ is not Hermitian and thus cannot represent an observable. In fact, $\hat{\mathbf{v}} \wedge \hat{\mathbf{B}}$ is not Hermitian either. It is only the combination of the two, which is of course the same as the rhs of Eq. (8.76) that is Hermitian and gives the correct form for the quantum version of the Lorentz force.

The last term on the rhs of Eq. (8.78) does have important physical consequences. In its absence the Lorentz force acts only locally, at the point indicated by $\mathbf{x} = (x, y, z)$. However, the last term in Eq. (8.78) is non-local, since it contains a derivative with respect to configuration space. This gives rise, for example, to the Aharonov-Bohm effect [42, 126].

8.5   THE MAXWELL EQUATIONS

In the previous section it was shown that by generalizing the momentum operator to

$$\hat{\mathbf{P}} = -i\nabla + \hat{\mathbf{Q}}$$

then the rate of change of momentum with respect to time was equal to what was recognizable as the Lorentz force and as a consequence, the existence of electric and magnetic fields was apparent. We can take this analysis further by noting that the derivative of $\hat{\mathbf{P}}$ with respect to time includes the derivative of $\hat{\mathbf{Q}}$ with resect to time. If we evaluate this independently we find[6]

$$\frac{d\hat{\mathbf{Q}}}{dt} = [\hat{\mathbf{Q}}, \hat{H}] = \hat{\mathbf{v}}.\nabla\hat{\mathbf{Q}} + \frac{i}{2}\nabla^2\hat{\mathbf{Q}}. \tag{8.79}$$

We can recognize the first term on the rhs of Eq. (8.79) as the contribution to the total derivative of $\hat{\mathbf{Q}}$ due to motion over inhomogeneities in its configuration space dependence. The final term in Eq. (8.79) plays the same role here as does the final term in Eq. (8.78), i.e., it keeps the rhs of the equation Hermitian, since $\hat{\mathbf{v}}.\nabla\hat{\mathbf{Q}}$ alone is not Hermitian. If $\hat{\mathbf{Q}}$ were explicitly time dependent then we would find

$$\frac{d\hat{\mathbf{Q}}}{dt} = \frac{\partial\hat{\mathbf{Q}}}{\partial t} + \hat{\mathbf{v}}.\nabla\hat{\mathbf{Q}} + \frac{i}{2}\nabla^2\hat{\mathbf{Q}}. \tag{8.80}$$

---

[6]In subscript notation we find

$$\frac{d\hat{Q}_i}{dt} = \frac{1}{2}(\hat{v}_j\hat{Q}_{i,j} + \hat{Q}_{i,j}\hat{v}_j) = \hat{v}_j\hat{Q}_{i,j} + \frac{1}{2}\hat{Q}_{i,jj}.$$

The terms after the first equals sign are difficult to write in vector notation, but suffix notation makes clear the origin of the final term.

Then we would have

$$\frac{d\hat{\mathbf{P}}}{dt} = \frac{\partial \hat{\mathbf{Q}}}{\partial t} + [\hat{\mathbf{P}}, \hat{H}]. \tag{8.81}$$

The Lorentz force equation would then become

$$\frac{d\hat{\mathbf{P}}}{dt} = \frac{\partial \hat{\mathbf{Q}}}{\partial t} - \nabla V + \hat{\mathbf{v}} \wedge \hat{\mathbf{B}} + i\frac{1}{2}\nabla \wedge \hat{\mathbf{B}}. \tag{8.82}$$

Now we need to redefine the electric field as

$$\hat{\mathbf{E}} = \frac{\partial \hat{\mathbf{Q}}}{\partial t} - \nabla V. \tag{8.83}$$

Eq. (8.83), together with $\hat{\mathbf{B}} = -\nabla \wedge \hat{\mathbf{Q}}$ then lead to

$$\nabla \cdot \hat{\mathbf{B}} = 0 \tag{8.84}$$

and

$$\nabla \wedge \hat{\mathbf{E}} = -\frac{\partial \hat{\mathbf{B}}}{\partial t}. \tag{8.85}$$

Eqs. (8.84) and (8.85) are immediately recognizable as two of Maxwell's equations. It is also quite clear from the definitions of the electric and magnetic fields that $-\hat{\mathbf{Q}}$ can be identified as the conventional vector potential in operator form, that is well-known in classical electromagnetic theory. The difference here is that it simply appears as a generalization of the canonical quantization relation between the components of momentum and the co-ordinates of configuration space, from which the electromagnetic field emerges, rather than part of the representation of a field already established from experimental observation and classical theory. It is quite remarkable that these have been derived ultimately from a representation of the number operator that is defined in the universal quantum equation, Eq. (1.1).

The method used above to derive the Maxwell equations, Eqs. (8.84) and (8.85), is similar to the method described by Dyson, which is attributed to Feynman [28]. Dyson also makes the point that Feynman argued that the other two Maxwell equations

$$\nabla \cdot \hat{\mathbf{E}} = \rho \tag{8.86}$$

and

$$\nabla \wedge \hat{\mathbf{B}} - \frac{\partial \hat{\mathbf{E}}}{\partial t} = \mathbf{J} \tag{8.87}$$

are simply definitions of ρ and $\mathbf{J}$ which specify, respectively, charge and current densities, if these exist. Then it is straightforward to check that the densities satisfy a continuity equation of the form

$$\frac{\partial \rho}{\partial t} + \nabla \cdot \mathbf{J} = 0. \tag{8.88}$$

8.6   ELECTROMAGNETIC WAVES AND PHOTONS

The Maxwell equations, Eqs. (8.84) to (8.87) support a variety of wave types. Here we are interested in dispersionless waves that occur when $\rho = \mathbf{J} = 0$. Then we get, from Eq. (8.86)

$$\nabla . \hat{\mathbf{E}} = \nabla . \frac{\partial \hat{\mathbf{Q}}}{\partial t} = \frac{\partial \nabla . \hat{\mathbf{Q}}}{\partial t} = 0. \tag{8.89}$$

The full time dependence of $\hat{\mathbf{Q}}$ can be maintained, while satisfying Eq. (8.89), if we let $\nabla . \hat{\mathbf{Q}} = 0$. This condition is referred to as the *Coulomb gauge*.

Substituting the definitions of $\hat{\mathbf{E}}$ and $\hat{\mathbf{B}}$, in terms of $\hat{\mathbf{Q}}$, into Eq. (8.87), leads to

$$\frac{\partial^2 \hat{\mathbf{Q}}}{\partial t^2} - \nabla^2 \hat{\mathbf{Q}} = 0, \tag{8.90}$$

where we have used

$$\nabla \wedge \nabla \wedge \hat{\mathbf{Q}} = \nabla (\nabla . \hat{\mathbf{Q}}) - \nabla^2 \hat{\mathbf{Q}} = -\nabla^2 \hat{\mathbf{Q}},$$

with the aid of the Coulomb gauge condition.

We can recognize Eq. (8.90) as the dispersionless wave equation which supports plane waves with a dependence on $\mathbf{x}$ and t that is proportional to a sinusoidal function with a phase argument of the form[7], $\mathbf{k}.\mathbf{x} - \omega t$, where $\mathbf{k}$ is a wavevector and ω is an angular frequency. These are of course electromagnetic waves that propagate in empty space. Here we choose a Hermitian solution to Eq. (8.90) of the form[8]

$$\hat{\mathbf{Q}} = \hat{\mathbf{Q}}(0) \sin(\mathbf{k}.\mathbf{x} - \omega t).$$

Substituting this solution into the wave equation, gives dispersion relation of the form

$$\omega^2 - \mathbf{k}^2 = 0. \tag{8.91}$$

Bearing in mind that the space and time co-ordinates are dimensionless here, then the wave propagation speed is

$$v_{ph} = \frac{\omega}{k},$$

where $k = \| \mathbf{k} \|$. Thus, with the dispersion relation, Eq. (8.91), the phase speed is just equal to unity. There is nothing strange in this. It is a consequence of using dimensionless variables. This point will be dealt with in more detail in Chapter 11.

From Eqs. (8.84), (8.85) and (8.89) we find

$$\mathbf{k}.\hat{\mathbf{E}} = \mathbf{k}.\hat{\mathbf{B}} = 0 \text{ and } \mathbf{k} \wedge \hat{\mathbf{E}} = \omega \hat{\mathbf{B}},$$

---

[7] The reason for choosing this form of the phase argument is dealt with in detail in Chapters 10 and 11.

[8] In principle we could use $\exp(i(\mathbf{k}.\mathbf{x} - \omega t))$ as the sinusoidal function, but this has both real and imaginary parts, so we cannot use it in conjunction with Hermitian operators like $\hat{\mathbf{Q}}$, since this would introduce anti-Hermitian parts. Consequently we only use solutions that are linear combinations of $\sin(\mathbf{k}.\mathbf{x} - \omega t)$ and $\cos(\mathbf{k}.\mathbf{x} - \omega t)$, with real coefficients.

which imply the electric field and magnetic field are both transverse to the direction of propagation and mutually perpendicular. These results, with the aid of the dispersion relation, Eq. (8.91), also imply that $\hat{\mathbf{E}}^2 = \hat{\mathbf{B}}^2$. From these relations it is straightforward to show that

$$\frac{\partial^2 \hat{\mathbf{Q}}}{\partial t^2} + \omega^2 \hat{\mathbf{Q}} = 0, \tag{8.92}$$

which means we can treat the electromagnetic waves as simple harmonic oscillators. Eq. (8.92) also implies that

$$\frac{\partial^2 \hat{\mathbf{E}}}{\partial t^2} + \omega^2 \hat{\mathbf{E}} = \frac{\partial^2 \hat{\mathbf{B}}}{\partial t^2} + \omega^2 \hat{\mathbf{B}} = 0. \tag{8.93}$$

Now consider a special case of an electromagnetic wave with a wave vector $\mathbf{k} = (k, 0, 0)$, an electric field $\hat{\mathbf{E}} = (0, \hat{E}, 0)$ and a magnetic field, $\hat{\mathbf{B}} = (0, 0, \hat{B})$. Imagine now that we confine the electromagnetic wave field to a finite volume in the form of a standing wave. The space occupied by the wave can be taken as a cube with edges parallel to the x-, y- and z-axes and linear dimensions, L. The standing wave structure is found by separating variables to find solutions to the wave equation, so we can take [44]

$$\hat{E} = \gamma \hat{U}(t) \cos(kx), \tag{8.94}$$

where γ is a real constant of proportionality and $\hat{U}(t)$ is a time dependent Hermitian operator. Eq. (8.85) then implies that $\hat{B}$ must be proportional to $\sin(kx)$, so we write

$$\hat{B} = \gamma \hat{V}(t) \sin(kx), \tag{8.95}$$

where $\hat{V}(t)$ is a time dependent Hermitian operator. The relationships between $\hat{E}$ and $\hat{B}$, from Eq. (8.85) now reduce to

$$\frac{\partial \hat{E}}{\partial x} = -\frac{\partial \hat{B}}{\partial t},$$

which, on substituting Eqs. (8.94) and (8.95) together with $\omega = k$, give

$$\hat{U} = \frac{\partial \hat{V}}{\partial t}. \tag{8.96}$$

Similarly, the relationship from Eq. (8.87) reduces to

$$-\frac{\partial \hat{B}}{\partial x} = \frac{\partial \hat{E}}{\partial t},$$

which gives

$$\hat{V} = -\frac{\partial \hat{U}}{\partial t}. \tag{8.97}$$

The relationships between $\hat{U}$ and $\hat{V}$ in Eqs. (8.96) and (8.97) are consistent with their being a pair of canonical variables just like those in Eqs. (3.42) and (3.43), or $\hat{x}$ and $\hat{p}$

in Eqs. (5.13) and (5.14). The implication is that we can associate a natural number operator, $\hat{N}$ with the field, just like we did with the harmonic oscillator in Sections 3.4 and 5.1, where

$$\frac{1}{2}(\hat{U}^2 + \hat{V}^2) = \hat{N} + \frac{1}{2}.$$

However, we would like the find an expression for $\hat{N}$ that involves $\hat{E}$ and $\hat{B}$ rather that $\hat{U}$ and $\hat{V}$. Unfortunately we cannot do this immediately since $\hat{E}$ and $\hat{B}$ do not form a canonical pair. The problem arises due to the factors $\sin(kx)$ and $\cos(kx)$ in $\hat{E}$ and $\hat{B}$. There is a way out of this difficulty, if we treat $\hat{E}$ and $\hat{B}$ as density amplitudes like the wave functions in quantum mechanics. Then we get

$$\int_0^L (\hat{E}^2 + \hat{B}^2)\mathrm{d}x = \gamma^2 \int_0^L (\hat{U}^2 \cos^2(kx) + \hat{V}^2 \sin^2(kx))\mathrm{d}x$$

$$= \frac{\gamma^2 L}{2}(\hat{U}^2 + \hat{V}^2) = \gamma^2 L(\hat{N} + \frac{1}{2}),$$

(8.98)

where we have taken L to be equal to a whole number of half wavelengths of the electromagnetic wave. So, we have succeeded in removing the x-dependence in the relation between the field amplitudes and $\hat{N}$, by integrating over x. We note also that the operators $\hat{U}$ and $\hat{V}$ are bosonic and so the value of the eigenvalues, n, of $\hat{N}$ is unlimited. We can interpret n as the number of *photons* of electromagnetic energy in the electromagnetic field.

It is a straightforward matter to construct the Hamiltonian for this system of n photons. It is exactly the same as that for the harmonic oscillator in Section 5.1, so in energy units we expect

$$\hat{H} = \hbar\omega(\hat{N} + \frac{1}{2}) = \frac{\hbar\omega}{\gamma^2 L} \int_0^L (\hat{E}^2 + \hat{B}^2)\mathrm{d}x.$$

(8.99)

Now the expression for the energy of the field confined to the volume L^3 that we would expect from classical theory is

$$\frac{L^2}{2} \int_0^L (\hat{E}^2 + \hat{B}^2)\mathrm{d}x,$$

so the expression in Eq. (8.99) can be brought into agreement with the classical definition of the energy density of the electromagnetic field if we take

$$\hbar\omega = \frac{\gamma^2 L^3}{2}.$$

Eq. (8.99) teaches us two important lessons about what we will call the *itemization of a field*, by which we mean using a field to represent a number of items, rather than a local amplitude like $\hat{A}$, that we introduced in Chapter 3. The first thing to note is that the process depends on interpreting the field amplitudes as the density amplitudes of the natural number operator and that these require integrating over a configuration space to get $\hat{N}$. The second point is that the energy density of the electromagnetic field is proportional to $\hat{E}^2 + \hat{B}^2$, which is consistent with the classical result [60], bearing in mind that here, the field amplitudes are dimensionless.

So, we have found that we can use fields distributed over configuration space in the form of operator density amplitudes to express the itemization of nature. We shall return to this new way of defining the number operator in Chapter 10, when we consider what we might learn about quantum fields in general from the use of field amplitude densities. To a certain extent this process is a kind of primitive blueprint for establishing a theory of quantum fields, but at this stage we should avoid calling it a *quantum field theory*.

8.7 GOLDILOCKS AND THE THREE DIMENSIONS

Finally, in this chapter, we address the general question of dimensionality. It is useful to now enquire if there is an isotropic four-dimensional oscillator which could be used to define a four-dimensional configuration space with $\mathbf{r} = (x_1, x_2, x_3, x_4)$ such that $r^2 = x_1^2 + x_2^2 + x_3^2 + x_4^2$. Following the arguments for the three-dimensional case, this would require there to be a complete complementary set of four components of linear momentum that would allow four-dimensional linear motion, and also four independent components of angular momentum that form a closed Lie group, so that we could guarantee four independent axes of rotation. There is no problem defining the linear momentum components, which are just the four differentials with respect to x_1, x_2, x_3 and x_4. Let us examine the possible components of angular momentum. We begin with a degenerate four-category system defined by

$$\begin{aligned} \hat{N} &= \hat{N}_1 + \hat{N}_2 + \hat{N}_3 + \hat{N}_4 \\ &= \hat{A}_1^\dagger \hat{A}_1 + \hat{A}_2^\dagger \hat{A}_2 + \hat{A}_3^\dagger \hat{A}_3 + \hat{A}_4^\dagger \hat{A}_4. \end{aligned} \tag{8.100}$$

Clearly, there are just four components of configuration space and linear momentum of the form

$$\hat{X}_i = \frac{1}{\sqrt{2}}(\hat{A}_i + \hat{A}_i^\dagger) \text{ and } \hat{P}_i = \frac{i}{\sqrt{2}}(\hat{A}_i^\dagger - \hat{A}_i),$$

where $i = 1, 2, 3$, or 4. Next we construct four components of angular moment in the form described by Eq. (8.4), but now the subscripts i, j and k are cyclic permutations of $1, 2, 3$ and 4. The four allowed components are then, $\hat{L}_1 = i(\hat{A}_3^\dagger \hat{A}_2 - \hat{A}_2^\dagger \hat{A}_3)$, $\hat{L}_2 = i(\hat{A}_4^\dagger \hat{A}_3 - \hat{A}_3^\dagger \hat{A}_4)$, $\hat{L}_3 = i(\hat{A}_1^\dagger \hat{A}_4 - \hat{A}_4^\dagger \hat{A}_1)$ and $\hat{L}_4 = i(\hat{A}_2^\dagger \hat{A}_1 - \hat{A}_1^\dagger \hat{A}_2)$. Now we require these four components to form a closed Lie group. It is a straightforward matter to check this. Let us look at $[\hat{L}_1, \hat{L}_2]$. We get

$$\begin{aligned} [\hat{L}_1, \hat{L}_2] &= -[\hat{A}_3^\dagger \hat{A}_2 - \hat{A}_2^\dagger \hat{A}_3, \hat{A}_4^\dagger \hat{A}_3 - \hat{A}_3^\dagger \hat{A}_4] \\ &= \hat{A}_4^\dagger \hat{A}_2 - \hat{A}_2^\dagger \hat{A}_4. \end{aligned} \tag{8.101}$$

We see immediately that $i(\hat{A}_4^\dagger \hat{A}_2 - \hat{A}_2^\dagger \hat{A}_4)$ is not one of the four components of angular momentum, and so their Lie group is not closed. It is easy to check that there are six possible combinations of the type in Eq. (8.101). It turns out to be impossible to find a set of components of angular momentum of the type in Eq. (8.101), that are just four in number, using the four sets of creation and annihilation operators in Eq. (8.100). So, the system does not have the required symmetry properties to define a four-dimensional configuration space.

So, a kind of *Goldilocks principle* is at work here. We saw in Chapter 7 that two categories yielded two components of both configuration space and linear momentum space but only one axis of rotation. Here, four categories have yielded four components of both configuration space and linear momentum space, but too many axes of rotation. Three categories are just right, with exactly three components of configuration space, linear momentum space and axes of rotation. Later, in Chapter 11, we shall see that it is possible to define a kind of four-dimensional space but it is a *space-time* that has *pseudo-Euclidean* rather than the purely Euclidean properties of configuration space, that is more properly described as a $3 + 1$ dimensional framework, rather than a four-dimensional one. A four-dimensional system gives the impression that one can interchange in a cyclic manner any four of the components, which is not the case with space-time. One can in fact only interchange the three configuration space components in a cyclic manner. The time component must remain as the $+1$ element of the system, and cannot be interchanged with a configuration space component.

It is interesting to note that in this case $3 + 1$ framework is consistent with the algebra of *quaternions*. This is an associative anti-commuting algebra with one real and three imaginary elements. It can be proved that it is not possible to have a self-consistent algebra of this kind with more than four elements [23, 108]. Quaternions were discovered by Hamilton in the nineteenth century, but fell out of fashion with the advent of conventional vector algebra. However, in the last few decades the power of quaternions in formulating fundamental physics is becoming increasingly recognized [108]. For example, Adler [1] has constructed an entirely quaternion version of quantum mechanics.

The fact that the components $\hat{L}_i$ form a closed three component Lie group of Hermitian operators is not only important in defining the number of dimensions of configuration space. The $\hat{L}_i$ are of course bosonic operators. Recall that we found a closed three component Lie group of fermionic Hermitian operators in Section 4.4, the $\hat{S}_i$s. This is no coincidence. Both sets of operators have rotational properties, but whereas the $\hat{L}_i$s exhibit rotation in configuration space, their fermionic cousins have no configuration space to rotate in. They rotate in the phase space, ξ that associated with the S-type equation involving the number operator, $\hat{N}$. The bosonic Hermitian operator, $\hat{U}$ and $\hat{V}$ do rotate through angle ξ in this case, but these have nothing to do with the components, $\hat{L}_i$, even though they are constructed from the bosonic components, $\hat{X}_j$ and $\hat{P}_k$. However, the two sets of rotational Hermitian operators, $\hat{L}_i$ and $\hat{S}_i$, can be brought together in a single concept, but we will only see that once special relativity has emerged. We will need to wait until Chapter 11 for that.

9 Interactions in multi-category systems

9.1 MULTIPLE-CATEGORY LINEAR TRANSFORMATIONS

In this section we begin with the multiple-category system of S independent categories that was introduced in Chapter 4, but then we will apply a unitary transformation to the annihilation operators in order to model interactions between the categories. This constitutes a generalization of the unitary transformation that was introduced in Chapter 6 to deal with the two-category case. That required a 2×2 matrix. Here we will require an $S \times S$ matrix transformation.

Consider a system of S independent categories, with a total number of items represented by a number operator, $\hat{N}$, that is given by

$$\hat{N} = \sum_{i=1}^{S} \hat{N}_i, \tag{9.1}$$

where the number of elements in the ith category, Cat$_i$, is represented by $\hat{N}_i = \hat{A}_i^\dagger \hat{A}_i$ and $\hat{A}_i$ is a natural number amplitude that can obey either Bosonic or Fermionic rules. The states of this system can be represented in occupation number formalism by an S-dimensional Fock state

$$|n_1, n_2, \ldots, n_i, \ldots n_S\rangle$$

where

$$\hat{N}_i |n_1, n_2, \ldots, n_i, \ldots n_S\rangle = n_i |n_1, n_2, \ldots, n_i, \ldots n_S\rangle.$$

The Hamiltonian, $\hat{\Omega}$, of the system is given by

$$\hat{\Omega} = \sum_{i=1}^{S} \omega_i \hat{N}_i = \sum_{i=1}^{S} \omega_i \hat{A}_i^\dagger \hat{A}_i, \tag{9.2}$$

where ω_i is the eigenfrequency of Cat$_i$. All of the $\hat{N}_i$s commute with $\hat{\Omega}$ and so are individually invariants. The system may be transformed by a linear transformation, according to

$$\hat{A}_i = \sum_j \gamma_{ij} \hat{C}_j, \tag{9.3}$$

where γ_{ij} represents the element of an $S \times S$ matrix (operator), $\hat{\Gamma}$.

The number operator in the new representation is then obtained by substituting Eq. (9.3) into Eq. (9.1) to give

$$\hat{N} = \sum_{i=1}^{S} \sum_{j,k} \gamma_{ik}^* \gamma_{ij} \hat{C}_k^\dagger \hat{C}_j. \tag{9.4}$$

DOI: 10.1201/9781003377504-9

Up to now the linear transformation matrix, $\hat{\Gamma}$, is undefined. We would now like to choose a linear transformation that results in

$$\hat{N} = \sum_{i=1}^{S} \hat{M}_i, \qquad (9.5)$$

where $\hat{M}_i = \hat{C}_i^\dagger \hat{C}_i$ is a new number operator. We can achieve this if we let

$$\gamma_{ik}^* = \gamma_{ki}^{-1}. \qquad (9.6)$$

In Eq. (9.6), γ_{ik}^* is and element of the matrix, $\hat{\Gamma}^{*T}$, the complex conjugate of the transpose of $\hat{\Gamma}$, while γ_{ki}^{-1} is an element of the operator, $\hat{\Gamma}^{-1}$, the inverse of $\hat{\Gamma}$, which by definition satisfies, $\hat{\Gamma}^{-1}\hat{\Gamma} = \hat{I}$, where $\hat{I}$ is the unit matrix with elements equal to δ_{ij}, the Kronecker delta. In component form this is just

$$\sum_j \gamma_{ij}^{-1} \gamma_{jk} = \delta_{ik}. \qquad (9.7)$$

Eq. (9.6) implies that $\hat{\Gamma}^{*T} = \hat{\Gamma}^{-1}$ which means that $\hat{\Gamma}$ is, by definition, a unitary matrix [53]. With these conditions, Eq. (9.4) becomes

$$\hat{N} = \sum_{i=1}^{S} \sum_{j,k}^{S} \gamma_{ki}^{-1} \gamma_{ij} \hat{C}_k^\dagger \hat{C}_j = \sum_{j,k}^{S} \delta_{kj} \hat{C}_k^\dagger \hat{C}_j = \sum_{j=1}^{S} \hat{C}_j^\dagger \hat{C}_j. \qquad (9.8)$$

The final part of Eq. (9.8) is now the same as Eq. (9.5) as required, so the two representations contain the same total number of items.

Now suppose that the $\hat{A}_i$s are bosonic, so we take

$$[\hat{A}_i, \hat{A}_j^\dagger] = \delta_{ij}. \qquad (9.9)$$

Given Eq. (9.3), the inverse relation is

$$\hat{C}_i = \sum_m \gamma_{im}^{-1} \hat{A}_m, \qquad (9.10)$$

then

$$[\hat{C}_i, \hat{C}_j^\dagger] = \sum_{m,n} \gamma_{im}^{-1} \gamma_{in}^{*-1} [\hat{A}_m, \hat{A}_n^\dagger] = \sum_{m,n} \gamma_{im}^{-1} \gamma_{jn}^{*-1} \delta_{nm} = \sum_m \gamma_{im}^{-1} \gamma_{jm}^{*-1}. \qquad (9.11)$$

Now we use the fact that if a matrix is unitary then so is its inverse [53], then $\gamma_{jm}^{*-1} = \gamma_{mj}$. Then Eq. (9.11) becomes

$$[\hat{C}_i, \hat{C}_j^\dagger] = \sum_m \gamma_{im}^{-1} \gamma_{mj} = \delta_{ij}, \qquad (9.12)$$

as required. Following a similar procedure we can show that

$$[\hat{A}_i, \hat{A}_j] = 0 \implies [\hat{C}_i, \hat{C}_j] = 0$$

and for Fermions
$$\{\hat{A}_i, \hat{A}_j^\dagger\} = 0 \implies \{\hat{C}_i, \hat{C}_j^\dagger\} = 0$$
and
$$\{\hat{A}_i, \hat{A}_j\} = 0 \implies \{\hat{C}_i, \hat{C}_j\} = 0.$$

Thus one finds that if $\hat{A}_i$ and $\hat{A}_i^\dagger$ are bosonic then so are $\hat{C}_i$ and $\hat{C}_i^\dagger$ and if $\hat{A}_i$ and $\hat{A}_i^\dagger$ are fermionic, then so are $\hat{C}_i$ and $\hat{C}_i^\dagger$. Note that the only way that the commutation properties of the brackets of either type is preserved by the unitary transformation is if it equal to zero or the Kronecker delta. This rules out parastatistical type relationships[1].

The eigenstates for the new natural number operators, $\hat{M}_i$, are represented in occupation number formalism by

$$|m_1, m_2, \ldots, m_i, \ldots, m_S\rangle,$$

where

$$\hat{M}_i|m_1, m_2, \ldots, m_i, \ldots, m_S\rangle = m_i|m_1, m_2, \ldots, m_i, \ldots, m_S\rangle, \qquad (9.13)$$

for either the bosonic or fermionic cases. The two cases give the same result also for

$$\hat{C}_i|m_1, m_2, \ldots, m_i, \ldots, m_S\rangle = \sqrt{m_i}|m_1, m_2, \ldots, m_i - 1, \ldots, m_S\rangle, \qquad (9.14)$$

but whereas

$$\hat{C}_i^\dagger|m_1, m_2, \ldots, m_i, \ldots, m_S\rangle = \sqrt{m_i + 1}|m_1, m_2, \ldots, m_i + 1, \ldots, m_S\rangle, \qquad (9.15)$$

for bosons, the fermion case requires

$$\hat{C}_i^\dagger|m_1, m_2, \ldots, m_i, \ldots, m_S\rangle = \sqrt{1 - m_i}|m_1, m_2, \ldots, m_i + 1, \ldots, m_S\rangle. \qquad (9.16)$$

The new representation of the Hamiltonian may be obtained, by substituting the transformed creation and annihilation operators into Eq. (9.2) and then separating the terms with the same subscripts from those with different ones. Then the Hamiltonian has the form

$$\hat{\Omega} = \sum_i W_i \hat{C}_i^\dagger \hat{C}_i + \sum_{jk} V_{jk} \hat{C}_j^\dagger \hat{C}_k, \qquad (9.17)$$

where $W_i = \sum_j \omega_j \gamma_{ij}^{-1} \gamma_{ji}$ and $V_{jk} = \sum_i \omega_i \gamma_{ji}^{-1} \gamma_{ik}$, when $j \neq k$.

Although the total number of items in the system is invariant when represented by either the sum over $\hat{N}_i$ in Eq. (9.1) or over $\hat{M}_i$ in Eq. (9.5), $\hat{M}_i$ no longer commutes with $\hat{\Omega}$ in Eq. (9.17), when the coefficients $V_{jk} \neq 0$, so $\frac{d\hat{M}_i}{dt} \neq 0$. However, it is straightforward to show that we still have $\frac{d\hat{N}}{dt} = 0$ for the system as a whole, as indicated above. So, we can still interpret Eqs. (9.5) as a system of N items, where

$$N = \sum_i^S n_i = \sum_i^S m_i. \qquad (9.18)$$

---

[1]Systems that are neither fermionic nor bosonic or are a mixture of the two are referred to as parastatistics [86]. Such systems are found in the theory of superconductivity, for example [66, 121].

Although the n_is are fixed in time, the m_is are time dependent like the case we found in Chapter 6, when there was an interaction term in the transformed Hamiltonian.

The population is still divided into S categories in both representations, but in the $\hat{M}_i$ representation, the sum over jk in Eq. (9.17) indicates that an item in Cat_k is transferred into Cat_j, since

$$\hat{C}_j^\dagger \hat{C}_k |m_1, \ldots, m_j, \ldots, m_k, \ldots\rangle$$
$$= \sqrt{(m_j + 1)m_k} |m_1, \ldots, m_j + 1, \ldots, m_k - 1, \ldots\rangle. \quad (9.19)$$

It is this exchange of items between categories that suggests we could use this picture to interpret the items as particles in a physical system. The different categories would indicate different states, of energy, say, or of momentum. This exchange of particles between states then constitutes a scattering process. It turns out that there is a close relationship between the above result and the generalization of the potential operator, $\hat{V} = V(\hat{X})$ in the Schrödinger equation in the earlier sections. It can be shown that $V_{jk} = \langle m_j | V | m_k \rangle$ [121], which means that the potential operator from the Schrödinger equation is directly related to the coefficients of the scattering term in the many-body Hamiltonian, through an inner product involving operating with the potential on the Fock space vectors. There is an important interpretation of what this amounts to. In the Schrödinger equation picture we are dealing with a single particle in an external potential. The implication of the connection to many-body scattering is that in fact this potential is as a result of interactions between the particle subject to the Schrödinger equation and a population of neighbouring particles that actually are the source of the potential.

It is also worth pointing out that the steps from Eq. (9.2) to Eq. (9.17) constitute a back-to-front derivation of particle interactions, since, in standard many-body physics, one usually starts with the Hamiltonian in the form in Eq. (9.17) (see ref. [121]). Reversing the transform by using the inverse of the matrix, γ_{ij}, then yields the Hamiltonian in the form of Eq. (9.2), from which the energy eigenvalues of the system can be determined. These are, of course, the ω_is in Eq. (9.2). This reverse transform is actually the usual Bogoliubov transformation that is well known in many-body quantum physics [121].

Finally, we note that there are two quite remarkable outcomes of the unitary transformation above. First, that it preserves both the boson and fermion rules, but works for nothing else. Second, the total population of elements, represented by $\hat{N}$ in Eq. (9.17) is invariant in either representation.

9.2 FERMION INTERACTIONS VIA BOSON EXCHANGE

The picture presented in Section 9.1 can be extended further in an interesting way by interpreting the constants V_{jk} as a background cause of the scattering process. We take the creation and annihilation operators in Eq. (9.17) to represent a fermionic system and then subject the V_{jk}s to a small oscillatory perturbation represented by a real variable, χ, such that $V_{jk}(\chi) = V_{jk}(0) + \chi V'_{jk}(0)$, where $V'_{jk}(0)$ is the first derivative of $V_{jk}(\chi)$ with respect to χ, at $\chi = 0$, and $V_{jk}(0)$ is the unperturbed background.

We then treat χ as an Hermitian operator $\hat{\chi}$, from which we can construct a bosonic creation and annihilation operator pair using $\hat{\chi}$ and $\frac{\partial}{\partial \hat{\chi}}$ (see Chapter 3). So, we can let $\hat{S} = \frac{1}{\sqrt{2}}(\hat{\chi} + \frac{\partial}{\partial \hat{\chi}})$, then $\hat{\chi} = \frac{1}{\sqrt{2}}(\hat{S} + \hat{S}^{\dagger})$, where $[\hat{S}, \hat{S}^{\dagger}] = 1$. Substituting this into Eq. (9.17) and tidying up the result a little yields

$$\hat{\Omega} = \sum_i W_i \hat{C}_i^{\dagger} \hat{C}_i + \omega_s \hat{S}^{\dagger} \hat{S} + \sum_{jk} V_{jk}(\hat{S}^{\dagger} + \hat{S}) \hat{C}_j^{\dagger} \hat{C}_k, \tag{9.20}$$

where the second term on the rhs has been introduced to represent the energy of the bosonic oscillations to preserve consistency. It can later be made to tend to zero (see below). Eq. (9.20) represents the Hamiltonian of an interacting fermionic system that is perturbed by a bosonic perturbation. The time dependence of $\hat{S}$ may be evaluated with the use of the H-type equation,

$$i\frac{d\hat{S}}{dt} = [\hat{S}, \hat{\Omega}] = \omega_s \hat{S} + \sum_{jk} V_{jk} \hat{C}_j^{\dagger} \hat{C}_k. \tag{9.21}$$

Thus the equation for the operator, $\hat{S}$, takes the form of a forced harmonic oscillator. If the bosonic perturbation is small, then the forcing term, $\sum_{jk} V_{jk} \hat{C}_j^{\dagger} \hat{C}_k$, may be approximated by making the assumption that the fermionic operators have their unperturbed time dependencies, i.e. $\hat{C}_k(t) = \hat{C}_k(0) \exp(-iW_k t)$, etc. Then the differential equation equation for $\hat{S}(t)$ may be integrated explicitly [43] to yield

$$\hat{S}(t) = -\sum_{jk} \frac{V_{jk} \hat{C}_j^{\dagger} \hat{C}_k}{W_j - W_k + \omega_s}, \tag{9.22}$$

where it has been assumed that $\hat{S}(-\infty) = 0$, which just means that the bosons have zero amplitude before they are excited. $\hat{S}^{\dagger}$ can be treated in a similar way and then one finds

$$\hat{S}^{\dagger}(t) = \sum_{jk} \frac{V_{jk} \hat{C}_j^{\dagger} \hat{C}_k}{W_j - W_k - \omega_s}. \tag{9.23}$$

Adding Eqs. (9.22) and (9.23) then yields

$$\hat{S}^{\dagger}(t) + \hat{S}(t) = \sum_{jk} \frac{V_{jk} \hat{C}_j^{\dagger} \hat{C}_k}{W_j - W_k - \omega_s} - \sum_{jk} \frac{V_{jk} \hat{C}_j^{\dagger} \hat{C}_k}{W_j - W_k + \omega_s}$$
$$= \sum_{jk} \frac{2\omega_s V_{jk} \hat{C}_j^{\dagger} \hat{C}_k}{(W_j - W_k)^2 - \omega_s^2}. \tag{9.24}$$

The above result may be substituted back into Eq. (9.21) to yield an equation of the form [70]

$$\hat{\Omega} = \sum_i W_i \hat{C}_i^{\dagger} \hat{C}_i + \sum_{jklm} V_{jklm} \hat{C}_j^{\dagger} \hat{C}_k^{\dagger} \hat{C}_l \hat{C}_m, \tag{9.25}$$

where

$$V_{jklm} = \frac{2\omega_s V_{jk} V_{lm}}{(W_j - W_k)^2 - \omega_s^2}. \tag{9.26}$$

The second sum on the right hand side of Eq. (9.25) represents pair interactions in which items of type l and m are scattered into types j and k. Remarkably, even with this more complicated form of Hamiltonian, the total number of items given by $\hat{N}$ in Eq. (9.5) is a constant of the motion and is unaffected by the interaction process.

The above results indicate that the bosonic fluctuations can be considered to be excited by the interacting fermions and also do not appear explicitly in the final form of the Hamiltonian, Eq. (9.25). In this case the bosons are referred to as virtual particles. It is instructive to see this by using diagrams to solve Eq. (9.20). This is depicted in Fig. 9.1. The resultant **c** in this diagram represents fermionic pair interaction via

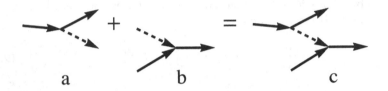

a	b	c

Figure 9.1 The solid arrows represent fermions and the dashed arrow bosons. (a) represents the first term in the sum in Eq. (9.20) where a boson is excited and (b) the second, where it is absorbed. (c) represents the sum of (a) and (b) by treating the bosons therein as a single linking line to represent the second sum in Eq. (9.25).

boson exchange and can be used to represent a variety of physically significant processes, such as the Coulomb interaction between two electrons in which a photon carries the electromagnetic force between them. The scattering coefficient, V_{jklm}, is given by $\langle m_l, m_m | V | m_j, m_k \rangle$, where the potential, V comes from the appropriate Schrödinger equation for the interaction. Eq. (9.25) is widely used in many-body physics [15, 121, 132].

9.3   BCS THEORY

A special case of Eq. (9.25) is of particular interest. Bardeen, Cooper and Schreiffer, in formulating their BCS theory of superconductivity [111], identified a special situation in the scattering coefficient, Eq. (9.26) when $W_j = W_k$. Superconductivity is an important and fascinating phenomenon in its own right, but it is beyond the scope of the present book to go into details of the condensed matter aspects of the theory. Interested readers should consult some of the many textbooks and monographs that deal with these aspects (see for example, refs.[15, 43, 121, 111, 132]). What is important here is that characteristics of BCS theory turn out to be of immense importance in fundamental particle physics as well as its applications in a condensed matter context. For the moment, we will examine the BCS mechanism in its many-body context in a simplified way. We first need to briefly introduce the notion of a *Fermi sea*.

9.3.1 THE FERMI SEA

The simplest model of an assembly of fermions is one in which they occupy a three-dimensional space enclosed by impenetrable walls. This is effectively a three-dimensional infinite square well potential, i.e., a three-dimensional version of the infinite square well we met in Chapter 5, that had energy levels proportional to $(n+1)^2$, where n is a natural number. The wave functions of this system comprise standing plane waves with energies equal to $\frac{(\hbar k)^2}{2m}$, where k is the wave number of the wave and m is the mass of the fermion. The lowest energies available to the fermions are thus associated with the longest wavelengths. These are of the order of the linear dimensions of the box containing the fermions. The idea now is that we begin adding fermions to the box in which the energy states just described are available for occupation. The first fermions to be placed in the box can occupy states with the lowest energy, but as more are added, they must be put in higher energy states because of the Pauli exclusion principle which only allows one fermion in a given state. So, when the last fermion is placed in the box it occupies a state with the highest energy. Because we have three dimensions, the system is degenerate and several states can have the same highest energy. The highest energy will obviously depend on how many fermions are contained in the box. The more fermions the higher the energy level needed to accommodate them. This highest energy level needed is called the *Fermi energy*, and the collection of fermions with energies up to the Fermi energy is called the *Fermi sea*.

This simple model is commonly used to explain the behaviour of electrons in metals. Then one considers the box to be a lattice of atoms. The fermions are then electrons given up by the atoms to form essentially a free electron gas. The superconducting behaviour that BCS theory deals with then comes about due to fluctuations in the positions of the atomic nuclei about their equilibrium locations in the regular lattice. These lattice vibrations are treated as harmonic oscillators just as we envisaged in the perturbation expansion of the background potential that we carried out with the aid of the bosonic operator $\hat{S}$ and its adjoint in Section 9.2. These bosonic perturbations that associated with lattice vibrations are termed *phonons*[2]. It is these that couple the fermions via phonon exchange as in Fig. 9.1. Next we will look at how this coupling can lead to an attractive force between the fermions.

9.3.2 FERMION-FERMION ATTRACTION

We first note that the way Eqs. (9.25) and (9.26) were obtained meant that V_{jk} and V_{lm} essentially have the same value. So, if we set them both equal to v, then V_{jklm} is just

$$V_{jklm} = \frac{2\omega_s v^2}{(W_j - W_k)^2 - \omega_s^2}.$$

Here, W_j and W_k are interpreted as being measured relative to the Fermi energy introduced above. So, if the two fermions have energies close to the Fermi energy,

---

[2]The term, phonon, is used in relation to sound waves as the term, photon, is used in relation to a light wave. Lattice vibrations are associated with thermal vibrations that propagate as sound waves.

then $W_j = W_k$ applies. Then, as realized by Bardeen, Cooper and Schreiffer, we get $V_{jklm} = -\frac{2v^2}{\omega_s}$, which is negative. This implies an attractive potential between the interacting fermions that tends to keep them together as a pair. In the limit as ω_s tends to zero[3], the interaction strength can be made vanishingly small, which indicates the pairing effect involves an instability and happens spontaneously. The instability is referred to as a *Cooper instability* and the resulting pairs are called Cooper pairs in the BCS theory of superconductors.

9.3.3 THE ENEGY GAP

With the aid of the *mean-field approximation*(see Appendix G) that is widely used in many-body quantum theory, the Hamiltonian can be reduced to the form [15, 132]

$$\hat{\Omega} = \sum_i W_i \hat{C}_i^\dagger \hat{C}_i - \Delta \sum_{jk} (\hat{C}_j^\dagger \hat{C}_k^\dagger + \hat{C}_k \hat{C}_j), \tag{9.27}$$

where[4] $\Delta = \frac{v^2}{\omega_s} \langle \hat{C}_k \hat{C}_j \rangle$. Notice that Eq. (9.27) has exactly the same form as Eq. (7.58), apart from the sign of the coefficient of the interaction term, but this has no effect on the outcome. Thus, the BCS mechanism involving an exchange of bosons between the fermions provides a physical explanation for the coupling between the fermionic creation and annihilation operators that we introduced earlier. The eigenvalues of the system governed by Eq. (9.27) may be found by following the same procedure that we used with Eq. (7.58). So, after applying the Heisenberg equation of motion to Eq. (9.27), as we did with Eq. (7.58), we can write the eigenvalue equations as

$$E\hat{C}_k = \omega \hat{C}_k + \Delta \hat{C}_j^\dagger, \tag{9.28}$$

and

$$E\hat{C}_j^\dagger = -\omega \hat{C}_j^\dagger + \Delta \hat{C}_k, \tag{9.29}$$

where E is the energy eigenvalue and $\omega = W_j = W_k$. These equations show that the annihilation of a fermion in state k is coupled to the excitation of a fermion in state j. Also note that in the absence of the coupling, $\Delta = 0$ and then $E = \omega$.

There is an alternative description of coupling process, which is that it can be described as being between the annihilation of a particle from state k, represented by $\hat{C}_k$ and the annihilation of a *hole*, or absence of a particle, from state, j, that is removed when a particle is created there. That is then represented by $\hat{C}_j^\dagger$. This is a subtle but highly significant point. Notice that in the two equations Eqs. (9.28) and (9.29) the coefficients involving ω on the rhs, are opposite in sign. These are the eigenvalues of the bare particles when $\Delta = 0$. However, with $\Delta \neq 0$, the eigenvalues, E on the lhs both have the same sign. This means that the loss of a particle from a state below the Fermi energy is positive, relative to the Fermi level, and has the same

---

[3]This condition also removes the term, $\omega_s \hat{S}$, from Eq. (9.21), that was originally added in an ad hoc fashion.

[4]$\langle \hat{C}_k \hat{C}_j \rangle$ represents an inner product involving the groundstate function of the BCS system. We can also take $\langle \hat{C}_j^\dagger \hat{C}_k^\dagger \rangle = \langle \hat{C}_k \hat{C}_j \rangle$ and then Δ is real. See ref.[15] for details.

sign as and energy of a state above the Fermi energy. Evaluating the determinant that results from Eqs. (9.28) and (9.29), one finds the energy eigenvalues satisfy

$$E^2 = \omega^2 + \Delta^2. \tag{9.30}$$

Just as with Eq. (7.64), we can put the energy eigenvalue equations in matrix form, i.e.

$$\hat{E} \begin{pmatrix} \hat{C}_k \\ \hat{C}_j^\dagger \end{pmatrix} = \begin{pmatrix} \omega & \Delta \\ \Delta & -\omega \end{pmatrix} \begin{pmatrix} \hat{C}_k \\ \hat{C}_j^\dagger \end{pmatrix}, \tag{9.31}$$

where

$$\hat{E} = \begin{pmatrix} E & 0 \\ 0 & E \end{pmatrix}. \tag{9.32}$$

We note, as in Eq. (7.66), that

$$\begin{pmatrix} \omega & \Delta \\ \Delta & -\omega \end{pmatrix} = \omega \begin{pmatrix} 1 & 0 \\ 0 & -1 \end{pmatrix} + \Delta \begin{pmatrix} 0 & 1 \\ 1 & 0 \end{pmatrix}. \tag{9.33}$$

The two 2×2 matrices on the right hand side of Eq. (9.33) can be recognized the conventional representations of the Pauli matrices, $\hat{\sigma}_3$ and $\hat{\sigma}_1$ respectively, that we first encountered in Chapter 4. Then, just like in Eq. (7.67), one finds

$$\hat{E}\phi = (\omega\hat{\sigma}_3 + \Delta\hat{\sigma}_1)\phi, \tag{9.34}$$

where

$$\phi = \begin{pmatrix} \hat{C}_k \\ \hat{C}_j^\dagger \end{pmatrix}. \tag{9.35}$$

The form of Eq. (9.34) has same as that of Dirac equation for a relativistic fermion in reduced dimensions [26, 67]. We will meet this in Chapters 11 and 12, where its dynamical significance will become clearer.

Eq. (9.30) is the so-called *Superconducting energy gap equation* that accounts for the superconducting state that is found in certain metals at extremely low temperatures [15, 43, 111]. It can be understood as follows. There are clearly two eigenvalue solutions to Eq. (9.30), given by

$$E = \pm\sqrt{\omega^2 + \Delta^2}. \tag{9.36}$$

In the absence of boson exchange between fermions, $\Delta = 0$, no Cooper pairs exist, and there is no gap between the positive and negative (relative to the Fermi energy) energy states. The effect of electron scattering via boson exchange gives rise to the energy gap of 2Δ between the upper and lower branches of the eigenvalue equation as illustrated in Fig. 9.2. It is the existence of the energy gap that gives rise to the superconducting property.

Finally, we note that the BCS mechanism has a number of features that lead Nambu [83] to see it as the basis of an explanation for how fermions acquire mass [35]. First there is the energy gap that has exactly the same form as that found in the

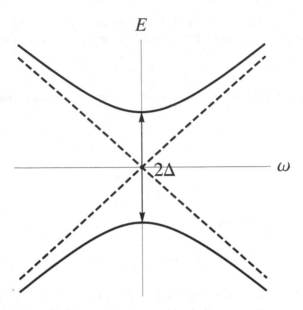

Figure 9.2 BCS superconducting energy gap. The ω axis corresponds to the Fermi surface. The solid curves correspond to the two eigenvalues in Eq. (9.36). The dashed lines correspond to the gapless case when $\Delta = 0$.

energy-momentum dispersion relation of the Dirac relativistic equation for fermions that arises as a consequence of the fermionic rest mass. Second, there is the appearance of the Pauli spin matrices in the spinor form of the superconducting energy equation, Eq. (9.34), that is entirely equivalent to the Dirac equation in reduced dimensions, although, as we shall see in Chapter 11, the coefficients need reinterpreting. Nambu also noticed that the particle/hole description of the coupling mechanism that leads to the energy gap and which explains the phenomenon of superconductivity is mirrored closely by the coupling between particles and antiparticles that is a key feature of the quantum field description of fermions. We shall explore some of these issues further in Chapters 11 and 12.

10 Field itemization

10.1 CONFIGURATION SPACE FOR FIELDS

The phase, ξ, that we introduced as part of natural number dynamics in Chapter 3, plays an absolutely key role in developing the dynamical theory that is the central theme of this book. We saw in Chapter 4, where we introduced multi-category systems, how the phase, ξ_i, on which the raising and lowering operators of each of the categories of a system functionally depend, needs to be parameterized in order to be able to both distinguish between the categories, but also to allow a universal scalar parameter to apply to the system as a whole and thus enable us to construct a single equation of motion for the whole system. Initially we chose the write $\xi_i = \omega_i t$, where ω_i serves as a label for the ith category and t is a universal parameter that applies to the system as a whole. We have seen that we can associate t with time and then ω_i behaves as an angular frequency that also serves as a scaled energy parameter. However, there is nothing to stop us parameterizing the phase with more than one universal parameter, and adding other labels for the categories. So we could, for example, generalize the phase, ξ_i to

$$\xi_i = \omega_i t + \kappa_i s, \tag{10.1}$$

where κ_i serves as a second category label, and s is a second systems-wide variable. For the rest of this chapter we only need to consider a single category and so we will drop the category label, i, from now on. Then we would find the annihilation operator, $\hat{A}$, for a system of n items, represented by the number operator $\hat{N}$, would have the form

$$\hat{A}(x,t) = \hat{A}(0,0)\exp(-i(\omega t + \kappa s)). \tag{10.2}$$

We are completely free to interpret this more general phase in any suitable way we choose. Having chosen to interpret t as time, we can choose to interpret s as, for example, a configuration space coordinate, so that the phase represents a propagating plane wave. A blueprint for this step is the discovery, in Chapter 8, of the electromagnetic field, that had the form of a propagating wave field. As we saw in Chapter 8, the phase of a wave propagating along the x-axis has the form $kx - \omega t$, so this suggests that we make the substitution[1], $\kappa s = -kx$, where k is the wave number, which is equal to 2π divided by the wavelength. Then, Eq. (10.2) becomes[2]

$$\hat{A}(x,t) = \hat{A}(0,0)\exp(i(kx - \omega t)). \tag{10.3}$$

---

[1]The reason for the negative sign in the substitution, $\kappa s = -kx$, will be discussed in detail in the next chapter.

[2]Notice that here we can use the complex version of the sinusoid, $\exp(i(kx - \omega t))$, rather than the real option, because, unlike in the case of the electric and magnetic fields in Chapter 8, which were Hermitian, $\hat{A}(x,t)$ is a mixture of Hermitian and anti-Hermitian parts.

DOI: 10.1201/9781003377504-10

Up to this point, the one-dimensional coordinate has been formally treated as an operator, $\hat{x}$, that is one of the Hermitian parts of the creation and annihilation operators, the other being the linear momentum operator, $\hat{p}$. Then, because of the commutation relation between $\hat{x}$ and $\hat{p}$, we have been able to represent $\hat{x}$ by the scalar variable x and $\hat{p}$ in terms of the derivative with respect to x, as explained in Chapter 5. The suggestion above that, instead of thinking of x as an operator, even though it can be represented by a scalar variable, we now treat it as a continuous systems variable, like time, completely changes its status. By introducing the co-ordinate, x, as a second systems-wide variable to parameterize the phase, ξ, we achieve two key results. First this puts the configuration space co-ordinate and time on an equal footing. This allows *space-time* and special relativity to be developed in a formal manner. Phase itself is a profoundly important invariant quantity in this regard and the invariance of phase in the form, $kx - \omega t$, has important consequences for the nature of space-time and relativistic quantum mechanics. These issues will be explored in detail in the next chapter. Secondly, the plane wave formulation allows us to consider wave fields that propagate over regions of space and this naturally brings into consideration quantities that may be expressed as densities that can then be integrated over space, as we found with the electromagnetic field in Chapter 8. Relativistic space-time and wave fields are essential ingredients of a quantum theory of fields [115, 125].

It is worth noting at this point that there is a natural pairing in the form of the phase in Eq. (10.3). Clearly, the pair (t, x) represents the systems-wide continuous variables. The complementary pair (ω, k) originated as category labels, but in the single category case, as here, can be considered as complementary variables in what is sometimes referred to as *reciprocal space*. This is in some ways an unfortunate description of the space that is represented by (ω, k), since it gives the impression that there are two separate spaces involved. The term reciprocal is only really a reference to the units used to measure a distance, x, in metres, say, and then a wave number, k, is in units of per metre. This just ensures that the product kx is dimensionless. Similarly, t has units of seconds, while ω has units of per second, so that ωt is dimensionless. As we shall see in the next chapter, (t, x) and (ω, k) can be teated as a pair of two-dimensional vectors. The phase can then be treated as a kind of scalar product involving these two vectors. Then (t, x) and (ω, k) need to share a common set of basis vectors in order that products that constitute the phase in Eq. (10.3) are meaningful. In this sense they belong to the same space. For now we just note that there is a duality between the two pairs that is the basis of a Fourier duality that stems from the sinusoidal character of the plane waves we are considering here. This Fourier duality will be developed in the next section. To avoid any confusion about space and reciprocal space, in what follows we will take t, x, ω and k as dimensionless parameters, unless otherwise stated.

In the next section we develop a new form of natural number dynamics that takes the electromagnetic field as a kind of blueprint, in as much as, we introduce densities over configuration space to represent natural numbers. This leads to a representation of itemization by field amplitudes, rather than just time dependent operators like the number operator amplitudes $\hat{A}$ and $\hat{A}^{\dagger}$. We will refer to this as *field itemization*. The fields involved will be referred to as *quantum fields*. However, this does not mean that we will be developing what is generally referred to as *quantum field theory* (QFT). That is a very well established theoretical technique that is based on Langrangian methods. It is beyond the scope of the present book (see comments in Section 13.2, of the epilogue). There are numerous books on quantum field theory, which the reader should consult for further details, see for example refs. [109, 115, 125].

10.2 ITEMS IN A FIELD: THE QUANTUM FIELD CONCEPT

It is now a relatively short step from the many-body formalism that has been developed in earlier chapters, to treat operators as functions of configuration space as well as time. It should be stressed again, that what follows in not standard quantum field theory. Here we just want to introduce the concept of the field amplitude to represent the items in a category, rather that the number operator amplitudes that we introduced in Chapter 3, and in so doing represent the items in a category by field itemization. This change in emphasis means that instead of operators such as that in Eq. (10.3), we use the density amplitude[3]

$$\hat{a}\exp\left(i(kx-\omega t)\right),$$

which represents a plane wave field in operator form. If we regard these plane waves as a Fourier spectrum, with amplitudes $\hat{a}(k) = \hat{a}\exp(-i\omega t)$, we can construct time and configuration dependent fields, $\hat{\phi}(x)$ and $\hat{\phi}^\dagger(x)$ from

$$\hat{\phi}(x) = \frac{1}{\sqrt{2\pi}}\int \hat{a}\exp(i(kx-\omega t))\mathrm{d}k = \frac{1}{\sqrt{2\pi}}\int \hat{a}(k)\exp(ikx)\mathrm{d}k, \qquad (10.4)$$

and

$$\hat{\phi}^\dagger(x) = \frac{1}{\sqrt{2\pi}}\int \hat{a}^\dagger\exp(-i(kx-\omega t))\mathrm{d}k = \frac{1}{\sqrt{2\pi}}\int \hat{a}^\dagger(k)\exp(-ikx)\mathrm{d}k. \qquad (10.5)$$

We recognize Eqs. (10.4) and (10.5) as Fourier transforms so $\hat{\phi}(x) \leftrightarrow \hat{a}(k)$ form a Fourier pair. The number operator becomes,

$$\begin{aligned}
\hat{N} &= \int \hat{\phi}^\dagger(x)\hat{\phi}(x)\mathrm{d}x \\
&= \frac{1}{2\pi}\iiint \hat{a}^\dagger(k')\hat{a}(k)\exp(i(k-k')x)\mathrm{d}k'\mathrm{d}k\mathrm{d}x \\
&= \int \hat{a}^\dagger(k)\hat{a}(k)\mathrm{d}k,
\end{aligned} \qquad (10.6)$$

where we have used

$$\int \exp(i(k-k')x)\mathrm{d}x = 2\pi\delta(k'-k).$$

---

[3]The frequency, ω is implicitly a function of k, since there is invariably some dispersion relation, $\omega = \omega(k)$, involved, as was the case with the electromagnetic field in Chapter 8.

What Eq. (10.6) tells us is that $\hat{\phi}^\dagger(x)\hat{\phi}(x)$ behaves as a particle density in configuration space which gives the particle number when integrated over that space. Also, $\hat{a}^\dagger(k)\hat{a}(k)$ can be considered as a particle density in k-space that gives the particle number when integrated over that space. This is an expression of Fourier duality between x and k that is part of the duality, $(\omega,k) \leftrightarrow (t,x)$. The equivalence of the densities in configuration space and k-space, from a mathematical point of view, is a consequence of Parsival's theorem of Fourier transforms that is expressed in Eq. (10.6).

The number operator in Eq. (10.6) is valid for both bosonic and fermionic fields, as long as these are confined to single category systems. Multi-category need to be treated somewhat differently. The commutation relations for the bosonic and fermionic annihilation and creation operators, respectively obey

$$[\hat{a}(k'),\hat{a}^\dagger(k)] = \delta(k' - k)$$

and

$$\{\hat{a}(k'),\hat{a}^\dagger(k)\} = \delta(k' - k). \tag{10.7}$$

For the corresponding configuration space functions we need

$$[\hat{\phi}(x'),\hat{\phi}^\dagger(x)] = \delta(x' - x)$$

and

$$\{\hat{\phi}(x'),\hat{\phi}^\dagger(x)\} = \delta(x' - x). \tag{10.8}$$

In addition we have

$$[\hat{\phi}(x'),\hat{\phi}(x)] = [\hat{\phi}^\dagger(x'),\hat{\phi}^\dagger(x)] = [\hat{a}(k'),\hat{a}(k)] = [\hat{a}^\dagger(k'),\hat{a}^\dagger(k)] = 0 \tag{10.9}$$

for bosonic fields, and

$$\{\hat{\phi}(x'),\hat{\phi}(x)\} = \{\hat{\phi}^\dagger(x'),\hat{\phi}^\dagger(x)\} = \{\hat{a}(k'),\hat{a}(k)\} = \{\hat{a}^\dagger(k'),\hat{a}^\dagger(k)\} = 0, \tag{10.10}$$

for fermionic fields. Then, for example, we have

$$\int [\hat{\phi}(x'),\hat{\phi}^\dagger(x)]dx = 1, \tag{10.11}$$

with corresponding results for the other brackets in Eqs. (10.7) and (10.8). Next we show how the Hamiltonian in field itemization may to be rewritten in terms of integrals with integrands involving $\hat{a}$ and $\hat{a}^\dagger$, and $\hat{\phi}$ and $\hat{\phi}^\dagger$.

10.3 THE HAMILTONIAN OF AN ITEMIZED FIELD

We can construct the Hamiltonian of this single category system by noting that the number density is in k-space is $\hat{a}^\dagger(k)\hat{a}(k)$. The energy of each item is ω and so the energy density should be $\omega\hat{a}^\dagger(k)\hat{a}(k)$. Thus we can expect the Hamiltonian, $\hat{\Omega}$ to be

$$\hat{\Omega} = \int \omega\hat{a}^\dagger(k)\hat{a}(k)dk. \tag{10.12}$$

Notice that the frequency, ω must be within the integral in Eq. (10.12), since it is in general a function of k. Now using Eq. (10.4), we can see that

$$\frac{\partial \hat{\phi}(x)}{\partial t} = \frac{1}{\sqrt{2\pi}} \int \hat{a} \frac{\partial \exp(i(kx - \omega t))}{\partial t} dk = -\frac{i}{\sqrt{2\pi}} \int \omega \hat{a}(k) \exp(ikx) dk. \quad (10.13)$$

Then we find

$$\hat{\Omega} = \frac{1}{2\pi} \int \hat{\phi}^\dagger(x) i \frac{\partial}{\partial t} \hat{\phi}(x) dx = \int \omega \hat{a}^\dagger(k) \hat{a}(k) dk. \quad (10.14)$$

10.4 THE PARTICLE/ANTI-PARTICLE HAMILTONIAN

It is worth noting that the field itemization formalism demonstrates its power in the case of fermionic fields that comprise particle and *anti-particle* fields as discussed in Chapter 9, in the context of BSC theory and as will be looked at again in Chapter 11, when we examine Nambu's explanation of fermionic mass in more detail [35]. The key point of the mechanism in both the BCS theory of superconductors and Nambu's theory of fermionic mass is the coupling between a creation operator and an annihilation operator, as in Eq. (9.27). To model this coupling, a single field may be constructed that combines a field $\hat{a}$ and a field $\hat{b}^\dagger$ as

$$\hat{\phi}(x) = \frac{1}{\sqrt{2\pi}} \int (\hat{a}(k) \exp(ikx) + \hat{b}^\dagger(k) \exp(-ikx)) dk, \quad (10.15)$$

and

$$\hat{\phi}^\dagger(x) = \frac{1}{\sqrt{2\pi}} \int (\hat{a}^\dagger(k) \exp(-ikx) + \hat{b}(k) \exp(ikx)) dk, \quad (10.16)$$

where $\hat{b}(k) = \hat{b} \exp(-i\omega t)$. If the field $\hat{a}$ is interpreted as the annihilation of a particle, then the field $\hat{b}^\dagger$ is not interpreted as the creation of a particle but as rather the annihilation of an anti-particle.

We then find

$$\int \hat{\phi}^\dagger(x) \hat{\phi}(x) dx = \frac{1}{2\pi} \iiint (\hat{a}^\dagger(k') \exp(-ik'x) + \hat{b}(k') \exp(ik'x))$$
$$\times (\hat{a}(k) \exp(ikx) + \hat{b}^\dagger(k) \exp(-ikx)) dk' dk dx$$
$$= \int \hat{a}^\dagger(k) \hat{a}(k) dk + \int \hat{b}(k) \hat{b}^\dagger(k) dk$$
$$= \hat{N}_a - \hat{N}_b + 1, \quad (10.17)$$

where we have used the integrated form of the fermionic anti-commutation rule

$$\int \hat{b}(k) \hat{b}^\dagger(k) dk = 1 - \int \hat{b}^\dagger(k) \hat{b}(k) dk, \quad (10.18)$$

with $\hat{N}_a$ and $\hat{N}_b$ being the number operators for particles and antiparticles respectively. Clearly, this does not give the total particle number, $\hat{N} = \hat{N}_a + \hat{N}_b$. Recall that we found that it was $\hat{N}_a - \hat{N}_b$ that was conserved when we treated coupled fermionic creation and annihilation operators in Section 7.5. However, if we note that Eqs. (10.6) and (10.7) are invariant to a new phase shift, ζ, such that, $\hat{a} \rightarrow \hat{a}\exp(-i\zeta)$ and if we similarly assume $\hat{b} \rightarrow \hat{b}\exp(-i\zeta)$, then we have

$$i\frac{\partial\hat{a}(k)\exp(i(kx-\zeta))}{\partial\zeta} = \hat{a}(k)\exp(i(kx-\zeta)) \tag{10.19}$$

and

$$i\frac{\partial\hat{b}^\dagger(k)\exp(i(\zeta-kx))}{\partial\zeta} = -\hat{b}^\dagger(k)\exp(i(\zeta-kx)). \tag{10.20}$$

Then

$$i\int\hat{\phi}^\dagger(x)\frac{\partial\hat{\phi}(x)}{\partial\zeta}dx = \iiint (\hat{a}^\dagger(k')\exp(i(\zeta-k'x))+\hat{b}(k')\exp(i(k'x-\zeta)))$$

$$\times (\hat{a}(k)\exp(i(kx-\zeta))-\hat{b}^\dagger(k)\exp(i(\zeta-kx)))dk'dkdx$$

$$= \int(\hat{a}^\dagger(k)\hat{a}(k)-\hat{b}(k)\hat{b}^\dagger(k))dk$$

$$= (\hat{N}_a+\hat{N}_b-1), \tag{10.21}$$

where we have used Eq. (10.18). Apart from the -1, Eq. (10.21) represents the total number of fermions.

The result above gives the impression that the operator $i\frac{\partial}{\partial\zeta}$ is acting like $i\frac{d}{d\xi}$ in Eq. (3.14), and so is equivalent to a number operator, $\hat{N}$. Although this interpretation works in the case of a single category field as in Eq. (10.6), in Eq. (10.21) we cannot simply put the lhs equal to $\hat{N}$ because it is not guaranteed to give a non-negative result. We can see this by operating on the vacuum state, $|0,0\rangle$. The result would be -1. We will come back to this point later, after treating the Hamiltonian.

Turning now to the Hamiltonian, we note that

$$\frac{\partial\hat{\phi}(x)}{\partial t} = \frac{1}{\sqrt{2\pi}}\int (\omega(\hat{a}(k)\exp(ikx)-\hat{b}^\dagger(k)\exp(-ikx)))dk, \tag{10.22}$$

where we have used

$$i\frac{\partial(\hat{a}(k)\exp(ikx))}{\partial t} = i\frac{\partial(\hat{a}\exp(i(kx-\omega t)))}{\partial t} \tag{10.23}$$

$$= \omega\hat{a}(k)\exp(-ikx)$$

and a corresponding result for $\hat{b}^\dagger(k)\exp(-ikx)$. Then we get

$$i\int\hat{\phi}^\dagger(x)\frac{\partial\hat{\phi}(x)}{\partial t}dx = \frac{1}{2\pi}\iiint (\hat{a}^\dagger(k')\exp(-ik'x)+\hat{b}(k')\exp(ik'x))$$

$$\times \omega(\hat{a}(k)\exp(ikx)-\hat{b}^\dagger(k)\exp(-ikx))dk'dkdx$$

$$= \int\omega(\hat{a}^\dagger(k)\hat{a}(k)-\hat{b}(k)\hat{b}^\dagger(k))dk$$

$$= \omega(\hat{N}_a+\hat{N}_b-1), \tag{10.24}$$

where again Eq. (10.18) has been applied. Eq. (10.24) agrees with the expected result for the Hamiltonian operator for a system containing $\hat{N}_a$ particles and $\hat{N}_b$ anti-particles, apart from the added constant of $-\omega$. We can see that this added constant is the vacuum state energy, when we operate on $|0,0\rangle$. This negative vacuum energy clearly has the same origin as the -1 in Eq. (10.21). It arises from the integration of $\delta(k'-k)$ in the anti-commutation relation in Eq. (10.7). It means that if we take Eq. (10.24) as the energy of the system, we cannot guarantee that it is non-negative. This issue is the same as that which arises in standard quantum field theory [125]. The approach taken there is that the vacuum energy can be ignored, since, it is argued, it is the energy against which the energy of a system is measured, rather like a calibration offset. So, the true energy of the system is taken as

$$\omega(\hat{N}_a + \hat{N}_b).$$

Here we have no reason to disagree with the experts in quantum field theory!

Looking back at Eq. (10.17), we can see that if charge exists in units of q, then the total charge is given by

$$q \int \hat{\phi}^{\dagger}(x)\hat{\phi}(x)\mathrm{d}x = q(\hat{N}_a - \hat{N}_b + 1), \qquad (10.25)$$

apart from the added constant of q, given that the particles and anti-particles carry charges with opposite signs. As in the case of the vacuum energy in Eq. (10.24), the vacuum charge can be ignored. However, it is interesting to note that, in the fermionic case, where, the occupation numbers can only have values of 0 or 1, it does ensure that the integral in Eq. (10.25) is non-negative.

Having established the basic framework for a theory of field itemization with one dimension of configuration space, it is a relatively straightforward matter to generalize it to one with three components of configuration space (see for example refs.[115, 109]). The relativistic aspects of the field theory are explored further in the next chapter by examining the invariant properties of the phase of a propagating plane wave field.

11 Phase invariance: The emergence of space-time and wave mechanics

11.1 REPRESENTING PHASE

We saw in the previous chapter that quantum fields could be pictured as propagating waves. The basic wave structure is a plane sinusoidal wave that varies in time and in one-dimension of configuration space as

$$\exp(-i\xi) = \exp(i(kx - \omega t)),$$

although the amplitude of these waves is represented by an operator, as in Eq. (10.4). The phase, $\xi = (\omega t - kx)$ in the above expression has the form it does so that points of fixed phase move with a positive speed for increasing x and increasing t, when both ω and k are positive quantities. To see this we note that the condition for following a fixed phase value is

$$\frac{d\xi}{dt} = \omega - k\frac{dx}{dt} = 0 \implies \frac{dx}{dt} = \frac{\omega}{k}. \tag{11.1}$$

The minus sign between the spatial and temporal parts of the phase ensures that, when ω and k are both positive, then a positive time step, dt, leads to a positive step, dx, along the x-axis, so the wave crests, say, move towards increasing values of x. The ratio of ω to k is called the *phase speed* of the wave. For the moment this is a speed in name only, since it must be remembered that x, t, k, and ω are dimensionless for now.

Highlighting the minus sign between the temporal and spatial parts of the phase may seem rather a trivial point, but it has important consequences, as shall see shortly. However, before we look at that, we need to check a related issue first. As is well-known the wave equation, such as Eq. (8.90) and the resulting dispersion relation, Eq. (8.91) admit two types of solution. One has a phase of the form, $\xi = (\omega t - kx)$, that we have just dealt with, but the other solution has a phase that is commonly written, $\xi = (\omega t + kx)$. Now since we are making the minus sign in the earlier form such a key issue in what follows, it might be thought that this second form of the phase needs to be treated differently. This is not the case. It is important to note that in both forms of the phase, ω and k are positive numbers. Thus when $\xi = (\omega t + kx)$, the speed of points of constant phase obey

$$\frac{dx}{dt} = -\frac{\omega}{k} = \frac{\omega}{-k},$$

DOI: 10.1201/9781003377504-11

which means that the points of constant phase travel towards decreasing values of x. The last step in the above expression implies that if we treat the wave number as the negative value, $-k = k'$, in this case, then $\xi = \omega t - k'x$. The two forms of phase then both appear with the minus sign between their temporal and spatial parts and represent a pair of plane waves travelling in opposite directions, with the sign of the wavenumber indicating the sense of the wave progression along the x-axis.

Before leaving this section, we will switch to the conventional way of writing phase and define the phase as $\theta = -\xi$, so from now on, the phase is

$$\theta = kx - \omega t. \tag{11.2}$$

Notice that there is still a minus sign between the spatial and temporal parts of the phase in Eq. (11.2), and the condition $\frac{d\theta}{dt} = 0$ again leads to Eq. (11.1).

11.1.1 PHASE AS A SCALAR PRODUCT

We can treat the pair, (x, t) as a vector in a two-dimensional vector space, and also the pair, (k, ω) as a vector in *the same* vector space. Then the phase in Eq. (11.2) can be treated as a scalar product of these two vectors. This is of course not a usual *Euclidean space*, because of the minus sign between the two terms in Eq. (11.2)[1]. Compare the phase in Eq. (11.2) with the phase in two dimensions of configuration space, given by

$$\theta = kx + ly, \tag{11.3}$$

where y is a second coordinate orthogonal to x and l is the corresponding component of the wave vector, (k, l). The phase in Eq. (11.3) is the familiar Euclidean product of (x, y) and (k, l). By contrast, the scalar production in Eq. (11.2) is termed *pseudo-Euclidean*.

Eqs. (11.2) and (11.3) are two particular examples of a more general definition of a scalar product of two vectors. This more general definition can be written in tensor notation as

$$g_{ij}u_iv_j, \tag{11.4}$$

where g_{ij} is referred to as the *metric tensor* of the space. The components of the two vectors are represented by u_i and v_j, where summation is understood to be over repeated indices. We can represent g_{ij} by the components of a square matrix, then Eq. (11.4) becomes $\mathbf{u}^T \hat{G} \mathbf{v}$, where $\mathbf{v}$ is a column vector, $\mathbf{u}^T$ is a row vector equal to the transpose of the column vector $\mathbf{u}$, and $\hat{G}$ is a square matrix that operates on the column vectors. In explicit component form, for a two-dimensional space, we have

$$\mathbf{u}^T \hat{G} \mathbf{v} = \begin{pmatrix} u_1, & u_2 \end{pmatrix} \begin{pmatrix} g_{11} & g_{12} \\ g_{21} & g_{22} \end{pmatrix} \begin{pmatrix} v_1 \\ v_2 \end{pmatrix} = g_{11}u_1v_1 + g_{12}u_1v_2 + g_{21}u_2v_1 + g_{22}u_2v_2. \tag{11.5}$$

If we let g_{ij} be $\hat{G}_E$ in the two-dimensional Euclidean case, then we have

$$\hat{G}_E = \begin{pmatrix} 1 & 0 \\ 0 & 1 \end{pmatrix}, \tag{11.6}$$

which is just the identity matrix.

---

[1]This rule is the same for both positive and negative values of k.

In the pseudo-Euclidean case, g_{ij} would be represented by

$$\hat{G}_P = \begin{pmatrix} 1 & 0 \\ 0 & -1 \end{pmatrix}. \tag{11.7}$$

Both the Euclidean and pseudo-Euclidean spaces above are termed *flat*[2] because the components of g_{ij} in each case are independent of u_i and v_j, for all i and j.

In what follows it is important to emphasize the need for a common vector space for both vectors in a scalar product. Conventionally, physicists regard the space of (x, y) as configuration space and that of (k, l) as a reciprocal space. This is somewhat misleading, because it gives the impression that these are two different spaces. This is because physicists have in mind that distance is measured in metres, for example, whereas the corresponding components of the wave vectors have units of radians per metre. This ensures that the product kx is dimensionless, as it must be to represent a phase, which is itself dimensionless. These units are simply descriptive labels and have nothing to do with the mathematical construction of a scalar product. The point is that in order for two vectors to be multiplied together using algebraic rules they must be related to the same set of orthogonal basis vectors and hence occupy the same space from a mathematical point of view. This is important to bear in mind in what follows. In order to make this point manifest we shall initially assume that all of the vector components are dimensionless and we can add units later, as necessary. In addition, we can define a pair of orthogonal unit basis vectors which can be used in both cases above. Writing the basis vectors as two-dimensional column vectors, e_1 and e_2, where

$$e_1 = \begin{pmatrix} 1 \\ 0 \end{pmatrix} \text{ and } e_2 = \begin{pmatrix} 0 \\ 1 \end{pmatrix}. \tag{11.8}$$

The corresponding row form of the basis vectors are the transposed the unit basis column vectors in Eq. (11.8), so we have for either the Euclidean or the pseudo-Euclidean space in two dimension, $e_1^T = (1, 0)$ and $e_2^T = (0, 1)$. Then

$$(u_1, u_2) = u_1 e_1^T + u_2 e_2^T, \tag{11.9}$$

where (u_1, u_2) represents any of (k, ω), (x, t), (k, l) or (x, y).

The Euclidean scalar product in Eq. (11.3) is[3]

$$kx + ly = (k, \quad l) \begin{pmatrix} 1 & 0 \\ 0 & 1 \end{pmatrix} \begin{pmatrix} x \\ y \end{pmatrix}. \tag{11.10}$$

The pseudo-Euclidean phase in Eq. (11.2) is

$$kx - \omega t = (k, \quad \omega) \begin{pmatrix} 1 & 0 \\ 0 & -1 \end{pmatrix} \begin{pmatrix} x \\ t \end{pmatrix}. \tag{11.11}$$

---

[2] Spaces in which the metric is a function of the vector components are called *Riemannian* or *pseudo-Riemannian* and correspond to curved spaces.

[3] Clearly we could drop the square matrix from the product in this case, but we leave it in here and in what follows in order to compare and contrast it with the pseudo-Euclidean case.

The Euclidean scalar products involving the two basis vectors are

$$(1, \ 0) \begin{pmatrix} 1 & 0 \\ 0 & 1 \end{pmatrix} \begin{pmatrix} 0 \\ 1 \end{pmatrix} = 0, \tag{11.12}$$

$$(1, \ 0) \begin{pmatrix} 1 & 0 \\ 0 & 1 \end{pmatrix} \begin{pmatrix} 1 \\ 0 \end{pmatrix} = 1, \tag{11.13}$$

and

$$(0, \ 1) \begin{pmatrix} 1 & 0 \\ 0 & 1 \end{pmatrix} \begin{pmatrix} 0 \\ 1 \end{pmatrix} = 1. \tag{11.14}$$

Eq. (11.12) indicates that the scalar product between the two basis vectors is zero, indicating their mutual orthogonality. Eqs. (11.13) and (11.14) show that the lengths of the two basis vectors in the Euclidean case are unity. We get something rather different in the pseudo-Euclidean case. There,

$$(1, \ 0) \begin{pmatrix} 1 & 0 \\ 0 & -1 \end{pmatrix} \begin{pmatrix} 0 \\ 1 \end{pmatrix} = 0, \tag{11.15}$$

$$(1, \ 0) \begin{pmatrix} 1 & 0 \\ 0 & -1 \end{pmatrix} \begin{pmatrix} 1 \\ 0 \end{pmatrix} = 1, \tag{11.16}$$

and

$$(0, \ 1) \begin{pmatrix} 1 & 0 \\ 0 & -1 \end{pmatrix} \begin{pmatrix} 0 \\ 1 \end{pmatrix} = -1. \tag{11.17}$$

So, whereas the orthogonality between the two basis vectors is still true and the length of e_1 is still unity, the *length* of e_2 is -1. This is typical of the pseudo-Euclidean spaces i.e., the measure that corresponds to a positive definite length in Euclidean spaces is not generally positive-definite in pseudo-Euclidean spaces. This property plays a crucial role in the implications of phase invariance in Euclidean and pseudo-Euclidean spaces, which we will investigate shortly. Next we look at changing the frame of reference of vectors in Euclidean and pseudo-Euclidean spaces. Changing the frame of reference is achieved through a linear transformation.

11.1.2 LINEAR TRANSFORMATION

It cannot be emphasized too strongly that the linear transformation we are going to apply, applies to the vector space itself and as a result the same transformation applies to all of the vectors in the space, that are used in the calculation of phase. This is not only essential for the mathematical structure, but it is also consistent with the conceptual issue we are dealing with. We are seeking to find the conditions that leave the phase invariant when we change our frame of reference. We should think of this in terms of changing the system of basis vectors which are used to define the properties of the vector space involved. This means that whatever transformation we use applies equally well to $(1,0)$ and $(0,1)$ as it does to (k,ω), (x,t), (k,l) and (x,y).

The linear transformation of the vectors can be represented by a 2×2 matrix of the form

$$\hat{T}_L = \begin{pmatrix} \alpha & \beta \\ \gamma & \delta \end{pmatrix}, \tag{11.18}$$

where α, β, γ and δ are constant coefficients yet to be determined. As we shall see, they take on different values for the pseudo-Euclidean case from those in the Euclidean case. Let us apply $\hat{T}_L$ to a general vector (u_1, u_2). Then

$$\begin{pmatrix} u_1' \\ u_2' \end{pmatrix} = \begin{pmatrix} \alpha & \beta \\ \gamma & \delta \end{pmatrix} \begin{pmatrix} u_1 \\ u_2 \end{pmatrix} = \begin{pmatrix} \alpha u_1 + \beta u_2 \\ \gamma u_1 + \delta u_2 \end{pmatrix}. \tag{11.19}$$

We can transform $(1,0)^T$ and $(0,1)^T$ using Eq. (11.18) as

$$\begin{pmatrix} \alpha & \beta \\ \gamma & \delta \end{pmatrix} \begin{pmatrix} 1 \\ 0 \end{pmatrix} = \begin{pmatrix} \alpha \\ \gamma \end{pmatrix}$$

and

$$\begin{pmatrix} \alpha & \beta \\ \gamma & \delta \end{pmatrix} \begin{pmatrix} 0 \\ 1 \end{pmatrix} = \begin{pmatrix} \beta \\ \delta \end{pmatrix}. \tag{11.20}$$

Notice that

$$\begin{pmatrix} u_1' \\ u_2' \end{pmatrix} = u_1 \begin{pmatrix} \alpha \\ \gamma \end{pmatrix} + u_2 \begin{pmatrix} \beta \\ \delta \end{pmatrix} = \begin{pmatrix} \alpha u_1 + \beta u_2 \\ \gamma u_1 + \delta u_2 \end{pmatrix}, \tag{11.21}$$

which shows that the transform of an arbitrary vector can be obtained from a linear superposition of the transforms of $(1,0)^T$ and $(0,1)^T$ and so must work for all of the individual two-dimensional vectors, i.e, both (k,l) and (x,y) and also (k,ω) and (x,t). This again emphasizes the point made earlier that changing the frame of reference through a linear transformation for a scalar product involves the same transformation for the two vectors in the scalar product. In addition to transforming column vectors as in Eq. (11.19) it is useful to know how to transform row vectors. This involves the transpose of the product of square matrix, $\hat{T}_L$ and the column vector. This is given by $(\hat{T}_L \mathbf{u})^T = \mathbf{u}^T \hat{T}_L^T$. So

$$(u_1', \ u_2') = (u_1, \ u_2) \begin{pmatrix} \alpha & \gamma \\ \beta & \delta \end{pmatrix} = (\alpha u_1 + \beta u_2, \ \gamma u_1 + \delta u_2). \tag{11.22}$$

It is clear that the final row vector in Eq. (11.22) is the transpose of final column vector in Eq. (11.21), as it should be.

We are now in a position to examine the effect of transforming the variables in Eq. (11.2) via a linear transformation while insisting that the phase remain invariant. The meaning of this transformation will become apparent shortly. The invariance of phase itself is readily justified. It is first and foremost a scalar quantity. If we think of the expression in Eq. (11.2) as representing the phase of a propagating wave then clearly a wave crest should remain a wave crest. Alternatively a complete cycle of phase remains a complete cycle. These features of phase cannot depend on our frame of reference. Notice also that phase invariance has nothing at all to do with any physical properties of a system. As Jackson [60] points out, elapsed phase is proportional to the number of wave crests that have passed the observer. Since this is merely a counting operation it must be independent of any frame of reference. The consequences of this frame invariance of phase is examined next.

11.1.3 PHASE INVARIANCE FOR EUCLIDEAN GEOMETRY

Let us begin by writing the phase in Eq. (11.4) in symbolic form as

$$\theta = \mathbf{u}^T \hat{G}_E \mathbf{v}, \tag{11.23}$$

where $\mathbf{u}^T$ represents (k, l) and $\mathbf{v}^T$ represents (x, y). Then the invariance of the phase, under the linear transformation, $\hat{T}_L$, implies

$$\mathbf{u}'^T \hat{G}_E \mathbf{v}' = \mathbf{u}^T \hat{T}_L^T \hat{G}_E \hat{T}_L \mathbf{v} = \mathbf{u}^T \hat{G}_E \mathbf{v}. \tag{11.24}$$

Eq. (11.24) implies

$$\hat{T}_L^T \hat{G}_E \hat{T}_L = \hat{G}_E. \tag{11.25}$$

Notice that Eq. (11.25) is independent of $\mathbf{u}$ and $\mathbf{v}$. This is to be expected since, as we have already argued, the transformation should be the same for all vectors in a particular space. In component form Eq. (11.25) is just

$$\begin{pmatrix} \alpha^2 + \gamma^2 & \alpha\beta + \gamma\delta \\ \alpha\beta + \gamma\delta & \beta^2 + \delta^2 \end{pmatrix} = \begin{pmatrix} 1 & 0 \\ 0 & 1 \end{pmatrix}, \tag{11.26}$$

and phase invarience implies

$$u_1' v_1' + u_2' v_2' = u_1 v_1 + u_2 v_2. \tag{11.27}$$

Eq. (11.26) implies

$$\alpha^2 + \gamma^2 = 1,$$
$$\beta^2 + \delta^2 = 1,$$

and

$$\alpha\beta + \gamma\delta = 0. \tag{11.28}$$

Eqs. (11.28) can then be solved in terms of α to give $\beta = \sqrt{1 - \alpha^2}$, $\gamma = -\sqrt{1 - \alpha^2}$, and $\delta = \alpha$. Because of these relationships we can use trigonometric functions to represent the components of the transformation matrix. We can write $\alpha = \delta = \cos \eta$, where η is some arbitrary angle. Then $\beta = -\gamma = \sin \eta$, and thus

$$\begin{pmatrix} u_1' \\ u_2' \end{pmatrix} = \begin{pmatrix} \cos \eta & \sin \eta \\ -\sin \eta & \cos \eta \end{pmatrix} \begin{pmatrix} u_1 \\ u_2 \end{pmatrix}, \tag{11.29}$$

which can be understood as a rotation of the vector, (u_1, u_2) through the angle, η. In symbolic form, Eq. (11.29) can be written

$$\mathbf{u}' = \hat{R}(\eta)\mathbf{u}, \tag{11.30}$$

...ere

$$\hat{R}(\eta) = \begin{pmatrix} \cos\eta & \sin\eta \\ -\sin\eta & \cos\eta \end{pmatrix}. \tag{11.31}$$

...e reverse of the transform in Eq. (11.30), is

$$\mathbf{u} = \hat{R}^{-1}(\eta)\mathbf{u}', \tag{11.32}$$

...ere $\hat{R}^{-1}(\eta)$ is the inverse of $\hat{R}(\eta)$, and is given by

$$\hat{R}^{-1}(\eta) = \begin{pmatrix} \cos\eta & -\sin\eta \\ \sin\eta & \cos\eta \end{pmatrix}, \tag{11.33}$$

...ich is just the reverse rotation to that in Eq. (11.31). Notice also that there is some ...bitrariness here. We could have defined the forward transform by the matrix in Eq. ...1.33) and the reverse by that in Eq. (11.31). Clearly, the starting point and end ...int of the transforms are interchangeable. It is simply a matter of choice which we ...gard as the forward and which the reverse transformation.

Obviously $(v_1, v_2) \rightarrow (v_1', v_2')$ corresponds to the same rotation as in Eq. (11.29). ..., under rotation, the phase in Eq. (11.23) and hence the specific form in Eq. (11.3) ...invariant. In addition the modulus of each vector is invariant, since Eq. (11.29) ...ds directly to

$$u_1'^2 + u_2'^2 = u_1^2 + u_2^2 = u_0^2, \tag{11.34}$$

...ere u_0 is the invariant length of (u_1, u_2). Similarly

$$v_1'^2 + v_2'^2 = v_1^2 + v_2^2 = v_0^2, \tag{11.35}$$

...ere v_0 is the invariant length of (v_1, v_2). $u_1^2 + u_2^2$ is referred to as the norm of the ...ctor, (u_1, u_2) and $v_1^2 + v_2^2$ the norm of (v_1, v_2). That these are positive definite is a ...y property of a Euclidean space, in addition to it being a flat space, as noted earlier. ...tice that we could have gotten to Eqs. (11.34) and (11.35) directly by noting that, ...Eq. (11.27) is true for any pair of vectors then it must be true for a scalar product of ..., u_2) with itself, hence Eq. (11.34) and likewise for the scalar product of (v_1, v_2) ...th itself, hence Eq. (11.35). In terms of the phase in Eq. (11.4), the invariants in ...dition to the phase itself are

$$k^2 + l^2 = k_0^2 \tag{11.36}$$

...d

$$x^2 + y^2 = r_0^2, \tag{11.37}$$

...ere k_0 and r_0 are the invariant quantities.

Notice finally, that from Eq. (11.29) we find $(1,0)$ and $(0,1)$ transform to ...s $\eta, -\sin\eta)$ and $(\sin\eta, \cos\eta)$. Then

$$(\cos\eta, \quad -\sin\eta) \begin{pmatrix} 1 & 0 \\ 0 & 1 \end{pmatrix} \begin{pmatrix} \sin\eta \\ \cos\eta \end{pmatrix} = 0, \tag{11.38}$$

$$(\cos\eta, \quad -\sin\eta) \begin{pmatrix} 1 & 0 \\ 0 & 1 \end{pmatrix} \begin{pmatrix} \cos\eta \\ -\sin\eta \end{pmatrix} = 1, \tag{11.39}$$

and

$$(\sin\eta, \quad \cos\eta) \begin{pmatrix} 1 & 0 \\ 0 & 1 \end{pmatrix} \begin{pmatrix} \sin\eta \\ \cos\eta \end{pmatrix} = 1, \tag{11.40}$$

which means that the orthogonality condition and the lengths of the basis vectors are invariant under the linear transformation in Eq. (11.29).

11.1.4 PHASE INVARIANCE IN PSEUDO-EUCLIDEAN GEOMETRY

We will now examine the effects of the linear transformation, $\hat{T}_L$ on the pseudo-Euclidean space that has the two-dimensional metric tensor represented by the matrix, $\hat{G}_P$. We will follow the same procedure as we did for the Euclidean space above. Now we represent the phase in Eq. (11.2) symbolically as

$$\theta = \mathbf{u}^T \hat{G}_P \mathbf{v}, \tag{11.41}$$

where now, $\mathbf{u}^T$ represents (k, ω) and $\mathbf{v}^T$ represents (x, t). Then the invariance of the phase, under the linear transformation, $\hat{T}_L$ now implies

$$\mathbf{u}'^T \hat{G}_P \mathbf{v}' = \mathbf{u}^T \hat{T}_L^T \hat{G}_P \hat{T}_L \mathbf{v} = \mathbf{u}^T \hat{G}_P \mathbf{v}. \tag{11.42}$$

Eq. (11.42) implies

$$\hat{T}_L^T \hat{G}_P \hat{T}_L = \hat{G}_P, \tag{11.43}$$

which again is independent of $\mathbf{u}$ and $\mathbf{v}$. In component form Eq. (11.43) is just

$$\begin{pmatrix} \alpha^2 - \gamma^2 & \alpha\beta - \gamma\delta \\ \alpha\beta - \gamma\delta & \beta^2 - \delta^2 \end{pmatrix} = \begin{pmatrix} 1 & 0 \\ 0 & -1 \end{pmatrix}, \tag{11.44}$$

and the phase invariance implies

$$u_1' v_1' - u_2' v_2' = u_1 v_1 - u_2 v_2. \tag{11.45}$$

Eq. (11.44) implies

$$\alpha^2 - \gamma^2 = 1,$$
$$\beta^2 - \delta^2 = -1,$$

and

$$\alpha\beta - \gamma\delta = 0. \tag{11.46}$$

Eqs. (11.46) can then be solved in terms of α to give $\beta = \gamma = \sqrt{\alpha^2 - 1}$, and $\delta = \alpha$. Because of these relationships we can use hyperbolic functions to represent the components of the transformation matrix. We can now write $\alpha = \delta = \cosh\eta$, where η is some arbitrary argument. Then we let[4] $\beta = \gamma = -\sinh\eta$, and thus

$$\begin{pmatrix} u_1' \\ u_2' \end{pmatrix} = \begin{pmatrix} \cosh\eta & -\sinh\eta \\ -\sinh\eta & \cosh\eta \end{pmatrix} \begin{pmatrix} u_1 \\ u_2 \end{pmatrix}. \tag{11.47}$$

---

[4]We could have chosen to write $\beta = \sinh\eta$, but it is a matter of convention to choose this option. The reason for this choice will become clear when we look at its application to the theory of special relativity later.

...mbolically

$$\mathbf{u}' = \hat{P}(\eta)\mathbf{u}, \tag{11.48}$$

...ere the pseudo-rotation, $\hat{P}(\eta)$ is

$$\hat{P}(\eta) = \begin{pmatrix} \cosh\eta & -\sinh\eta \\ -\sinh\eta & \cosh\eta \end{pmatrix}. \tag{11.49}$$

...e inverse of the transform in Eq. (11.48) is

$$\mathbf{u} = \hat{P}^{-1}(\eta)\mathbf{u}', \tag{11.50}$$

...ere $\hat{P}^{-1}(\eta)$ is the inverse of $\hat{P}(\eta)$. It is given by

$$\hat{P}^{-1}(\eta) = \begin{pmatrix} \cosh\eta & \sinh\eta \\ \sinh\eta & \cosh\eta \end{pmatrix}. \tag{11.51}$$

Obviously $(v_1, v_2) \to (v_1', v_2')$ corresponds to the same transformation, $\hat{P}(\eta)$. So, ...der pseudo-rotation, the phase in Eq. (11.41) and hence the specific form in Eq. ...1.2) is invariant. In addition the modulus of each vector is invariant, since Eq. ...1.45) leads directly to

$$u_2'^2 - u_1'^2 = u_2^2 - u_1^2 = u_0^2, \tag{11.52}$$

...ere u_0 is the invariant length of (u_1, u_2). Similarly

$$v_2'^2 - v_1'^2 = v_2^2 - v_1^2 = v_0^2, \tag{11.53}$$

...ere v_0 is the invariant length of (v_1, v_2). Unlike the Euclidean case, here in the ...eudo-Euclidean case these lengths are clearly not positive definite. Notice that we ...uld have got to Eqs. (11.52) and (11.53) directly by noting that, if Eq. (11.45) is ...e for any pair of vectors then it must be true for a scalar product of (u_1, u_2) with ...elf, hence Eq. (11.52) and likewise for the scalar product of (v_1, v_2) with itself, ...nce Eq. (11.53).

Notice finally, that from Eq. (11.47) we find $(1, 0)$ and $(0, 1)$ transform to ...shη, sinη) and $(\sinh\eta, \cosh\eta)$. Then

$$(\cosh\eta, \quad \sinh\eta) \begin{pmatrix} 1 & 0 \\ 0 & -1 \end{pmatrix} \begin{pmatrix} \sinh\eta \\ \cosh\eta \end{pmatrix} = 0, \tag{11.54}$$

$$(\cosh\eta, \quad \sinh\eta) \begin{pmatrix} 1 & 0 \\ 0 & -1 \end{pmatrix} \begin{pmatrix} \cosh\eta \\ \sinh\eta \end{pmatrix} = 1, \tag{11.55}$$

$$(\sinh\eta, \quad \cosh\eta) \begin{pmatrix} 1 & 0 \\ 0 & -1 \end{pmatrix} \begin{pmatrix} \sinh\eta \\ \cosh\eta \end{pmatrix} = -1, \tag{11.56}$$

...ich means that the orthogonality condition and the lengths of the basis vectors are ...ariant under the linear transformation in Eq. (11.49).

Now of course, the whole purpose of this exercise was to look at the consequences of a linear transformation on the phase in Eq. (11.2). We are now in a position to state the result, simply by letting $(u_1, u_2) = (k, \omega)$ and $(v_1, v_2) = (x, t)$. Thus
we have

$$\begin{pmatrix} k' \\ \omega' \end{pmatrix} = \begin{pmatrix} \cosh \eta & -\sinh \eta \\ -\sinh \eta & \cosh \eta \end{pmatrix} \begin{pmatrix} k \\ \omega \end{pmatrix} \tag{11.57}$$

and

$$\begin{pmatrix} x' \\ t' \end{pmatrix} = \begin{pmatrix} \cosh \eta & -\sinh \eta \\ -\sinh \eta & \cosh \eta \end{pmatrix} \begin{pmatrix} x \\ t \end{pmatrix}. \tag{11.58}$$

Then Eqs. (11.52) and (11.53) become

$$\omega'^2 - k'^2 = \omega^2 - k^2 = \omega_0^2, \tag{11.59}$$

where ω_0 is an invariant, independent of the frame of reference. Similarly, with (x, t),
we get

$$t'^2 - x'^2 = t^2 - x^2 = s_0^2, \tag{11.60}$$

where s_0 is also an invariant, independent of the frame of reference.

Notice that, unlike the Euclidean case, Eqs. (11.36) and (11.37), where the invariants are positive definite or zero, in the pseudo-Euclidean case, the invariants are not
positive definite but are equal to the difference between two squares. In the form in
Eqs. (11.58) and (11.59), we have chosen to make ω^2 larger than k^2 and t^2 larger than
x^2. We could have chosen to reverse either or both of these conditions. We will deal
with the reason for the choice in Eqs. (11.59) and (11.60) in the following section.

11.2 SPECIAL RELATIVITY

In the previous section we obtained an invariance relation involving wave parameters,
k and ω in Eq. (11.59) and another involving space and time co-ordinates, x ant t in
Eq. (11.60), from the invariance of phase, Eq. (11.2), under a linear transformation.
The requirement that phase be invariant constrains the linear transforms to the form
in the matrix $\hat{P}_\eta$, in Eq. (11.49). Anyone familiar with the theory of *special relativity*
will recognize Eq. (11.60) as the equation in dimensionless form for the invariant
interval, s_0 that results from the Lorentz transformation of Einstein's theory. The
traditional way of introducing special relativity begins with Einstein's principle of
equivalence, i.e., that the speed of light is the same in all inertial frames. As is well-
known, this approach is soundly based on the experimental evidence from the famous
Michelson-Morley experiment. However, as acknowledged by Einstein himself [30],
there has been some criticism of the approach for an over-reliance on light signals. It
is not the intention here to raise such criticisms of the method itself, which has served
generations of physics students for over a century, but it is worth pointing out that
the association with light signals can lead to some more general wave implications
being missed. These implications are dealt with below.

As is well-known, Einstein obtained the null form of Eq. (11.60), i.e., for $s_0 = 0$
on the basis that the speed of light was the same in all inertial frames [30]. Switching
from one inertial frame to another was a concrete way of describing what we have

lled, thus far, a change of frame of reference. Einstein required further arguments
get to the more general form with $s_0 \neq 0$.

In the present work, our approach is to develop physical theory without assum-
g physical processes beforehand. The arguments above are based entirely on phase
variance, which is a property of abstract wave patterns. The advantage of this ap-
oach is that not only do we get directly the space-time invariance condition in its
n-null form, but also a wave dispersion relation, Eq. (11.59) which has profound
plications for the origins of wave mechanics, as we shall see. So, the phase in-
riance arguments kills two birds with one stone. First we will deal with special
ativity.

As pointed out above, we can interpret what we have called a change in the frame
reference under the action of a linear transformation as a change in inertial frame
the case of a pseudo-Euclidean space, which corresponds to the form of phase in
. (11.2). Thus the transform $(x,t) \rightarrow (x',t')$ corresponds to Eq. (11.58). Thus

$$x' = x\cosh\eta - t\sinh\eta. \tag{11.61}$$

is well-known, inertial frames are frames that move at constant relative velocity.
will assume that the primed frame is moving at a constant speed, v along the
axis. Let us look at the point in the primed frame for which $x' = 0$, i.e., the origin
the primed frame. Then, with $x' = 0$, Eq. (11.61) gives

$$\frac{x}{t} = \tanh\eta = v, \tag{11.62}$$

m which we can deduce that $\cosh\eta = \frac{1}{\sqrt{1-v^2}}$ and $\sinh\eta = \frac{v}{\sqrt{1-v^2}}$.

Now with $x = vt$ from Eq. (11.62), substituted into the expression for s_0 in Eq.
.60) we get

$$s_0 = t\sqrt{1-v^2}. \tag{11.63}$$

real s_0, there is an upper limit on $|v|$ of 1.

Recalling that thus far, x and t are in dimensionless form, in order to compare
se results with the standard form of the equations of special relativity, we can
nporarily put x in length units, and t in time units. To achieve this, we use a con-
sion factor between them in the form of a velocity in units of length per unit time.
we call this conversion factor c, then $t \rightarrow ct$ and $v \rightarrow v/c$. Then,

$$s_0 = t\sqrt{1 - \frac{v^2}{c^2}}. \tag{11.64}$$

w we see that c corresponds to the upper limit to velocity in any inertial frame.
nust therefore be a universal constant. We know of course that this happens to be
speed of light in a vacuum, but we did not need to know that in order to derive
above results. We will see later that c is not so much associated specifically with
ht but any massless object. We note also that introducing c does not modify Eq.

(11.62), but the transformation in velocity form, which we relabel, $\hat{P}_v$, becomes

$$\begin{pmatrix} x' \\ ct' \end{pmatrix} = \begin{pmatrix} \dfrac{1}{\sqrt{1-v^2/c^2}} & -\dfrac{v/c}{\sqrt{1-v^2/c^2}} \\ -\dfrac{v/c}{\sqrt{1-v^2/c^2}} & \dfrac{1}{\sqrt{1-v^2/c^2}} \end{pmatrix} \begin{pmatrix} x \\ ct \end{pmatrix}, \tag{11.65}$$

which is the standard form for the Lorentz transformation in special relativity.

As has been explained at length, the linear transformation, $\hat{P}_v$, that transforms x and t also transforms k and ω, so

$$\begin{pmatrix} k' \\ \dfrac{\omega'}{c} \end{pmatrix} = \begin{pmatrix} \dfrac{1}{\sqrt{1-v^2/c^2}} & -\dfrac{v/c}{\sqrt{1-v^2/c^2}} \\ -\dfrac{v/c}{\sqrt{1-v^2/c^2}} & \dfrac{1}{\sqrt{1-v^2/c^2}} \end{pmatrix} \begin{pmatrix} k \\ \dfrac{\omega}{c} \end{pmatrix}, \tag{11.66}$$

Then we have

$$\omega'^2 - k'^2 c^2 = \omega^2 - k^2 c^2 = \omega_0^2. \tag{11.67}$$

11.3　RELATIVISTIC WAVE MECHANICS: THE DIRAC EQUATION

The phase invariance approach above has allowed us to recover the special theory of relativity without recourse to light signals, thus freeing the conceptual basis from electromagnetic phenomena and allowing a more universal interpretation in terms of the intrinsic properties of spaces. However, the real bonus of the method is that it brings into play a wave dispersion relation in the form of an invariant frequency from which emerges relativistic wave mechanics in a way that is free from any preconceived physical notions.

We note that the conserved quantity, involving k and ω in Eq. (11.59), lead to[5]

$$\omega^2 = k^2 + \omega_0^2. \tag{11.68}$$

Eq. (11.68) represents a dispersion relation for waves with (angular) frequency ω and wave number k. We can then envisage a wave field defined in terms of a scalar wave function $\psi(x,t)$, constructed from a spectrum of plane waves with frequencies $\omega(k)$ that are functions of k, as specified in Eq. (11.68). $\psi(x,t)$ is then defined by the Fourier integral [101] (also see Appendix C)

$$\psi(x,t) = \frac{1}{\sqrt{2\pi}} \int_{-\infty}^{\infty} F(k)\exp(i(kx - \omega(k)t))\mathrm{d}k. \tag{11.69}$$

If the integral is finite, then the field will form a localized wave pulse. At $t = 0$

$$\psi(x,0) = \frac{1}{\sqrt{2\pi}} \int_{-\infty}^{\infty} F(k)\exp(ikx)\mathrm{d}k, \tag{11.70}$$

---

[5]Here we are reverting to dimensionless form.

so the inverse Fourier transform of Eq. (11.70) gives the k-spectrum

$$F(k) = \frac{1}{\sqrt{2\pi}} \int_{-\infty}^{\infty} \psi(x,0) \exp(-ikx) dx. \tag{11.71}$$

This is the k-spectrum of the wave field. The spectral power $F(k)^* F(k)$ does not change as the wave field propagates. To see this we note that $F(k) \exp(-i\omega(k)t)) = F(k,t)$ is the Fourier amplitude of $\psi(x,t)$, since

$$\psi(x,t) = \frac{1}{\sqrt{2\pi}} \int_{-\infty}^{\infty} F(k,t) \exp(i(kx)) dk. \tag{11.72}$$

The spectral power at any instant is

$$F^*(k,t)F(k,t) = F^*(k) \exp(i\omega(k)t))F(k) \exp(-i\omega(k)t)) = F^*(k)F(k),$$

and so does not change with time. All that happens during propagation is that, in general, plane wave components of the spectrum travel with different phase speeds. This leads to the change of relative phase between the different spectral components and this caused the shape of the pulse to change. As shown in Appendix C, the mean position of the pulse propagates not at the mean phase speed in Eq. (11.1), i.e.,

$$\frac{\omega(k)}{k} = \frac{\sqrt{k^2 + \omega_0^2}}{k}, \tag{11.73}$$

but rather at the mean group speed, v_g, given by

$$v_g = \frac{\partial \omega}{\partial k} = \frac{k}{\sqrt{k^2 + \omega_0^2}} = \frac{k}{\omega}. \tag{11.74}$$

We can formally partially differentiate Eq. (11.72) with respect to t and get

$$\frac{\partial \psi(x,t)}{\partial t} = -i \frac{1}{\sqrt{2\pi}} \int_{-\infty}^{\infty} \omega(k)F(k) \exp(i(kx - \omega(k)t)) dk$$

$$= -i \frac{1}{\sqrt{2\pi}} \int_{-\infty}^{\infty} \sqrt{k^2 + \omega_0^2} F(k) \exp(i(kx - \omega(k)t)) dk. \tag{11.75}$$

Since $F(k) \exp(-i\omega(k)t) = F(k,t)$ is the Fourier transform of $\psi(x,t)$, then, using standard Fourier theory we can write

$$\frac{\partial^p \psi(x,t)}{\partial x^p} = \frac{1}{\sqrt{2\pi}} \int_{-\infty}^{\infty} (ik)^p F(k) \exp(i(kx - \omega(k)t)) dk, \tag{11.76}$$

where p is a natural number. This means that partial derivatives of $\psi(x,t)$ with respect to x are generated by powers of k, multiplied by $F(k)$. The problem we have is that $\sqrt{k^2 + \omega_0^2}$ is not a simple power of k, so we cannot convert the integral in Eq. (11.76) into an pth order differential with respect to x. One way out of the problem is to see that the second derivative of $\psi(x,t)$ with respect to t does make things easier, since

$$
\begin{aligned}
\frac{\partial^2 \psi(x,t)}{\partial t^2} &= -\frac{1}{\sqrt{2\pi}} \int_{-\infty}^{\infty} F(k) \exp(i(kx - \omega(k)t)) \omega^2(k) dk \\
&= -\frac{1}{\sqrt{2\pi}} \int_{-\infty}^{\infty} F(k) \exp(i(kx - \omega(k)t)) (k^2 + \omega_0^2) dk \\
&= -\omega_0^2 \psi(x,t) - \frac{\partial^2 \psi(x,t)}{\partial x^2}.
\end{aligned}
\tag{11.77}
$$

Eq. (11.77) is just the Klein-Gordon equation [82, 101, 109].

An alternative way out is to assume, $\omega_0 \gg k$, then to a good approximation, $\sqrt{k^2 + \omega_0^2} \approx \omega_0 + \frac{k^2}{2\omega_0}$. Then we get

$$
\begin{aligned}
\frac{\partial \psi(x,t)}{\partial t} &= -i \frac{1}{\sqrt{2\pi}} \int_{-\infty}^{\infty} F(k) \exp(i(kx - \omega(k)t)) \sqrt{k^2 + \omega_0^2} dk \\
&= -i \frac{1}{\sqrt{2\pi}} \int_{-\infty}^{\infty} F(k) \exp(i(kx - \omega(k)t)) (\omega_0 + \frac{k^2}{2\omega_0})) dk \\
&= -i(\omega_0 \psi(x,t) + \frac{1}{2\omega_0} \frac{\partial^2 \psi(x,t)}{\partial x^2}).
\end{aligned}
\tag{11.78}
$$

Writing $\tilde{\psi}(x,t) = \psi(x,t) \exp(i\omega_0 t)$, then we get

$$
i \frac{\partial \tilde{\psi}(x,t)}{\partial t} = \frac{1}{2\omega_0} \frac{\partial^2 \tilde{\psi}(x,t)}{\partial x^2}.
\tag{11.79}
$$

This is a scaled form of the Schrödinger equation for a free particle that we met in Chapter 5.

Another alternative solution to the difficulty of $\sqrt{k^2 + \omega_0^2}$ not being a simple power of k is to use the result that the square root can be put into linear form by using anti-commuting operators such that we developed in Section 4.5. Then

$$
\sqrt{k^2 + \omega_0^2} = \hat{\sigma}_i k + \hat{\sigma}_j \omega_0,
\tag{11.80}
$$

where $\hat{\sigma}_i$ and $\hat{\sigma}_j$ are any pair of the Pauli spinors with $i \neq j$. Then, $\hat{\sigma}_i \hat{\sigma}_j + \hat{\sigma}_j \hat{\sigma}_i = 0$ and $\hat{\sigma}_i^2 = \hat{\sigma}_j^2 = 1$. Then we get

$$\frac{\partial \psi(x,t)}{\partial t} = -i \frac{1}{\sqrt{2\pi}} \int_{-\infty}^{\infty} \sqrt{k^2 + \omega_0^2} F(k) \exp(i(kx - \omega(k)t)) dk$$

$$= -i \frac{1}{\sqrt{2\pi}} \int_{-\infty}^{\infty} (\hat{\sigma}_i k + \hat{\sigma}_j \omega_0) F(k) \exp(i(kx - \omega(k)t)) dk. \tag{11.81}$$

With a little rearranging, Eq. (11.81) is just

$$i \frac{\partial \psi(x,t)}{\partial t} = -i \hat{\sigma}_i \frac{\partial \psi(x,t)}{\partial x} + \hat{\sigma}_j \omega_0 \psi(x,t). \tag{11.82}$$

We can recognize Eq. (11.82) as the Dirac equation of the relativistic electron in scaled form for one spatial dimension and time[6]. So, not only has phase invariance given us Einstein's theory of special relativity, it has also thrown up relativistic wave mechanics.

Eq. (11.82) can be written in terms of the Hamiltonian operator $\hat{\Omega}_1$ and the momentum operator $\hat{P}_x$ so

$$\hat{\Omega}_1 = \hat{\sigma}_i \hat{P}_x + \hat{\sigma}_j \omega_0. \tag{11.83}$$

The plane wave function, $\exp(i(kx - \omega t))$ is an eigenfunction of both $\hat{\Omega}$ and $\hat{P}_x$, with respective eigenvalues of ω and k. This means that ω and k are, respectively, the scalar energy and linear momentum eigenvalues. In the next section we examine the relationship between them from a kinematic viewpoint.

11.4 PHASE KINEMATICS

Phase kinematics refers to the dynamical properties of wave fields, particularly when they are viewed in terms of localized wave groups or pulses, that can be deduced from their dispersion properties, rather than from the equations of motion that govern their physical amplitudes. Particular attention will be focussed on the invariant quantities that were derived in Section 11.1.

We begin by writing the phase, using Eq. (11.63), in the form

$$\theta = (kv - \omega)t = s_0 \frac{(kv - \omega)}{\sqrt{1 - v^2}}. \tag{11.84}$$

Since both θ and s_0 are invariant, then we must have

$$\frac{\omega - kv}{\sqrt{1 - v^2}} = \omega_{00}, \tag{11.85}$$

where ω_{00} is also an invariant.

---

[6]This is commonly referred to as the $1 + 1$ dimensional case [67, 26].

We now wish to examine the situation in which the wave pulse is treated as an *object* moving at the speed v relative to the rest frame. So, we look for the condition for the group speed to be equal to v. From Eq. (11.74) we can see that this occurs when

$$v = \frac{k}{\omega} \tag{11.86}$$

Substituting Eq. (11.86) into Eq. (11.85) gives

$$\frac{\omega^2 - k^2}{\sqrt{\omega^2 - k^2}} = \sqrt{\omega^2 - k^2} = \omega_{00}, \tag{11.87}$$

from which we get

$$\omega^2 = k^2 + \omega_{00}^2. \tag{11.88}$$

It is then easy to see that we must have $\omega_{00} = \omega_0$, and so

$$\frac{\omega - kv}{\sqrt{1 - v^2}} = \omega_0, \tag{11.89}$$

$$k = \frac{\omega_0 v}{\sqrt{1 - v^2}}, \tag{11.90}$$

and

$$\omega = \frac{\omega_0}{\sqrt{1 - v^2}}. \tag{11.91}$$

Eq. (11.89) is equivalent to Eq. (11.85), but Eqs. (11.90) and (11.91) give us something new. We can regard ω_0 as the rest energy of the pulse, i.e., $\omega = \omega_0$ when $v = 0$, and then k as the relativistic momentum and ω as the relativistic energy associated with the moving wave field pulse.

Multiplying the variables on the lhs of Eqs. (11.90) and (11.91) by $\hbar$ puts the energy and one-dimensional momentum in standard form, with energy, $E = \hbar\omega$ and momentum, $p_x = \hbar k$. Multiplying Eqs. (11.68) by $\hbar^2$ and replacing v with $\frac{v}{c}$ then yields

$$E^2 = p_x^2 c^2 + m_0^2 c^4 \tag{11.92}$$

where $m_0 c^2 = \hbar\omega_0$. m_0 is of course the invariant rest mass and Eq. (11.92) is the well-known relativistic energy-momentum relation for a particle in one dimension.

Multiplying Eq. (11.83) by $\hbar$ and rescaling the velocity as with Eq. (11.92) then yields

$$\hat{H}_1 = \hat{\sigma}_i \hat{P}_x c + \hat{\sigma}_j m_0 c^2, \tag{11.93}$$

where $\hat{H}_1 = \hbar\hat{\Omega}_1$ is the one-dimensional Dirac Hamiltonian that applies to massive fermions[7].

---

[7]Compare Eq. (11.93) with Eq. (9.34) for the energy in BCS theory.

11.5 3+1 DIMENSIONS

The above results for a single spatial dimension can easily be extended to three spatial dimensions. We begin by writing the phase in $3+1$ dimensions as

$$\theta = \mathbf{k}.\mathbf{x} - \omega t, \tag{11.94}$$

where $\mathbf{k} = (k_x, k_y, k_z)$ and $\mathbf{x} = (x, y, z)$. We can assume that the relative velocity of the two frames of reference are orientated so the x is parallel to x', y is parallel to y' and z is parallel to z'. The scalar product of $\mathbf{k}$ and $\mathbf{x}$ in Eq. (11.95) is Euclidean, i.e.,

$$\mathbf{k}.\mathbf{x} = k_x x + k_y y + k_z z. \tag{11.95}$$

The relative motion between the frames is directed along the positive x axis. Then linear transformation, $\hat{G}_P$, will now take the form of a 4×4 matrix, given by

$$\hat{G}_P = \begin{pmatrix} 1 & 0 & 0 & 0 \\ 0 & 1 & 0 & 0 \\ 0 & 0 & 1 & 0 \\ 0 & 0 & 0 & -1 \end{pmatrix}, \tag{11.96}$$

so that

$$\mathbf{k}.\mathbf{x} - \omega t = \begin{pmatrix} k_x, & k_y, & k_z, & \omega \end{pmatrix} \begin{pmatrix} 1 & 0 & 0 & 0 \\ 0 & 1 & 0 & 0 \\ 0 & 0 & 1 & 0 \\ 0 & 0 & 0 & -1 \end{pmatrix} \begin{pmatrix} x \\ y \\ z \\ t \end{pmatrix}. \tag{11.97}$$

Also, since no relative motion takes place along the y and z axes, then, $k_y' = k_y$, $k_z' = k_z$, $y' = y$, $z' = z$, and

$$s_0^2 = t^2 - \mathbf{x}^2 = t'^2 - \mathbf{x}'^2 \tag{11.98}$$

and

$$s_0 = t \sqrt{1 - \mathbf{v}^2}, \tag{11.99}$$

where $\mathbf{v} = (dx/dt, dy/dt, dz/dt)$. The corresponding invariant quantity in $(\mathbf{k}, \omega)$ space is

$$\omega_0^2 = \omega^2 - \mathbf{k}^2 = \omega'^2 - \mathbf{k}'^2. \tag{11.100}$$

From Eq. (11.100) we find $\omega = \sqrt{\mathbf{k}^2 + \omega_0^2}$, which we wish to linearize as we did with Eq. (11.80). However, we now have four components to deal with, that will need four independent anti-commuting operators. This means we cannot use the 2×2 Pauli matrices because there are only three of these. Instead we write

$$\omega = \sqrt{\mathbf{k}^2 + \omega_0^2} = \hat{\alpha}.\mathbf{k} + \hat{\beta}\omega_0 \tag{11.101}$$

where $\hat{\alpha} = (\hat{\alpha}_x, \hat{\alpha}_y, \hat{\alpha}_z)$. $\hat{\alpha}_x$, $\hat{\alpha}_y$, $\hat{\alpha}_z$, and $\hat{\beta}$ are four independent anti-commuting operators. They are conventionally represented by a set of 4×4 Dirac matrices [118].

They obey the same anti-commutation rules as the Pauli matrices, i.e., $\{\hat{\alpha}_i, \hat{\alpha}_j\} = 2\delta_{ij}$, $\{\hat{\beta}, \hat{\alpha}_i\} = 0$ and $\hat{\beta}^2 = 1$.

We now need a wave field in three spatial dimensions, which takes the form

$$\frac{\partial \psi(\mathbf{x},t)}{\partial t} = -i\frac{1}{\sqrt{2\pi}} \int\limits_{-\infty}^{\infty} \sqrt{\mathbf{k}^2 + \omega_0^2} F(\mathbf{k}) \exp(i(\mathbf{k}.\mathbf{x} - \omega(\mathbf{k})t)) \mathrm{d}^3\mathbf{k}$$

(11.102)

$$= -i\frac{1}{\sqrt{2\pi}} \int\limits_{-\infty}^{\infty} (\hat{\alpha}.\mathbf{k} + \hat{\beta}\omega_0) F(\mathbf{k}) \exp(i(\mathbf{k}.\mathbf{x} - \omega(\mathbf{k})t)) \mathrm{d}k,$$

which leads to

$$i\frac{\partial \psi(\mathbf{x},t)}{\partial t} = -i\hat{\alpha}.\frac{\partial \psi(\mathbf{x},t)}{\partial \mathbf{x}} + \hat{\beta}\omega_0 \psi(x,t).$$

(11.103)

As with the $1+1$ dimensional case, the wave equation in Eq. (11.103) is easily put in Hamiltonian form. Then

$$\hat{\Omega}_3 = \hat{\alpha}.\hat{\mathbf{P}} + \hat{\beta}\omega_0,$$

(11.104)

where $\hat{\mathbf{P}} = (\hat{P}_x, \hat{P}_y, \hat{P}_z)$. The plane wave function $\exp(i(\mathbf{k}.\mathbf{x} - \omega(\mathbf{k})t))$ is an eigenfunction of both $\hat{\Omega}_3$ and $\hat{\mathbf{P}}$, with eigenvalues, respectively, of ω and $\mathbf{k}$, so ω and $\mathbf{k}$ represent the energy and momentum of the $3+1$ case. The kinematic relationship between them can again be obtained by considering the invariants under the linear transformation between reference frames, as follows.

From Eq. (11.94)

$$\theta = (\mathbf{k}.\mathbf{v} - \omega)t = s_0 \frac{(\mathbf{k}.\mathbf{v} - \omega)}{\sqrt{1 - \mathbf{v}^2}},$$

(11.105)

and we can now let

$$\frac{\omega - \mathbf{k}.\mathbf{v}}{\sqrt{1 - \mathbf{v}^2}} = \omega_{00}.$$

(11.106)

The dispersion relation is now,

$$\omega^2 = \mathbf{k}^2 + \omega_0^2$$

(11.107)

then the group velocity, $\mathbf{v}_g$ is

$$\mathbf{v}_g = \frac{\partial \omega}{\partial \mathbf{k}} = \frac{\mathbf{k}}{\omega}.$$

(11.108)

In this case we now want $\mathbf{v}_g = \mathbf{v}$ and so

$$\mathbf{v} = \frac{\mathbf{k}}{\omega}.$$

(11.109)

Substituting Eq. (11.109) into Eq. (11.106) yields

$$\frac{\omega^2 - \mathbf{k}^2}{\sqrt{\omega^2 - \mathbf{k}^2}} = \omega_{00},$$

(11.110)

from which we get

$$\omega^2 = \mathbf{k}^2 + \omega_{00}^2.$$

(11.111)

Comparing Eqs. (11.107) with (11.111), we can conclude that $\omega_{00} = \omega_0$, as in the one-dimensional case. So then, we have

$$\mathbf{k} = \frac{\omega_0 \mathbf{v}}{\sqrt{1 - \mathbf{v}^2}},$$ (11.112)

and

$$\omega = \frac{\omega_0}{\sqrt{1 - \mathbf{v}^2}}.$$ (11.113)

Now $\mathbf{k}$ and ω represent the relativistic momentum and energy for pulse kinetics in $3 + 1$ dimensions that now replace their one-dimensional counterparts, Eqs. (11.89) to (11.91).

Finally, putting Eqs. (11.107) and (11.104) gives us

$$(\hbar\omega)^2 = E^2 = \mathbf{p}^2 c^2 + m_0^2 c^4$$ (11.114)

and

$$\hbar\hat{\Omega}_3 = \hat{H}_3 = \hat{\alpha}.\hat{\mathbf{p}}c + \hat{\beta}m_0 c^2.$$ (11.115)

We recognize Eq. (11.114) as Einstein's energy-momentum equation of special relativity, and Eq. (11.115) as Dirac's equation of the relativistic electron.

11.6 THE ORIGIN OF FERMIONIC MASS

Having now established the form of the relativistic Hamiltonian operator and its energy eigenvalues and recognized that it is identical to the relativistic Dirac equation, we can begin to explore its physical significance. The energy eigenvalue of the Hamiltonian in the one-dimensional case in Eq. (11.92) takes the form of a pair energy equations that satisfy

$$E = \pm \sqrt{p_x^2 c^2 + m_0^2 c^4}.$$ (11.116)

These two solutions form two branches of an energy-momentum dispersion relation that is plotted in Fig. 11.1. We can immediately recognize the similarities between the relativistic energy-momentum relationship from Eq. (11.116) and the energy eigenvalues for BCS superconducting theory from Eq. (9.37) that were displayed in Fig. 9.2. Furthermore, the spinor nature of the superconductor equation, Eq. (9.34) has the same form as the one-dimensional Dirac equation, Eq. (11.83)[8]. What is also clear is the energy gap in the two sets of curves in Figs. 9.2 and 11.1. In the case of superconductors, it is the energy of 2Δ between electrons above and below the Fermi energy level that is responsible for the superconducting effect. It corresponds to a minimum energy needed to excite electrons from the so called *Fermi sea* and the conduction band. In the Dirac equation the gap of $2m_0c^2$ is the minimum energy for pair production of a positron and an electron. It is this feature that attracted Nambu's

---

[8]The particular components of the Pauli spinors involved to not have to correspond exactly. Recall that they can be interchanged anyway.

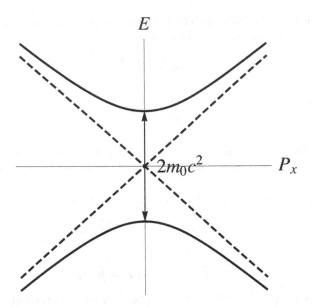

Figure 11.1 The fermionic energy gap. The solid curves represent the two relations between E and p_x from Eq. (11.116). The dashed curves correspond to the case when the rest mass, m_0 is zero. The energy gap here is $2m_0c^2$.

attention to the BCS theory of superconductivity, as an explanation of the existence of mass in fermions [35].

A key feature of the BCS mechanism, is the coupling between the annihilation of an electron in one state and the excitation of one in another state, that is evident in the BCS coupling equations, Eqs. (9.28) and (9.29). This is the same coupling we found using a linear transformation of fermion creation and annihilation operators in Section 7.5. Nambu recognized that, in the context of Dirac's theory of electrons, this is one way of thinking about the coupling between the annihilation of an electron in a negative energy state and the excitation of its anti-particle i.e. a positron in a positive energy state. The P_x axis in Fig. 11.1 corresponds to $E = 0$. This corresponds to the Fermi energy in BCS theory. In particle physics, $E = 0$ is referred to as *the vacuum*, from which particles are excited. This means that the lower curve corresponds to negative energy states. However, we do not see negative energy particles in Nambu's picture, since whenever an electron is excited in positive energy state then a positron is excited, which correspond to a loss of negative energy. Hence the positron appears with positive energy too. This process is well-known as *pair production*.

Nambu also pointed out the fact that Eqs. (9.28) and (9.29) could written in spinor form, indicated a close connection with the Dirac equation. Given the same functional dependence of their energy eigenvalues, the role of the energy gap in both phenomena, and the appearance of spinors in both energy equations, Nambu supposed that the same basic mechanism operates in both superconductivity and in the process by which fermions acquire mass. One can then suppose, by analogy with the superconducting energy gap, that the electron mass term arises due to electron scattering via boson exchange. However, whereas in the case of superconductors, the boson scattering is involves phonons, in the case of Dirac electrons, we now associate the scattering that gives rise to fermion mass with Higgs bosons [115].

This analogy that Nambu draws between the BCS theory of superconductivity and the theory of the origin of mass in fermions emphasizes the role played by the collective behaviour of the fermions in each case. It means that isolated electrons, which are usually referred to as *bare electrons* in this context, in the absence of the coupling via boson scattering, would have zero mass and would travel at the speed of light. These bare particles correspond to the dashed lines in Fig. 11.1. We will explore this energy gap again, in the context of a dynamical model of electron-positron pair production, in Chapter 12.

For a more detailed and rigorous discussion of the mathematical basis of the mechanisms involved in Nambu's explanation of the origin of fermion mass, a reading of Nambu's Nobel lecture [83] is strongly recommended. Of particular interest is an anecdotal description of how Nambu got his idea after attending a seminar by Schrieffer on superconductivity. Nambu makes clear how important the cross-fertilization between the two apparently different phenomena was in facilitating the development of his new theory and that starting in a new place may sometimes bring new insights and fresh understanding [35].

11.7 ANGULAR MOMENTUM REVISITED: THE EMERGENCE OF SPIN

We have seen that the angular momentum vector, $\hat{\mathbf{L}} = (\hat{L}_x, \hat{L}_y, \hat{L}_z)$, commutes with the spherically symmetric Hamiltonians in Chapter 8, including in the special case where the potential is absent. This implies that the angular momentum vector is a constant of the motion under these circumstances. The Hamiltonians in Chapter 8 are essentially non-relativistic. Quantities that are of particular significance in physics are those that are conserved during changing conditions, so there is especial significance given to operators that commute with the Hamiltonian operator of a system, since this implies such quantities do not change in time. In this section we wish to check whether the angular momentum vector is conserved for the potential-free relativistic Dirac equation. The full three-dimensional case is rather expensive algebraically. It is dealt with in detail in standard texts on relativistic quantum mechanics (see for example refs. [109, 118]). Here it is sufficient to check a system of just two dimensions of configuration space with their corresponding components of linear momentum, from which one component of angular momentum can be constructed perpendicular to the plane of the linear momentum components. This has the great advantage that we can represent the anti-commuting operators by Pauli spinors. Let the momentum component be $\hat{L}_k$

$$\hat{L}_k = \hat{X}_i \hat{P}_j - \hat{X}_j \hat{P}_i, \qquad (11.117)$$

where i, j and k are all different, but they must, in that sequence, form a right handed set. We can first check $\hat{L}_k$ in expression Eq. (11.117) commutes with the free non-relativistic case with a Hamiltonian of the form

$$\hat{\Omega}_2 = \frac{1}{2}(\hat{P}_i^2 + \hat{P}_j^2). \qquad (11.118)$$

We find

$$[\hat{L}_k, \hat{\Omega}_2] = i(\hat{P}_i\hat{P}_j - \hat{P}_j\hat{P}_i) = 0, \qquad (11.119)$$

as expected, since $[\hat{P}_i, \hat{P}_j] = 0$.

Now let us construct a two-dimensional Dirac equation. We can ignore the term involving the rest mass since it plays no part in the result, so we just choose

$$\hat{\Omega}_2 = \hat{\sigma}_i\hat{P}_i + \hat{\sigma}_j\hat{P}_j. \qquad (11.120)$$

Now

$$[\hat{L}_k, \hat{\Omega}_2] = [\hat{X}_i\hat{P}_j - \hat{X}_j\hat{P}_i, \hat{\sigma}_i\hat{P}_i + \hat{\sigma}_j\hat{P}_j] = i(\hat{\sigma}_i\hat{P}_j - \hat{\sigma}_j\hat{P}_i), \qquad (11.121)$$

which is not zero. So, the angular momentum around an axis perpendicular to the $(\hat{X}_i, \hat{X}_j)$ plane is not conserved.

The problem is solved by constructing a new angular momentum component,

$$\hat{J}_k = \hat{L}_k + \frac{1}{2}\hat{\sigma}_k, \qquad (11.122)$$

where $[\hat{\sigma}_i, \hat{\sigma}_j] = i2\hat{\sigma}_k$. Then we find

$$[\hat{J}_k, \hat{\Omega}_2] = 0. \qquad (11.123)$$

This is a crucial result. The implication is clear, that $\frac{1}{2}\hat{\sigma}_k$ is a component of an angular momentum. It is of course one of the components of intrinsic spin of the electron. Unlike $\hat{L}_k$, which is not an intrinsic element of the Dirac equation, $\frac{1}{2}\hat{\sigma}_k$ is an essential feature of the Dirac equation itself.

A further point of note is that the Pauli spinors are related to the three fermionic Hermitian operators with components, $\hat{S}_i$, that we came across is Section 4.4, where we found

$$\hat{S}_i = \frac{1}{2}\hat{\sigma}_k.$$

There are two important points to note about this. First, there are three independent components of $\hat{S}_i$ just as there are three independent components of $\hat{L}_i$. So, each component of angular momentum has a matching component of spin. This is neither arbitrary nor a coincidence. It is a consequence of the Goldilocks principle of the three dimensions that we discovered in Chapter 8. Secondly, the eigenvalues of each of the components $\hat{S}_i$ is $\pm\frac{1}{2}$. Now the eigenvalues of a single component of angular momentum has an integer value. Recall that we found that the eigenvalues of the components of the $\hat{J}_i$ operators that we looked at in Chapters 7 and 8 could have integer or half integer values. The connection to intrinsic spin provides an important interpretation of that result.

Obviously, the full picture of the three-dimensional angular moment needs the $3 + 1$ dimensional Dirac equation. Then the $3 + 1$ dimensional Dirac Hamiltonian,

$\hat{\Omega}_3$, commutes with all the components of the total angular momentum vector[9], $\hat{\mathbf{J}} = (\hat{J}_x, \hat{J}_y, \hat{J}_z)$. The total angular momentum is just the sum of the orbital angular momentum and the spin angular momentum, so

$$\hat{\mathbf{J}} = \hat{\mathbf{L}} + \hat{\mathbf{S}}. \tag{11.124}$$

Here we can see the importance that we have precisely three components of $\hat{\mathbf{L}}$ and three components of $\hat{\mathbf{S}}$ to make the three components of $\hat{\mathbf{J}}$ in a consistent way. It is the underlying Lie symmetry of both sets of components that gives us this.

11.8 INHOMOGENEOUS PROPAGATION AND THE MEANING OF ξ

Recall that the system variable ξ was introduced as part of the definition of the natural number operator in Eq. (3.14). There it was not parameterized in any way. Later the need for a system-wide time variable, t, led to the simple parameterization with $\xi = \omega t$. Later still when fields were introduced, the parameterization was broadened to include three configuration space co-ordinates to give

$$\xi = \omega t - \mathbf{k}.\mathbf{x},$$

which is equivalent to Eq. (11.94), recalling that $\theta = -\xi$. Now, Eq. (11.94) and the above equation are just a linear parameterization of ξ. However, we are not forced to confine attention to linear parameterization. We can consider something more general and just take ξ and θ as functions of x, y, z, and t, and so

$$\xi(x,y,z,t) = -\theta(x,y,z,t).$$

Then we can define $\mathbf{k}$ and ω by

$$\mathbf{k} = -\nabla\xi \text{ and } \omega = \frac{\partial\xi}{\partial t}. \tag{11.125}$$

Eqs. (11.125) are standard definitions in classical wave kinematics [128], but note they are equivalent to operating respectively with $\hat{\mathbf{P}}$ and $\hat{\Omega}$ on a plane wave a constant amplitude, Ψ, and an inhomogeneous phase, ξ, of the form, $\Psi\exp(-i\xi)$, such that

$$\mathbf{k}\Psi\exp(-i\xi) = \hat{\mathbf{P}}\Psi\exp(-i\xi) = -i\nabla(\Psi\exp(-i\xi)) = -(\nabla\xi)\Psi\exp(-i\xi)$$

and

$$\omega\Psi\exp(-i\xi) = \hat{\Omega}\Psi\exp(-i\xi) = i\frac{\partial\Psi\exp(-i\xi)}{\partial t} = \left(\frac{\partial\xi}{\partial t}\right)\Psi\exp(-i\xi).$$

Dividing these two chains of equations by $\Psi\exp(-i\xi)$ leads to Eqs. (11.125).

---

[9]See ref.[118] for details on the relationship between $\hat{\mathbf{J}}$ and the three-dimensional Dirac equation, Eq. (11.115).

Now, both $\mathbf{k}$ and ω in general will be functions of x, y, z, and t. In addition, ω will depend on the components of $\mathbf{k}$ through a dispersion relation, so we can let

$$\omega = f_\omega(k_x, k_y, k_z, x, y, z, t), \tag{11.126}$$

where $\mathbf{k} = (k_x, k_y, k_z)$. Notice that Eqs. (11.125) are also consistent with Eq. (11.94) in the case of linear parameterization.

Two results follow immediately from Eqs. (11.125). They are

$$\nabla \wedge \mathbf{k} = 0 \tag{11.127}$$

and

$$\frac{\partial \mathbf{k}}{\partial t} = -\frac{\partial \nabla \xi}{\partial t} = -\nabla \frac{\partial \xi}{\partial t} = -\nabla \omega. \tag{11.128}$$

Following Whitham [128], we evaluate Eq. (11.128) for a single component of $\mathbf{k}$. Let this be k_i, where i represents any of the three components. Then, with the aid of Eq. (11.126)

$$\frac{\partial k_i}{\partial t} = -\frac{\partial f_\omega}{\partial x_i} - \sum_j \frac{\partial f_\omega}{\partial x_j} \frac{\partial k_j}{\partial x_i} = -\frac{\partial f_\omega}{\partial x_i} - \sum_j \frac{\partial f_\omega}{\partial x_j} \frac{\partial k_i}{\partial x_j}, \tag{11.129}$$

where Eq. (11.127) has been utilized as

$$\frac{\partial k_j}{\partial x_i} = \frac{\partial k_i}{\partial x_j}.$$

So we get, after a little rearranging

$$\frac{d\mathbf{k}}{dt} = \frac{\partial \mathbf{k}}{\partial t} + \mathbf{v}.\nabla \mathbf{k} = -\nabla f_\omega, \tag{11.130}$$

where

$$\mathbf{v} = \frac{\partial f_\omega}{\partial \mathbf{k}},$$

which we recognize as the group velocity. This means that we can interpret, $\frac{d\mathbf{k}}{dt}$, in Eq. (11.130) as the total derivative of $\mathbf{k}$ in a frame moving with the group velocity. Recalling that $f_\omega = \omega$, then multiplying Eq. (11.130) by $\hbar$ leads to

$$\frac{d\hbar \mathbf{k}}{dt} = -\nabla(\hbar f_\omega), \tag{11.131}$$

which is Hamilton's equation. What this shows is that when we generalize the parameterization of ξ, the Hamiltonian function becomes dependent on the location in configuration space. This is the meaning of inhomogeneous propagation. It is consistent with there being a potential that is a function of the co-ordinates. In the homogeneous case that applied in Section 11.5, we had $\nabla(\hbar\omega) = 0$, and so the linear momentum was a constant in time.

Let us define a function, $\mathscr{L}$, as

$$\mathscr{L} = -\frac{d\xi}{dt} = -\frac{\partial \xi}{\partial t} - \mathbf{v}.\nabla \xi, \tag{11.132}$$

so, since $\xi = \xi(x, y, z, t)$, then

$$\mathscr{L} = \mathscr{L}(v_x, v_y, v_z, x, y, z, t), \tag{11.133}$$

where $\mathbf{v} = (v_x, v_y, v_z)$. Substituting, Eqs. (11.125) into Eq. (11.132) then gives $\mathscr{L} = \mathbf{v}.\mathbf{k} - \omega$ and so

$$\omega = \mathbf{v}.\mathbf{k} - \mathscr{L}. \tag{11.134}$$

Since $\hbar\omega$ represents energy, and $\hbar\mathbf{k}$ represents linear momentum, then we can recognize $\hbar\mathscr{L}$ as the *Lagrangian function*. Also

$$-\xi = \int \mathscr{L} dt,$$

so $-\hbar\xi$ is just the *action* as defined in standard mechanics.

The contents of Eqs. (11.126), (11.133) and (11.134) can be treated using standard methods involving a Legendre transformation [41] (also see Appendix H) to show that

$$\frac{\partial \mathscr{L}}{\partial \mathbf{v}} = \mathbf{k}, \quad \frac{\partial \mathscr{L}}{\partial t} = -\frac{\partial f_\omega}{\partial t} \quad \text{and} \quad \frac{\partial \mathscr{L}}{\partial \mathbf{x}} = -\frac{\partial f_\omega}{\partial \mathbf{x}}. \tag{11.135}$$

The first relation in Eqs. (11.135) is the standard definition of canonical momentum. The results in Eqs. (11.135), together with Eq. (11.131) yield

$$\frac{d}{dt}\frac{\partial \mathscr{L}}{\partial \mathbf{v}} - \frac{\partial \mathscr{L}}{\partial \mathbf{x}} = 0, \tag{11.136}$$

which is the Langrange equation of motion. This is of course entirely equivalent to the Hamilton equation, Eq. (11.131).

It is interesting to see, from Eq. (11.134), that the Lagrangian function is identical to the negative of the Doppler shifted frequency. We also note that, for the linear parameterization of ξ, in Section 11.5, then combining Eqs. (11.112) and (11.113) leads to

$$\hbar\mathscr{L} = \hbar(\mathbf{v}.\mathbf{k} - \omega) = \hbar\left(\frac{\omega_0 v^2}{\sqrt{1-v^2}} - \frac{\omega_0}{\sqrt{1-v^2}}\right) = -\hbar\omega_0\sqrt{1-v^2}, \tag{11.137}$$

which is the correct relativistic Langrangian for a free particle, with rest energy $\hbar\omega_0 = m_0 c^2$. In addition, if we use the expression $\mathscr{L} = \mathbf{k}.\mathbf{v} - \omega$ together with Eq. (11.101), then

$$\hbar\mathscr{L} = \hbar(\mathbf{k}.\mathbf{v} - \hat{\alpha}.\mathbf{k} - \hat{\beta}\omega_0), \tag{11.138}$$

so

$$\frac{\partial \hbar\mathscr{L}}{\partial \mathbf{v}} = \hbar\mathbf{k},$$

which is the correct canonical momentum.

We can generalize the above result by returning to the case when the phase is more generally inhomogeneous. Then we can add a function of the coordinates to $\mathscr{L}$ in Eq. (11.138), so that $\hbar\mathscr{L} = \hbar(\mathbf{k}.\mathbf{v} - \hat{\alpha}.\mathbf{k} - \hat{\beta}\omega_0) - V(x,y,z)$. Then it is easy to check that

$$\hbar\omega = \hbar(\mathbf{k}.\mathbf{v} - \mathscr{L}) = \hbar(\hat{\alpha}.\mathbf{k} + \hat{\beta}\omega_0) + V(x,y,z), \qquad (11.139)$$

which is the Hamiltonian of a relativistic particle in a potential, $V(x,y,z)$ [81].

Lagrangian methods play an important role in the development of quantum field theory, but that is best approached by means of a technique involving path integrals [33, 71, 115, 135], rather than the simple phase method outlined above (Also see comments in the epilogue).

11.9 NON-RELATIVISTIC LIMIT

We are now in a position to understand why the analysis in Chapter 5 lead to non-relativistic quantum mechanics and the Schrödinger equation rather than the Dirac equation. We have seen that the relativistic result comes from parameterizing the phase, ξ, as $\xi = \omega t - kx$. Then we insisted on the invariance of this phase under the linear transformation of the pseudo-Euclidean vectors, (x,t) and (k,ω). This required a single 2×2 matrix. In Chapter 5, we parameterized ξ as, $\xi = \omega t$. In this case we need to transform the scalars, t and ω with a 1×1 matrix, i.e., a simple scalar. Let this be γ, so that $t' = \gamma t$ and $\omega' = \gamma\omega$. Then phase invariance leads to

$$\omega' t' = \gamma^2 \omega t = \omega t.$$

So we can conclude that $\gamma = \pm 1$. The negative result adds nothing extra and we can take $\omega' = \omega$ and $t' = t$. If we compare this to the relativistic results in Eqs. (11.65) and (11.66), we can see that they are equivalent to the condition that $\frac{v}{c} \to 0$. So, as we already know, the relativistic case comes about for low relative speeds or by taking the upper limiting speed, c, to infinity. The important thing here is that we have seen that this is the same as only parameterizing ξ with time, as we did in Chapter 5. As a result we then got the Schrödinger equation, which is therefore the non-relativistic form of quantum mechanics.

12 Natural number dynamics II: Time-dependent population models

The intention, in this chapter, is to illustrate something of the versatility of natural number dynamics as a means of modelling time dependent populations of items that occur in systems of interacting categories and to demonstrate the universality of *the quantum nature of things* that flavours the whole of this book. The applicability of the natural number dynamics to systems outside those that are normally treated in physics, is made possible, to a certain degree, by the way natural number dynamics was introduced in Chapter 3, without any reference to physical systems, nor with any reliance on preconceived physical notions. The examples we will look at come from a variety of scientific fields including ecology, and molecular biology as well as physics. The examples from ecology and molecular biology involve bosonic natural number operators. The example from ecology is a type of predator-prey model and the molecular biology example involves an original model of cell division [106]. The application of the number operator in an extra-physical context is a very recent development of the use of quantum techniques. Quantum population dynamics in a social science context that has been pioneered by Bagarello and his co-workers [5, 6, 8, 7] has received a lot of attention recently.

Mathematical modelling of populations of living things has a history probably dating back to Fibonacci in the thirteenth century [4]. An early example of the use of differential calculus in population dynamics can be found in a paper submitted to the Academy of Sciences in Paris, in 1760, by Daniel Bernouilli, that presented an analysis of the effects of inoculation on deaths due to smallpox [4]. Mathematical models incorporating differential equations have been in common use to investigate the dynamical behaviour of systems involving interacting populations of living things, ever since Lotka [75] and Volterra [123] introduced their model of predator-prey competition in the 1920s. However, these models are based on continuous scalar variables. The continuum approach does lead to simplifications, since we can use continuous, scalar-valued functions and ordinary differential calculus, that was developed by Leibniz and Newton to deal with finding the rates of change of smooth mathematical functions, for rates of change of such populations. This approximation is often justified by arguing that if one is only interested in averages, as is usually the case in population models, then real numbers and not just counting numbers, are justifiable in most cases, especially when large populations are involved. Then, also, the minimum change in population number, being one, is a small fraction of the population as a whole, so any errors incurred should be small. However, it remains unclear whether modelling the average is the same thing as averaging a model, in the case of natural number valued populations, especially when population numbers are not large [106]. The nature of these models is often heuristic and it is usually taken for granted that the number continuum on the real number line can be used to model systems of discrete, countable entities like people, animals, plants, bacteria, and cells. Ecological systems [21], the spread of epidemics [59], and cancer cell population growth [20, 58] are just a few examples of what has been modelled in this way. The models that have been used in to predict the infection rates in the recent COVID-19 pandemic have all been based on the calculus of continuous variables to predict population numbers [40, 96].

With these considerations in mind, we are going to explore some simple models that contain populations of items that will be modelled using the number operator method that has been developed in earlier chapters. This ensures that population numbers remain natural numbers, even in dynamical situations in multiple category systems, especially when interactions are involved, so that items move between categories, as we saw in Chapter 6. The common factor in all of the examples in this chapter is that they are systems that contain, for the most part, two categories of items that can be accounted for by means of natural numbers and natural number operators, whether these items be photons or biological cells. These items then undergo interactions of various kinds that can be modelled using creation and annihilation operators. These serve to move items between the categories and so affect population numbers within categories. These models are referred to as *exchange models*.

We also include two examples from physics. The first of these relies on bosonic operators and comes from quantum optics. This involves the parametric amplification of photon populations by means of a high powered laser, which is a well-known piece of physics. The final example in this chapter that is also from physics, involves fermionic operators. It deals with the interactions between between fermions and their anti-particles. It is a rather instructive example that illustrates how the coupling between electrons and positrons, which, as we have seen in earlier chapters, accounts for their mass, can be seen as vacuum oscillations in which a particle/anti-particle pair is continually created and annihilated. The frequency of this population number oscillation turns out to be the well-known *zitterbewegung* frequency [81, 52, 108]. This simple little model is remarkably revealing about the nature of the electron-positron system and provides an interesting picture of fundamental particles, not as microscopic lumps of matter, but as harmonic excitations. Of particular interest here is that the oscillation is seen in the expectation value of the natural number operator associated with the particles and not in the field, so it is unnecessary to worry about whether the particles are singular points or strings.

We begin by reviewing the basic model that was introduced in Chapter 6, that will serve as a blueprint for more complicated models later in the chapter.

12.1   ONE-FOR-ONE EXCHANGE

Consider a system with a Hamiltonian of the form

$$\hat{\Omega} = U\hat{A}^\dagger\hat{A} + W\hat{B}^\dagger\hat{B} + V(\hat{A}^\dagger\hat{B} + \hat{B}^\dagger\hat{A}), \tag{12.1}$$

where U, W and V are constant coefficients, and $\hat{A}$ and $\hat{B}$ are two independent bosonic annihilation operators with their corresponding creation operators, $\hat{A}^\dagger$ and $\hat{B}^\dagger$. These obey the standard form of commutation relations for bosonic systems, i.e.

$$[\hat{A},\hat{A}^\dagger] = [\hat{B},\hat{B}^\dagger] = 1,$$

and

$$[\hat{A},\hat{B}] = [\hat{A},\hat{B}^\dagger] = [\hat{A}^\dagger,\hat{B}] = [\hat{A}^\dagger,\hat{B}^\dagger] = 0.$$

$\hat{\Omega}$ in Eq. (12.1) has exactly the same form as the Hamiltonian in Eq. (6.14) that was obtained by a unitary transformation of a Hamiltonian for a pair of non-interacting categories, so the origin of this model, as is the case for all of the models in this chapter, can be traced directly back to the universal quantum equation, Eq. (1.1). In Eq. (6.14) and again here in Eq. (12.1) the Hamiltonian represents the interaction between two populations that are represented by the number operators, $\hat{N} = \hat{A}^\dagger\hat{A}$ and $\hat{M} = \hat{B}^\dagger\hat{B}$. We will refer to the category of items associated with $\hat{A}$ as Cat$_A$ and those associated with $\hat{B}$ as Cat$_B$. The interacting character of the Hamiltonian comes exclusively from the term $V(\hat{A}^\dagger\hat{B}+\hat{B}^\dagger\hat{A})$.

It is useful to separate the Hamiltonian into a non-interacting part, $\hat{\Omega}_0$, where

$$\hat{\Omega}_0 = U\hat{N}+W\hat{M}, \tag{12.2}$$

and an interacting part, $\hat{\Omega}_I$, given by

$$\hat{\Omega}_I = V(\hat{A}^\dagger\hat{B}+\hat{B}^\dagger\hat{A}). \tag{12.3}$$

The interaction Hamiltonian, $\hat{\Omega}_I$, exhibits the one-for-one exchange behaviour in that, as we have seen previously, the first term on the rhs of Eq. (12.3), $\hat{B}$ removes *one* item from Cat$_B$ and then $\hat{A}^\dagger$ puts *one* item in Cat$_A$. The second term on the rhs of Eq. (12.3) does the reverse.

Hamiltonians with the interaction term like that in Eq. (12.1) lead to time dependent population numbers. We will follow the number operator approach to calculating the time dependent population numbers of the categories in this system, that was covered in Section 6.2.4, as an alternative to the eigenfrequency method. The reason for using this method is that it can be directly applied to more complicated exchange models that we will deal with later in this chapter.

That the interaction term is the direct cause of population number variations may be seen by calculating the time derivatives of the population numbers as represented by the number operators, $\hat{N}$ and $\hat{M}$, using the Heisenberg equation of motion. Then we find

$$i\frac{d\hat{N}}{dt} = [\hat{N},\hat{\Omega}] = V(\hat{A}^\dagger\hat{B} - \hat{B}^\dagger\hat{A}) \tag{12.4}$$

and

$$i\frac{d\hat{M}}{dt} = [\hat{M},\hat{\Omega}] = -V(\hat{A}^\dagger\hat{B} - \hat{B}^\dagger\hat{A}). \tag{12.5}$$

We can see that the two number operators are both independent of time when $V = 0$, which indicates that the interaction term, i.e., $\hat{\Omega}_I$ is responsible for the time dependence of the populations. Adding Eqs. (12.4) and (12.5) leads to

$$\frac{d(\hat{N} + \hat{M})}{dt} = 0. \tag{12.6}$$

So, although $\hat{N}$ and $\hat{M}$ vary individually when $V \neq 0$, the total population does not change with time.

If we now substitute the result from Eq. (12.4) back into the Heisenberg equation of motion, than, after a little manipulation

$$\begin{aligned}
\frac{d^2 \hat{N}}{dt^2} &= (U - W)V(\hat{A}^\dagger \hat{B} + \hat{B}^\dagger \hat{A}) - 2V^2(\hat{N} - \hat{M}) \\
&= (U - W)(\Omega - U\hat{N} - W\hat{M}) - 2V^2(\hat{N} - \hat{M}),
\end{aligned} \tag{12.7}$$

where we have used $V(\hat{A}^\dagger \hat{B} + \hat{A}^\dagger \hat{B}) = \hat{\Omega} - (U\hat{N} + W\hat{M})$. Further progress in solving the differential equation above is only possible by operating on an Fock state. It must be recalled that in the H-representation of a system, while the operators are time dependent, the eigenstates are not time dependent. As explained in Chapter 6, an appropriate choice of eigenstate for our two category system, when population numbers are time dependent, is the eigenstate of the operators at $t = 0$, which we designate $|n(0), m(0)\rangle$, such that

$$\hat{N}(0)|n(0), m(0)\rangle = n(0)|n(0), m(0)\rangle,$$

$$\hat{M}(0)|n(0), m(0)\rangle = m(0)|n(0), m(0)\rangle, \tag{12.8}$$

where $\hat{N}(0)$ indicates the operator $\hat{N}(t)$ at $t = 0$. Then we can define time dependent expectation values by [7]

$$\tilde{n}(t) = \langle n(0), m(0)|\hat{N}(t)|n(0), m(0)\rangle,$$

$$\tilde{m}(t) = \langle n(0), m(0)|\hat{M}(t)|n(0), m(0)\rangle. \tag{12.9}$$

Because the sum, $\hat{N}(t) + \hat{M}(t)$ is independent of time we can write $\tilde{m}(t) = n(0) + m(0) - \tilde{n}(t)$. We also note that $\hat{\Omega}$ must be invariant, since it obviously commutes with itself, so we have

$$\langle \hat{\Omega} \rangle = \langle \hat{\Omega}(0) \rangle = Un(0) + Wm(0), \tag{12.10}$$

since, given that $\hat{A}(0)|n(0), m(0)\rangle = \sqrt{n(0)}|n(0) - 1, m(0)\rangle$, etc., then

$$\langle n(0), m(0)|\hat{A}^\dagger(0)\hat{B}(0)|n(0), m(0)\rangle = 0,$$

$$\langle n(0), m(0)|\hat{B}^\dagger(0)\hat{A}(0)|n(0), m(0)\rangle = 0. \tag{12.11}$$

Finally, noting, from Section 2.8, that

$$\langle \frac{d^2 \hat{N}}{dt^2} \rangle = \frac{d^2 \langle \hat{N} \rangle}{dt^2} = \frac{d^2 \tilde{n}(t)}{dt^2} \tag{12.12}$$

then we get a differential equation for $\tilde{n}(t)$ of

$$\frac{d^2\tilde{n}(t)}{dt^2} + \omega^2\tilde{n}(t) = \omega^2 n(0) - 2V^2(n(0) - m(0)), \qquad (12.13)$$

where $\omega^2 = (U - W)^2 + 4V^2$.

Eq. (12.13) is a second order differential equation in $\tilde{n}(t)$ and requires two boundary conditions. The first is just, $\tilde{n}(t) = n(0)$ at $t = 0$. The second, involving the value of $\frac{d\tilde{n}(t)}{dt}$ is obtained from

$$\left.\frac{d\tilde{n}(t)}{dt}\right|_{t=0} = -iV\langle n(0), m(0)|\hat{A}^\dagger(0)\hat{B}(0) - \hat{B}^\dagger(0)\hat{A}(0)|n(0), m(0)\rangle = 0. \qquad (12.14)$$

With these two boundary conditions we obtain

$$\tilde{n}(t) = n(0) + 2(n(0) - m(0))\left(\frac{V}{\omega}\right)^2 (1 - \cos(\omega t)), \qquad (12.15)$$

which is identical in form to the solution we found in Eq. (6.42) by means of the eigenvalue equations for the coupled annihilation operators of the two-category system. It is interesting to note that the population variations of the two categories in this closed system, where the sum of the two populations is invariant, take the form of a mean plus a sinusoidal variation about the mean. In the case of Eq. (12.15) the mean is

$$n(0) - 2\left(\frac{V}{\omega}\right)^2 (n(0) - m(0))$$

and the amplitude of the oscillation is

$$2\left(\frac{V}{\omega}\right)^2 |n(0) - m(0)|,$$

from which we can see that the ratio of V to ω determines how big the fluctuation amplitude is. As we have seen above, the frequency increases with increasing V, but decreases with increases in the difference, $|U - W|$. So the difference in the two frequencies in the Hamiltonian acts like inertia, i.e., reluctance to change, whereas, V drives change. However, we can also see that the fluctuations disappear if $n(0) = m(0)$, and the maximum amplitude for a given set of initial conditions occurs when $|U - W| = 0$, because this maximizes the ratio of V to ω. An example of the time dependence of the populations of items for the exchange model associated with a Hamiltonian of the type in Eq. (12.1) is illustrated in Fig.(12.1).

The type of response we have found in the simple exchange model above, in which a fluctuation about a mean is observed is typical of even more complicated closed systems, as we shall see later. Indeed, the Hamiltonian, Eq. (12.1) and the time dependent population variations that it leads to serve as a blueprint for modelling interacting populations in a variety of situations, and it will be the basis for exploring interacting population models in what follows. There are classical systems which exhibit such behaviour. One in particular is quite close conceptually and

that is the well-known predator-prey model that was developed in the early twentieth century by Lotka [75] and Volterra [123]. The classical Lotka-Volterra model of
predator-prey interaction envisages a prey population that grows at the expense of
a prey population. In the operator exchange model above, we can interpret the one-
to-one exchange Hamiltonian, Eq. (12.1), as the survival of one member of the prey
population at the expense of the loss of one member of the predated population, plus
the reverse of this process. The result is the fluctuation of the two populations that
we see in Fig. 12.1. In the next section we have a brief look at the Lotka-Volterra
model as a comparison.

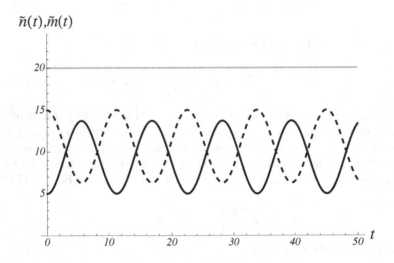

Figure 12.1   An example of the solution, Eq. (12.15). The dashed curve represents $\tilde{n}(t)$ and
the dark solid curve represents $\tilde{m}(t)$ as functions of t. The grey horizontal line represents the
constant sum, $\tilde{n}(t) + \tilde{m}(t)$. The initial values of $\tilde{n}(t)$ and $\tilde{m}(t)$ are respectively, 15 and 5. The
values of the constant parameters for the Hamiltonian in Eq. (12.1) are $U = 0.3$, $W = 0.1$ and
$V = 0.26$.

12.2   THE LOTKA-VOLTERRA EQUATIONS

The Lotka-Volterra differential equations for the time dependence of a prey population, $x(t)$ and a predator population, $y(t)$, take the form

$$\frac{dx(t)}{dt} = fx(t) - gy(t)x(t),$$

$$\frac{dy(t)}{dt} = -hy(t) + kx(t)y(t), \tag{12.16}$$

where f, g, h and k are constant parameters. The obvious difference between the
Lotka-Volterra equations and Eq. (12.1) is that the time dependent variables, $x(t)$

and $y(t)$, are scalar valued functions from the start. They are of course, by their nature, continuous variables and so one is making the assumption that the discrete nature of population numbers is not relevant. The simple argument for this form of the equations is based on exponential growth and decay. It is assumed that the linear growth rate, f, of the prey population, which is the birth rate less a smaller death rate, is modified in the presence of the predator population, which increases the death rate in direct proportion to the predator population. On the other hand the birth rate of the predator population is increased in direct proportion to the prey population. It is well known that in a certain part of that parameter space, the populations are bound and cyclic.

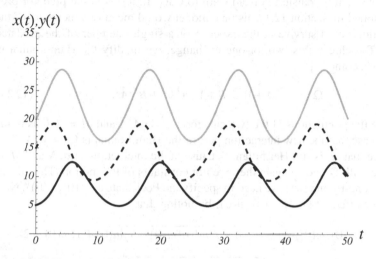

Figure 12.2 Variation of the prey, $x(t)$, (dashed curve) and predator, $y(t)$, (solid curve) populations as functions of t. The sum of the two populations is indicated by the grey curve. The initial value are $x(0) = 15$ and $y(0) = 5$. The constants in Eqs. (12.16) are $f = 0.35$, $h = 0.59$ and $g = k = 0.043$.

An example of this kind of solution is shown in Fig. 12.2. Here the initial values $x(0)$ and $y(0)$ of the interacting populations have been chosen to match the initial populations of the quantum system illustrated in Fig. 12.1, and the constant parameters for Eqs. (12.13) have been adjusted so that the frequency of oscillation and amplitudes are similar to those in Fig. 12.1.

Although the fluctuations in Figs. 12.1 and 12.2 are exactly periodic and of similar frequency and amplitudes, there are some striking differences. First we notice that the sum of the two populations, $x(t) + y(t)$ in Fig. 12.2 does not remain constant. One of the reasons for this is that the phase difference between the two populations in not π, whereas the populations in Fig. 12.1 are in exact anti-phase. Also the variations of the curves in Fig. 12.2, unlike those in Fig. 12.1, are not exactly sinusoidal. There is a noticeable sharpening of the peaks compared to the troughs in Fig. 12.2. The curves in Fig. 12.1 are exactly sinusoidal because they represent linear differential equations like Eq. (12.13). Eqs. (12.16) are clearly non-linear. Nonlinear behaviour in which the total population numbers are not conserved is also seen in two-category systems that are modelled by natural number operators, when the exchange term is not one-to-one. Such a model, with a two-for-one exchange term is examined next.

12.3 TWO-FOR-ONE EXCHANGE

In this section we will examine the consequences of an exchange processes between two populations of items that is *two-for-one* rather than one-for-one. By two-for-one we mean that two items will be transferred from Cat_A to Cat_B, while at the same time one item is transferred from Cat_B to Cat_A. In terms of the predator-prey model mentioned in Section 12.1, this new model would mean two members of the prey population could survive at the expense of a single member of the predated population. To achieve this two-for-one exchange, we modify the Hamiltonian in Eq. (12.1) to become [7]

$$\hat{\Omega} = U\hat{A}^\dagger\hat{A} + W\hat{B}^\dagger\hat{B} + V(\hat{A}^\dagger\hat{B}^2 + \hat{B}^{\dagger 2}\hat{A}). \tag{12.17}$$

We still have the populations of the two categories $\hat{N} = \hat{A}^\dagger\hat{A}$ and $\hat{M} = \hat{B}^\dagger\hat{B}$ as in the one-for-one case, but the new interaction part of the Hamiltonian is $\hat{\Omega}_I = V(\hat{A}^\dagger\hat{B}^2 + \hat{B}^{\dagger 2}\hat{A})$. Notice that $\hat{\Omega}_I$ is still Hermitian. Because of the interaction term, $\hat{N}$ and $\hat{M}$ do not commute with $\hat{\Omega}$ and so again they are not constants of the motion. This means that, as in the one-to-one case, we need to specify the Fock state as $|n(0), m(0)\rangle$. Now we can see the effect of $\hat{\Omega}_I$ on $|n(0), m(0)\rangle$ by noting that

$$\hat{A}^\dagger(0)\hat{B}^2(0)|n(0), m(0)\rangle = \sqrt{(n(0)+1)m(0)(m(0)-1)}|n(0)+1, m(0)-2\rangle,$$

$$\hat{B}^{\dagger 2}(0)\hat{A}(0)|n(0), m(0)\rangle = \sqrt{n(0)(m(0)+1)(m(0)+2)}|n(0)-1, m(0)+2\rangle. \tag{12.18}$$

So the first term in $\hat{\Omega}_I$ takes two items from Cat_B and puts one in Cat_A. The second term does the reverse. Then

$$i\frac{d\hat{N}}{dt} = [\hat{N}, \hat{\Omega}] = V(\hat{A}^\dagger\hat{B}^2 - \hat{B}^{\dagger 2}\hat{A}) \tag{12.19}$$

and

$$i\frac{d\hat{M}}{dt} = [\hat{M}, \hat{\Omega}] = -2V(\hat{A}^\dagger\hat{B}^2 - \hat{B}^{\dagger 2}\hat{A}). \tag{12.20}$$

Now we find that not only are $\hat{N}$ and $\hat{M}$ are not constants of the motion, but neither is $\hat{N} + \hat{M}$. In fact, after adding Eqs. (12.19) and (12.20), we have

$$\frac{d(2\hat{N}+\hat{M})}{dt} = 0, \tag{12.21}$$

so the new constant of the motion, K_0, is

$$K_0 = 2\hat{N}(t) + \hat{M}(t) = 2\hat{N}(0) + \hat{M}(0). \tag{12.22}$$

The second derivative of $\hat{N}$ with respect to time is

$$\frac{d^2\hat{N}}{dt^2} = (U - 2W)V(\hat{A}^\dagger\hat{B}^2 + \hat{B}^{\dagger 2}\hat{A}) - 2V^2(\hat{N} + 4\hat{N}\hat{M} - \hat{M}(\hat{M} - 1))$$

$$= (U - 2W)(\Omega - U\hat{N} - W\hat{M}) - 2V^2(\hat{N} + 4\hat{N}\hat{M} - \hat{M}(\hat{M} - 1)), \tag{12.23}$$

where we have used $V(\hat{A}^\dagger\hat{B}^2 + \hat{B}^{\dagger 2}\hat{A}) = \hat{\Omega} - (U\hat{N} + W\hat{M})$. The equation for the time dependent expectation values, after eliminating $\hat{M}$ with the aid of Eq. (12.22), becomes

$$\frac{d^2\tilde{n}(t)}{dt^2} + \omega^2\tilde{n}(t) = (U - 2W)^2 n(0) + 2V^2 K_0(K_0 - 1) + 24V^2\widetilde{n^2(t)}, \tag{12.24}$$

where $\omega^2 = (U - 2W)^2 + 16V^2 K_0$ and $\widetilde{n^2(t)} = \langle\hat{N}^2\rangle$.

Now we can immediately see a difficulty in solving the differential equation, Eq. (12.24) for $\tilde{n}(t)$. The last term on the rhs contains $\widetilde{n^2(t)}$. This is not generally equal to $\tilde{n}^2(t)$ for reasons similar to those which arose in the discussion concerning quantum noise in Section 6.2.3. We really need to generate a second differential equation for $\widetilde{n^2(t)}$. However, this would contain expectation values of even higher order. We can obtain a useful and instructive solution to Eq. (12.24) by utilizing this so-called *semi-classical approximation* [87, 88] widely used in quantum optics. Then, employing what is referred to as a first level of approximation [87], operators are replaced by their time dependent expectation values. Thus we assume $\widetilde{n^2(t)} \approx \tilde{n}^2(t)$. This still leaves a nonlinear differential equation, but at least this is solvable by standard numerical techniques. The resulting differential equation is second order and needs two boundary conditions. These are $\tilde{n}(0) = n(0)$ and

$$\left.\frac{d\tilde{n}(t)}{dt}\right|_{t=0} = -iV\langle n(0), m(0)|\hat{A}^\dagger(0)\hat{B}^2(0) - \hat{B}^{\dagger 2}(0)\hat{A}(0)|n(0), m(0)\rangle = 0. \tag{12.25}$$

Once a solution for $\tilde{n}(t)$ is found, then $\tilde{m}(t)$ is obtained simply from the conservation rule, Eq. (12.22).

An example of the kind of solution one obtains from this approximation is shown in Fig. 12.3. The values of $n(0)$ and $m(0)$ have been chosen to be that same as those in Fig. 12.1, to give some indication of the effect of the nonlinearity on the system. The values of U, W and V here, have been chosen so that the value of the frequency, ω is similar to that in the one-for-one case. Comparing Figs. 12.1 and 12.3, the most obvious difference is that the total population is no longer a conserved quantity, because of the new conservation rule Eq. (12.22). The fluctuation amplitudes are similar to those in Fig. 12.1, and are still in anti-phase, but the amplitudes of the two categories are no longer equal. The amplitude of $\tilde{n}(t)$ is noticeably smaller than that of $\tilde{m}(t)$. In fact, because of conservation rule, Eq. (12.22), we can expect the peak-to-peak amplitudes, Δm and Δn, to obey $\Delta m = 2\Delta n$.

12.4 PARAMETRIC AMPLIFICATION IN QUANTUM OPTICS

Here we are going to investigate the properties of the bosonic Hamiltonian in Eq. (6.74) that we obtained in Chapter 6 by linearly transforming a two-category system

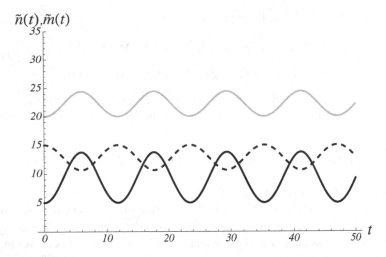

Figure 12.3 The dashed curve represents $\tilde{n}(t)$ and the dark solid curve represents $\tilde{m}(t)$ as functions of t. The grey curve represents the constant sum, $\tilde{n}(t) + \tilde{m}(t)$. The initial values of $\tilde{n}(t)$ and $\tilde{m}(t)$ are respectively, 15 and 5. The values of the constant parameters for the Hamiltonian in Eq. (12.1) are $U = 1.22$, $W = 0.32$ and $V = 0.03$.

by coupling a creation operator and an annihilation operator of the original system. This lead to a Hamiltonian of the form

$$\hat{\Omega} = U\hat{A}^{\dagger}\hat{A} + W\hat{B}^{\dagger}\hat{B} + V(\hat{A}^{\dagger}\hat{B}^{\dagger} + \hat{B}\hat{A}). \tag{12.26}$$

This only differs from Eq. (6.74) by the constant μ_{gh} being neglected here, since this has no effect on the dynamics of the system. Again, we have a two categories, Cat$_A$ with number operator, $\hat{N} = \hat{A}^{\dagger}\hat{A}$, and Cat$_B$ with number operator, $\hat{M} = \hat{B}^{\dagger}\hat{B}$.

We can again generate a second order differential using exactly the same approach as in Section 12.1. The first step is to find the time derivatives of the two number operators. This leads to

$$i\frac{d\hat{N}}{dt} = [\hat{N}, \hat{\Omega}] = V(\hat{A}^{\dagger}\hat{B}^{\dagger} - \hat{B}\hat{A}) \tag{12.27}$$

and

$$i\frac{d\hat{M}}{dt} = [\hat{M}, \hat{\Omega}] = V(\hat{A}^{\dagger}\hat{B}^{\dagger} - \hat{B}\hat{A}). \tag{12.28}$$

This leads to

$$\frac{d(\hat{N} - \hat{M})}{dt} = 0. \tag{12.29}$$

This is a different result than that found in Section 12.1. Here, although $\hat{N}$ and $\hat{M}$ still vary individually when $V \neq 0$, the total population is not constrained at all with time. If fact it is the difference between the two populations that is conserved, so

$$\hat{N}(t) - \hat{M}(t) = \hat{N}(0) - \hat{M}(0). \tag{12.30}$$

This has a major effect on the dynamics of the system. We can proceed to find the equation for the time dependent expectation values of the populations. This takes the form

$$\frac{d^2 \tilde{n}(t)}{dt^2} + \omega^2 \tilde{n}(t) = \omega^2 n(0) + 2V^2(n(0) + m(0) + 1). \tag{12.31}$$

where $\omega^2 = (U + W)^2 - 4V^2$. The boundary conditions for Eq. (12.31) are the same as those in Section 12.1. However, it is the nature of ω^2 that sets the agenda for the behaviour of the system. It differs considerably from that in Section 12.1. There it was positive-definite, which meant that ω was always real. Here we can see that ω^2 is only positive as long as $4V^2 < (U + W)^2$. This result is the same as that found in Section 6.3.2, using the eigenvalue equation method. If this condition is met then the solution to Eq. (12.31) is

$$\tilde{n}(t) = n(0) + 2\left(\frac{V}{\omega}\right)^2 (n(0) + m(0) + 1)(1 - \cos(\omega t)). \tag{12.32}$$

$\tilde{m}(t)$ is then obtained from Eq. (12.30) which gives

$$\tilde{m}(t) = \tilde{n}(t) - n(0) + m(0). \tag{12.33}$$

An example of the solution to Eq. (12.32) is shown in Fig. 12.4.

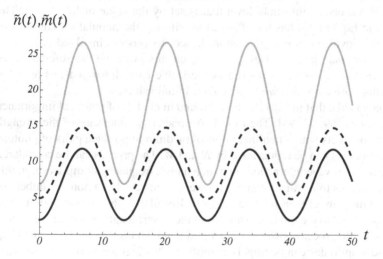

Figure 12.4 An example of the solution in Eq. (12.32). The dashed curve represents $\tilde{n}(t)$ and the dark solid curve represents $\tilde{m}(t)$ as functions of t. The grey curve represents the sum, $\tilde{n}(t) + \tilde{m}(t)$. The initial values of $\tilde{n}(t)$ and $\tilde{m}(t)$ are respectively, 5 and 2. The values of the constant parameters for the Hamiltonian in Eq. (12.26) are $U = 0.45$, $W = 0.25$ and $V = 0.26$.

The time dependent expectation values of the two populations illustrated in Fig. 12.4 both oscillate sinusoidally as they have done in the examples in Sections 12.1 and 12.2, but here they are in phase rather than anti-phase. This is due to the creation operators of the two categories being paired, as are the annihilation operators. It is also clear that the sum of the two populations is no longer conserved.

A more interesting situation arises, not least from a practical point of view, when $n(0) = m(0) = 0$, i.e., when initially there are no product photons. Then we get, from Eq. (12.32)

$$\tilde{n}(t) = \tilde{m}(t) = 2 \left(\frac{V}{\omega} \right)^2 (1 - \cos(\omega t)),$$

which shows that the populations of the two product photons rises from zero at $t = 0$ and reaches a maximum amplitude of $4 \left(\frac{V}{\omega} \right)^2$, when $\omega t = \pi$, and thereafter oscillates between zero and the maximum. Clearly, the maximum amplitude increases as the value of ω approaches zero, i.e., as the value of $(U + W)^2$ approaches $4V^2$. However, since ω approaches zero at the same time, then the rise in the populations happens at a slower and slower rate. If the condition $4V^2 > (U + W)^2$, is satisfied, then ω^2 becomes negative and instability ensues. Writing $\gamma^2 = 4V^2 - (U + W)^2$, then when the unstable condition applies, the solution to Eq. (12.31) becomes

$$\tilde{n}(t) = n(0) + 2 \left(\frac{V}{\gamma} \right)^2 (n(0) + m(0) + 1)(\cosh(\gamma t) - 1), \qquad (12.34)$$

with $\tilde{m}(t)$ again being evaluated from Eq. (12.33). The solution in Eq. (12.34) grows without bounds due to the instability. The instability occurs when the driver for the interaction, V, exceeds a threshold level that is set by the value of $|(U + W)|$. The Hamiltonian in Eq. (12.26) has the effect of amplifying the population numbers of both categories involved in the interaction, hence the process involved is referred to as *parametric amplification*. Devices based on this principle are referred to as *parametric amplifiers*. Notice that even in the stable case illustrated in Fig. 12.4, both populations are amplified well above their initial values.

Hamiltonians like that in Eq. (12.26) are found in models of practical importance in quantum optics [89, 92, 94]. Then the V parameter is a measure of the strength of a laser that interacts with a nonlinear optical medium to produce pairs of photons with populations that can be represented by $\hat{N}$ and $\hat{M}$. The process can be considered as a nonlinear three-wave interaction, in which a high frequency pump wave, in this case a laser, couples to a pair of waves of lower frequency. The photon numbers of the waves is a measure of their relative intensity. Recall that photon population numbers are proportional to the modulus squared of the electromagnetic field as shown in Sections 8.6. In the process power is transferred from the laser to the Cat_A and Cat_B waves whose amplitudes can be hugely amplified. This is referred to as a *parametric interaction*, which can lead to a *parametric instability*. Such processes have been studied in a variety of circumstances, particularly when high power electromagnetic waves interact with plasma. Some of the most interesting experiments into parametric instability have involved the use of very high power radio waves transmitted into the Earth's naturally occurring ionospheric plasma [100].

The unbounded growth in the photon population that we see in Eq. (12.34) is of course unrealistic in real experiments. There will always be a limit in practice. The reason why the Hamiltonian in Eq. (12.26) leads to the unbounded growth of the photon population is that it fails to take account of the depletion of the laser as it feeds energy into the product photons. By modifying the Hamiltonian, this depletion effect can be taken into account. Thus if instead of Eq. (12.26) we have a Hamiltonian of the form

$$\hat{\Omega} = U\hat{A}^{\dagger}\hat{A} + W\hat{B}^{\dagger}\hat{B} + S\hat{C}^{\dagger}\hat{C} + V(\hat{A}^{\dagger}\hat{B}^{\dagger}\hat{C} + \hat{C}^{\dagger}\hat{B}\hat{A}), \tag{12.35}$$

where $\hat{C}$ and its adjoint are annihilation and creation operators for the laser photons. The interaction term in the Hamiltonian is easily interpreted. The first part involving $\hat{A}^{\dagger}\hat{B}^{\dagger}\hat{C}$, simply represents the gain of photons of Cat_A and Cat_B, for the loss of a laser, Cat_C photon. The second part of the interaction Hamiltonian that involves $\hat{C}^{\dagger}\hat{B}\hat{A}$ has the reverse effect and preserves the Hermiticity of the Hamiltonian. Also included in the Hamiltonian is a term, $S\hat{C}^{\dagger}\hat{C}$ which simply accounts for the contribution to the Hamiltonian of the number of photons associated with the laser. S is a constant that plays the same role as U and W do for the Cat_A and Cat_B populations. The effect of including the Cat_C population explicitly to represent the laser is as follows. If the number operator for the laser is $\hat{O} = \hat{C}^{\dagger}\hat{C}$, then we find

$$i\frac{d\hat{N}}{dt} = [\hat{N}, \hat{\Omega}] = V(\hat{A}^{\dagger}\hat{B}^{\dagger}\hat{C} - \hat{C}^{\dagger}\hat{B}\hat{A}), \tag{12.36}$$

$$i\frac{d\hat{M}}{dt} = [\hat{M}, \hat{\Omega}] = V(\hat{A}^{\dagger}\hat{B}^{\dagger}\hat{C} - \hat{C}^{\dagger}\hat{B}\hat{A}). \tag{12.37}$$

and

$$i\frac{d\hat{O}}{dt} = [\hat{O}, \hat{\Omega}] = -V(\hat{A}^{\dagger}\hat{B}^{\dagger}\hat{C} - \hat{C}^{\dagger}\hat{B}\hat{A}). \tag{12.38}$$

Subtracting Eq. (12.37) from Eq. (12.36) again leads to Eq. (12.30), indicating that the difference in the populations of Cat_A and Cat_B photons is conserved, but in addition we now have

$$\frac{d(\hat{O} + \hat{N})}{dt} = 0 \tag{12.39}$$

and

$$\frac{d(\hat{O} + \hat{M})}{dt} = 0. \tag{12.40}$$

Eqs. (12.39) and (12.40) imply that the populations of the Cat_A and Cat_B populations is now limited. Adding Eqs. (12.39) and (12.40) leads to

$$2\hat{O}(t) + \hat{N}(t) + \hat{M}(t) = 2\hat{O}(0) + \hat{N}(0) + \hat{M}(0), \tag{12.41}$$

implying that the sum of the Cat_A and Cat_B populations can never exceed twice the laser, Cat_C population. An approximation to the Hamiltonian that is common used in quantum optics, and is referred to as the degenerate case, assumes that the Cat_B populations can be treated as the Cat_A population in such a way that we get two Cat_A photons from each Cat_C laser photon, so the conversion process can be represented by the Hamiltonian

$$\hat{\Omega} = 2U\hat{A}^{\dagger}\hat{A} + S\hat{C}^{\dagger}\hat{C} + V(\hat{A}^{\dagger 2}\hat{C} + \hat{C}^{\dagger}\hat{A}^{2}). \tag{12.42}$$

We can immediately recognize the Hamiltonian in Eq. (12.42) as being identical in form to the two-for-one exchange Hamiltonian in Eq. (12.17). This means that the differential Eq. (12.24) can be used the calculate the time dependent expectation values of the photon populations in the present case. An example of this calculation is shown in Fig. 12.5, where we have made the same approximation as in Eq. (12.24) with regard to neglecting quantum noise. This result clearly indicates that the depletion of the laser limits the growth of the instability and sets an upper limit to the population of product photons.

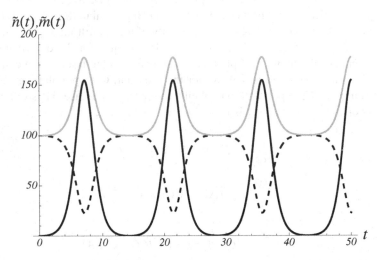

Figure 12.5 An example of the photon populations for a system governed by the Hamiltonian in Eq. (12.42). The size of the photon population of the laser pump is represented by the dashed curve. The population of the photons produced is represented by the dark curve. The grey curve represents the total population of photons. The initial laser population is 100 and that of the product photons is 0. It is clear that although there is a huge amplification effect of the produced photons, their population is limited by the depletion of the laser intensity.

12.5 HIGHER ORDER INTERACTIONS

So far we have seen how interaction between pairs of categories can be modelled using creation and annihilation operators to transfer items from one category to another. Three different types of interaction have been used so far. These are the one-to-one exchange, the two-to-one exchange and the type we used to model parametric amplification in Section 12.4. These have lead to various constants of the motion involving linear combinations of the pair of number operators involved. It is not difficult to imagine that interaction models of this type can be constructed with, for example, three-for-one or three-for-two exchanges. The interaction Hamiltonian with the most general pair exchange models of this type, that still retains the Hermitian property, has the form

$$\hat{\Omega}_I = V(\hat{A}^{\dagger p}\hat{B}^{\dagger q}\hat{A}^r\hat{B}^s + \hat{B}^{\dagger s}\hat{A}^{\dagger r}\hat{B}^q\hat{A}^p), \qquad (12.43)$$

where p, q, r and s are positive integers. $\hat{\Omega}_I$ represents asymmetrical exchange between a pair of categories, Cat_A and Cat_B, represented respectively by number operators, $\hat{N} = \hat{A}^\dagger \hat{A}$ and $\hat{M} = \hat{B}^\dagger \hat{B}$. Bearing in mind that when $V \neq 0$, the number operators will be time dependent, as we have seen in the cases of interaction that were treated earlier in this chapter, so, as before, we use a system state vector at a fixed time, represented by $|n(0), m(0)\rangle$. We then note that the effect of the first part of $\hat{\Omega}_I$ on $|n(0), m(0)\rangle$ can be found by operating with $\hat{A}^{\dagger p}(0)\hat{B}^{\dagger q}(0)\hat{A}^r(0)\hat{B}^s(0)$. This then produces the state $|n(0) + p - r, m(0) + q - s\rangle$, as long as $n(0) - r \geq 0$ and $m(0) - s \geq 0$. The total population, $\hat{N} + \hat{M}$ is not generally conserved during the interaction. Instead, there is a new conservation rule,

$$\frac{\mathrm{d}}{\mathrm{d}t}((s-q)\hat{N} + (p-r)\hat{M}) = 0. \qquad (12.44)$$

Notice that, the total population is constant if $p - r = s - q$. The special case we have seen so far correspond to, (a) the one-for-one interaction, where $p = s = 1$ and $q = r = 0$, which leads to conservation rule, Eq. (12.6); (b) the two-for-one case, where $p = 1$, $s = 2$ and $q = r = 0$, leads to the conservation rule, Eq. (12.21); (c) the interaction term in Eq. (12.26), where $p = q = 1$ and $r = s = 0$, leads to the conservation rule in Eq. (12.29).

In the next section we will utilize a version of the general interaction Hamiltonian to investigate an important process in molecular biology, namely the dynamics of cell division.

12.6   CELL DIVISION AND CELL POPULATION DYNAMICS

12.6.1   A SIMPLIFIED PICTURE OF CELL DIVISION

Cell division is an important and complicated process that is a central topic of microbiological and medical research. In particular it is central to the question of how and why cancers arise. In this section we will attempt to model cell division in a simple way by focussing on the basic features that can be captured using the creation and annihilation operator formalism that has been applied earlier in this chapter to the interaction of categories of items. In doing this, no attempt is made to reproduce the details of the microbiological mechanisms involved. The basic picture of cell division is simply that a single cell splits into two cells [54]. We then describe this process as the replacement of a single cell by a pair of cells. This is essentially a two-for-one exchange, but with a single category, Cat_C, of cells. We can devise an interaction Hamiltonian to represent this mechanism that will contain the sequence $\hat{C}^{\dagger 2}\hat{C}$. Here, the annihilation operator $\hat{C}$ represents the disappearance of one cell, and the pair of creation operators (i.e., $\hat{C}^{\dagger 2}$) represents the appearance of two cells of the same type in place of the original one. This is simply a case of, where there was one, there are now two. The number of cells present is then represented by $\hat{N} = \hat{C}^\dagger \hat{C}$. This picture can be made more realistic by including a trigger mechanism that stimulates the cell division. The trigger mechanism consists of a population of stimulating agents that take part in the interaction process. The population of these stimulating agents of category, Cat$_S$, can be represented by a second number operator, $\hat{M} = \hat{S}^\dagger \hat{S}$, with annihilation operator, $\hat{S}$ and creation operator, $\hat{S}^\dagger$. The material in this section is based largely on reference [106], where the use of number operators to model the cell division process was first introduced.

We will next outline two different scenarios of the cell division process. In the first the stimulus is exhausted in some way in the division process and in the second, we will assume that the stimulating agent acts as a catalyst so that it remains after cell division has taken place. As we shall see, the first of these models closely resembles normal cell division (NCD) that is associated with stable cell populations, whereas the second turns out to be able to mimic oncogenic cell division (OCD), that is associated with the unbounded growth of cell populations that indicates the onset of a cancer. Stimulation of normal cell division depends on the presence of an extrinsic factor, such as a growth factor, which can be quickly diluted out or degraded thereby removing the stimulus. In contrast, stimulation of cancer cell division is driven by an intrinsic factor, such as an activated oncogene, which is regenerated upon DNA replication and passes from one generation to the next; hence it acts as a catalyst and is never lost [46, 127].

12.6.2   NORMAL CELL DIVISION (NCD) MODEL: STABLE POPULATION

In this first model, the NCD model, it is supposed that the stimulus, whose presence is necessary to trigger cell division, is exhausted in some way by its participation in the process. Then the combined contribution to the interaction Hamiltonian of the

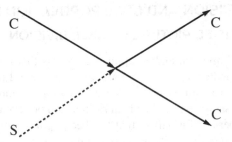

Figure 12.6   Schematic of normal cell division in which a cell (C) and a stimulating agent (S) enter the process and two cells result.

cell division and stimulus is $\hat{C}^{\dagger 2}\hat{C}\hat{S}$. The cell division process described by this set of operators is shown schematically in Fig. 12.6. Note that the direction of the arrows in Fig. 12.6 give an indication of time evolution. Adding the time reversed form of this, to preserve the self-adjoint property, the interaction Hamiltonian is then

$$\hat{\Omega}_I = V(\hat{C}^{\dagger 2}\hat{C}\hat{S} + \hat{S}^\dagger \hat{C}^\dagger \hat{C}^2), \tag{12.45}$$

where V is the cell division rate in the presence of the stimulus, and where the usual bosonic commutation relations apply. Notice that this is an example of the general form of interaction Hamiltonian in Eq. (12.43), with $p = 2$, $q = 0$, $r = s = 1$. The resulting Hamiltonian for the whole process is then

$$\hat{\Omega} = U\hat{N} + W\hat{M} + V(\hat{C}^{\dagger 2}\hat{C}\hat{S} + \hat{S}^{\dagger}\hat{C}^{\dagger}\hat{C}^2), \qquad (12.46)$$

Here U and W are the intrinsic rates associated with $\hat{C}$ and $\hat{S}$, respectively, when there is no interaction. By intrinsic we mean that in the absence of any interaction, then we would find $\hat{C}(t) = \hat{C}(0)\exp(-iUt)$, and $\hat{S}(t) = \hat{S}(0)\exp(-iWt)$. The Hamiltonian in Eq. (12.46) is very similar to the one we encountered in Section 12.4 for the quantum optics modelling of parametric amplifiers.

Applying the Heisenberg equation to the population operators, we get

$$i\frac{d\hat{N}}{dt} = [\hat{N},\hat{\Omega}] = V(\hat{C}^{\dagger 2}\hat{C}\hat{S} - \hat{S}^{\dagger}\hat{C}^{\dagger}\hat{C}^2) \qquad (12.47)$$

and

$$i\frac{d\hat{M}}{dt} = [\hat{M},\hat{\Omega}] = -V(\hat{C}^{\dagger 2}\hat{C}\hat{S} - \hat{S}^{\dagger}\hat{C}^{\dagger}\hat{C}^2) \qquad (12.48)$$

from which it is clear that

$$\frac{d(\hat{N}+\hat{M})}{dt} = 0.$$

This is exactly what is to be expected from Eqs. (12.43) and (12.44), with $p = 2$, $r = s = 1$, $q = 0$. So, in spite of the more complicated form of the interaction, the combined populations of the dividing cells and the stimuli is fixed, and we can conclude that the cell population is bound. The state of the system is then one of quasi-equilibrium where the individual population numbers of the participating categories may fluctuate, but only in such a way as to maintain the total population. This is enough for us to know that no unbounded growth in the cell population can occur in this model. However, it is worth looking a little further into the character of the equilibrium. A further application of the Heisenberg equation to Eq. (12.47), together with the population conservation rule yields

$$\frac{d^2\hat{N}}{dt^2} + \omega^2\hat{N} - 6V^2(1+N_0)\hat{N}^2 + 8V^2\hat{N}^3 = (U-W)(\hat{\Omega}+WN_0), \qquad (12.49)$$

where $N_0 = n(0) + m(0)$ and $\omega^2 = (U-W)^2 + 2V^2(1+N_0)$. The nonlinear terms in the operator equation present considerable difficulties, as has been explained earlier in Section 12.3. Here again, we use an approximate solution to the corresponding equation for the time dependent expectations by taking $\overline{n^2(t)} \approx \tilde{n}^2(t)$. In addition, because of the cubic term in Eq. (12.49), we also need $\overline{n^3(t)} \approx \tilde{n}^3(t)$. The resulting differential equation is then

$$\frac{d^2\tilde{n}(t)}{dt^2} + \omega^2\tilde{n}(t) - 6V^2(1+N_0)\tilde{n}^2(t) + 8V^2\tilde{n}^3(t) = (U-W)^2 n(0), \qquad (12.50)$$

where, $\tilde{n}(t) = \langle \hat{C}^{\dagger}(t)\hat{C}(t)\rangle$ and where the time independent Fock state is again taken as $|n(0), m(0)\rangle$.

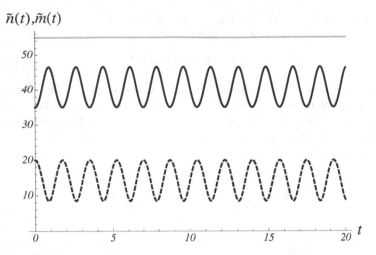

Figure 12.7 Approximate solution for the normal cell division (NCD) model. The cell and stimuli populations are shown as functions of (scaled) time, by numerical integration of Eq. (12.50). The solid curve represents the cell population and the dashed curve the trigger population. The horizontal grey line is the sum of the two populations.

An example of the numerical integration of Eq. (12.50) is shown in Fig. 12.7. The result shows a fluctuating pair of populations, with something very close to sinusoidal form on constant means, rather like the linear cases in Section 12.1. However, it is interesting to note, that, unlike in the linear cases, the fluctuation amplitudes here are not very sensitive to the interaction rate, V. There is actually greater sensitivity to the ratio of $n(0)$ to N_0. The amplitude of the Cat_C cell population is sensitive to the strength of the population of the Cat_S stimulating agents. Fig. 12.8 shows are similar result to that in Fig. 12.7, but with a lower value of $m(0)$. This leads to a lower fluctuation amplitude in the Cat_C population. This is opposite to what we saw in the simple linear equation in Section 12.1, where a lowering of the value of $m(0)$ would have increased the fluctuation level because of its linear increase with the increasing difference in the initial populations of the two categories, as is seen in Eq. (12.15).

12.6.3 ONCOGENIC CELL DIVISION (OCD) MODEL: EXPLOSIVE POPULATION GROWTH

In the second model, the division process is as before, but here the stimulus is a catalyst and so needs to be involved in such a way that it survives division process. This is achieved by using the form $\hat{C}^{\dagger 2}\hat{S}^{\dagger}\hat{C}\hat{S}$ in the interaction term in the Hamiltonian, which then has the self-adjoint form

$$\hat{\Omega}_I = V(\hat{C}^{\dagger 2}\hat{S}^{\dagger}\hat{C}\hat{S} + \hat{S}^{\dagger}\hat{C}^{\dagger}\hat{S}\hat{C}^2). \tag{12.51}$$

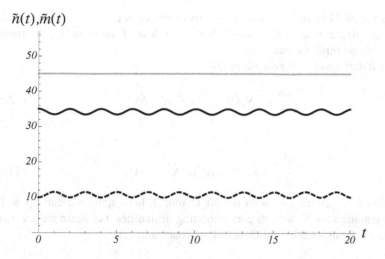

$\tilde{n}(t), \tilde{m}(t)$

Figure 12.8   As Fig. 12.7, but with a smaller population of stimulating agents.

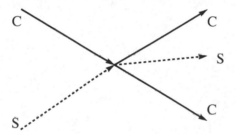

Figure 12.9   Schematic of oncogenic cell division in which a cell (C) and a stimulating agent (S) enter the process and two cells result, but now the stimulating agent survives the process.

This is again of the form in Eq. (12.43), but here, $p = 2, q = 1, r = 1$ and $s = 1$. Since $r = s$ here, then the conservation rule cannot involve $\hat{N}$. It simply consists of

$$\frac{d\hat{M}}{dt} = 0, \tag{12.52}$$

which means that $\hat{M}$, its eigenvalues and its expectation values are independent of time, so we can use $|n(0)\rangle$ as the time independent Fock state for the system and treat $\hat{M}$ as a scalar constant m. The cell division process corresponding to the interaction Hamiltonian is shown schematically in Fig. 12.9. The total Hamiltonian of the system is then

$$\hat{\Omega} = U\hat{N} + Vm(\hat{C}^\dagger \hat{N} + \hat{N}\hat{C}), \tag{12.53}$$

where we have dropped a constant term of Wm, since it has no effect on the equations of motion. We can interpret the time reversed term in Eq. (12.53) in a similar manner to that for the NCD model, i.e., the change from two cells to one is a representation of cell death. Since, in the OCD model there is no loss of the stimulus, the reverse process needs no further explanation.

The first derivative of $\hat{N}$ now becomes

$$i\frac{d\hat{N}}{dt} = [\hat{N}, \hat{\Omega}] = Vm(\hat{C}^\dagger\hat{N} - \hat{N}\hat{C}) \qquad (12.54)$$

and then

$$\frac{d^2\hat{N}}{dt^2} + \omega^2\hat{N} - 6V^2m^2\hat{N}^2 = U\hat{\Omega}, \qquad (12.55)$$

where $\omega^2 = U^2 + 2V^2m^2$. As with the NCD model, here again we end up with a nonlinear equation for $\hat{N}$, with its corresponding difficulties, but again we can make good progress by using the semi-classical approach and get

$$\frac{d^2\tilde{n}(t)}{dt^2} + \omega^2\tilde{n}(t) - 6V^2m^2\tilde{n}^2(t) = U^2n(0). \qquad (12.56)$$

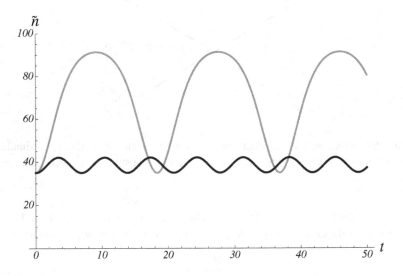

Figure 12.10 Solutions for dividing cell population as a function of (scaled) time for the OCD model, Eq. (12.56). For the grey curve, $\mu = 0.31$, which is close to the condition for instability. For the black curve, $\mu = 0.2$, which is well below the condition for instability.

The solutions to Eq. (12.56) are very different from those for the NCD model, as we shall see. The key features of $\tilde{n}(t)$ in the OCD model depend, essentially, on the parameter combination $\mu^2n(0)$, where $\mu = \frac{mV}{U}$. When $\mu^2n(0)$ is small, $\tilde{n}(t)$ is an almost sinusoidal fluctuation superimposed on a mean value a little above $n(0)$, as in the black curve in Fig. 12.10. As $\mu^2n(0)$ increases the amplitude increases and becomes less sinusoidal. Also the frequency decreases with increasing values of $\mu^2 n(0)$. These effects can clearly be seen in Fig. 12.10, where the grey curve in Fig. 12.10, corresponds a solution for a higher value of $\mu^2 n(0)$. Eventually, with further increases in the value of $\mu^2 n(0)$, the frequency goes to zero and the dividing cell population becomes unstable. This process is best illustrated by a population number phase space diagram. This is obtained by integrating Eq. (12.56) to give

$$\frac{d\tilde{n}(t)}{dt} = \pm \sqrt{2U^2 n(0)\tilde{n}(t) - \omega^2 \tilde{n}^2(t) + 4V^2 m^2 \tilde{n}^3(t) + K_0}, \qquad (12.57)$$

where K_0 is a constant of integration.

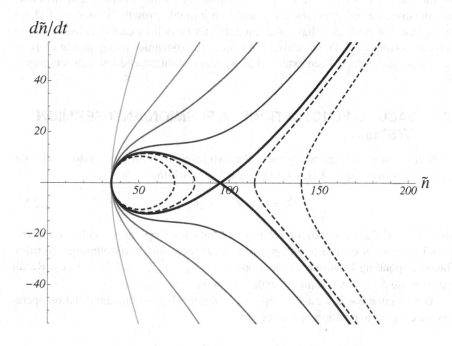

Figure 12.11   The curves, calculated from Eq. (12.57), represent the population phase space trajectories for different values of μ for the OCD model, with $n(0) = 35$. The smaller the value of μ, the darker the curve (the range used is $0.03 < \mu < 0.1$). The black curve is the separatrix between the regions where bound states exist and where they do not. The separatrix corresponds a value of $\mu = 0.0333211$. The closed dashed curves correspond to bound states, whereas the open curves represent unbound states.

Fig. 12.11 shows population phase space curves of $\frac{d\tilde{n}(t)}{dt}$ against $\tilde{n}(t)$, from Eq. (12.57), plotted for different values of μ, with K_0 chosen to make $\frac{d\tilde{n}(t)}{dt} = 0$ at $\tilde{n}(t) = n(0)$, which in this case is chosen to be 35, to be consistent with Fig. 12.10. The closed part of the dashed curves in Fig 12.11 correspond to stable states where the cell population gyrates around between an upper and lower bound (where the closed curves cross the $\tilde{n}$ axis). The lower bound is $n(0)$. The bold curve is the separatrix between the regions of bound and unbound solutions. It corresponds to the threshold value of[1] $\mu = 0.0333211$. As the value of μ is increased beyond threshold, there are no longer closed curves and the phase space curves show that the population values are capable of reaching levels well above their threshold values[2]. This result is consistent with the existence of an instability that is triggered in the above threshold conditions.

We can see from this model that a given initial cell population will maintain a quasi-stationary state with time, i.e., cell division will be going on, stimulated by the presence of a population of stimuli, as long as that population remains below a certain threshold. However, if the stimulating population numbers exceed that threshold, then the dividing cell population will undergo unstable growth. Thus the OCD has appropriate characteristics that make it suitable for modelling cancer cells, in spite of its obvious simplicity. The modelling of cancer cell dynamics using quantum operator models has recently been extended using more sophisticated analysis techniques [9].

12.7 VACUUM FLUCTUATIONS: A FERMION/ANTI-FERMION SYSTEM

We begin with the degenerate form of the two category fermionic system, with the Hamiltonian from Eq. (6.82), with $\omega_a = \omega_b = \omega$, such that

$$\hat{\Omega} = \omega(\hat{C}_a^\dagger \hat{C}_a + \hat{C}_b^\dagger \hat{C}_b), \tag{12.58}$$

where $\hat{C}_a$ and $\hat{C}_b$ are a pair of fermionic annihilation operators with their corresponding creation operators, obeying the usual fermionic anti-commutation rules. The corresponding fermionic number operators, $\hat{N}_a = \hat{C}_a^\dagger \hat{C}_a$ and $\hat{N}_b = \hat{C}_b^\dagger \hat{C}_b$. Recall that, because $\hat{N}_a$ and $\hat{N}_b$ commute with $\hat{\Omega}$ then they are invariants.

As in Section 6.4.2, we again apply the rotation, $\hat{R}(\gamma)$ so that annihilation operators are coupled to creation operators and get

$$\begin{pmatrix} \hat{C}_i \\ \hat{C}_j^\dagger \end{pmatrix} = \hat{R}(\gamma) \begin{pmatrix} \hat{C}_a \\ \hat{C}_b^\dagger \end{pmatrix} = \begin{pmatrix} \cos\gamma & -\sin\gamma \\ \sin\gamma & \cos\gamma \end{pmatrix} \begin{pmatrix} \hat{C}_a \\ \hat{C}_b^\dagger \end{pmatrix}, \tag{12.59}$$

where, as we saw in Section 6.4.2, $\hat{C}_i$ and $\hat{C}_j$ and their corresponding creation opereators obey fermionic anti-commutation rules. The Hamiltonian then becomes

$$\hat{\Omega} = p(\hat{C}_i^\dagger \hat{C}_i + \hat{C}_j^\dagger \hat{C}_j) + m(\hat{C}_j^\dagger \hat{C}_i^\dagger + \hat{C}_i \hat{C}_j) + \mu, \tag{12.60}$$

---

[1] The fact that this number is quoted to so many significant figures is an indication of how sensitive the equilibrium condition is to the value of μ.

[2] It may appear that with this mathematical framework the values that $\tilde{n}(t)$ could reach infinite levels. They can certainly reach levels that are much larger than $n(0)$, but, in practice, in instabilities of this kind there is always a limit imposed by external factors, such as, the system occupying a finite volume and having a finite energy source.

where

$$p = \omega\cos(2\gamma) \quad \text{and} \quad m = \omega\sin(2\gamma). \tag{12.61}$$

We also have

$$\mu = 2\omega\sin^2\gamma,$$

but we can neglect this constant in what follows, since it has no effect on the dynamics of the system. The nature of the coupling between the creation and annihilation operators that are associated with the Hamiltonian in Eq. (12.60) can be seen immediately by applying the Heisenberg equation of motion to them. This yields

$$i\frac{d\hat{C}_i}{dt} = [\hat{C}_i, \hat{\Omega}] = p\hat{C}_i - m\hat{C}_j^\dagger, \tag{12.62}$$

and

$$i\frac{d\hat{C}_j^\dagger}{dt} = [\hat{C}_j^\dagger, \hat{\Omega}] = -p\hat{C}_j^\dagger - m\hat{C}_i. \tag{12.63}$$

Eqs. (12.62) and (12.63) constitute a closed pair of equations from which the eigenvalues may be deduced. Following the same steps as for the degenerate case in Section 7.5, we note that the Hamiltonian can be written in spinor form as

$$\hat{\Omega}\phi = (p\hat{\sigma}_3 + m\hat{\sigma}_1)\phi, \tag{12.64}$$

where

$$\phi = \begin{pmatrix} \hat{C}_i \\ \hat{C}_j^\dagger \end{pmatrix}. \tag{12.65}$$

The eigenvalues of Eq. (12.64) are then

$$\omega = \pm\sqrt{p^2 + m^2}. \tag{12.66}$$

Notice that the eigenvalues in Eq. (12.66) are consistent with Eqs. (12.61), since, from Eq. (12.61) we have

$$p^2 + m^2 = \omega^2\cos^2(2\gamma) + \omega^2\sin^2(2\gamma) = \omega^2. \tag{12.67}$$

So, Eqs. (12.62) to (12.64) confirm that we are dealing with fermions obeying a $1+1$ dimensional Dirac equation[3].

Our next step is to find equations for the time dependent number operators, $\hat{N}_i = \hat{C}_i^\dagger\hat{C}_i$ and $\hat{N}_j = \hat{C}_j^\dagger\hat{C}_j$, neither of which, in general, commutes with $\hat{\Omega}$. Applying the Heisenberg equation of motion, we find

$$i\frac{d\hat{N}_i}{dt} = [\hat{N}_i, \hat{\Omega}] = m(\hat{C}_j^\dagger\hat{C}_i^\dagger - \hat{C}_i\hat{C}_j), \tag{12.68}$$

and

$$i\frac{d\hat{N}_j}{dt} = [\hat{N}_j, \hat{\Omega}] = m(\hat{C}_j^\dagger\hat{C}_i^\dagger - \hat{C}_i\hat{C}_j), \tag{12.69}$$

---

[3]This means one dimension of configuration space and time.

which confirms that

$$\frac{d(\hat{N}_i - \hat{N}_j)}{dt} = 0, \tag{12.70}$$

so $\hat{N}_i - \hat{N}_j$ is invariant, but individually, neither $\hat{N}_i$ nor $\hat{N}_j$ is.

Applying the Heisenberg equation of motion to Eq. (12.68) yields

$$\begin{aligned}
\frac{d^2 \hat{N}_i}{dt^2} &= 2pm(\hat{C}_j^\dagger \hat{C}_i^\dagger + \hat{C}_i \hat{C}_j) + 2m^2(1 - \hat{N}_i - \hat{N}_j) \\
&= 2p(\hat{\Omega} - p(\hat{N}_i + \hat{N}_j)) + 2m^2(1 - \hat{N}_i - \hat{N}_j) \\
&= 2p\hat{\Omega} - 2(p^2 + m^2)(\hat{N}_i + \hat{N}_j) + 2m^2.
\end{aligned} \tag{12.71}$$

The next task is to generate time dependent expectation values for $\hat{N}_i$ and $\hat{N}_j$. We use the same procedure as in Section 6.2. Recall that we need a time-independent Fock state. An appropriate one in this case is $|n_i(0), n_j(0)\rangle$, where $n_i(0)$ is the an eigenvalue of $\hat{N}_i$ at $t = 0$, and $n_j(0)$ is the an eigenvalue of $\hat{N}_j$ at $t = 0$. Then we define a time dependent expectation value of $\hat{N}_i(t)$ as $\tilde{n}_i(t) = \langle \hat{N}_i(t) \rangle$, where $\langle \hat{N}_i(t) \rangle = \langle n_i(0), n_j(0) | \hat{N}_i(t) | n_i(0), n_j(0) \rangle$, with a corresponding definition for $\tilde{n}_j(t)$. However, because $\hat{N}_i - \hat{N}_j$ is invariant, then we have

$$\tilde{n}_i(t) - \tilde{n}_j(t) = n_i(0) - n_j(0). \tag{12.72}$$

Also, since $\hat{\Omega}$ commutes with itself and must be invariant, then $\langle \hat{\Omega}(t) \rangle = \langle \hat{\Omega}(0) \rangle$, so

$$\langle \hat{\Omega}(t) \rangle = p(n_i(0) + n_j(0)). \tag{12.73}$$

Taking the expectation value of Eq. (12.71), with the aid of Eqs. (12.72) and (12.73), we get

$$\frac{d^2 \tilde{n}_i(t)}{dt^2} + \omega_n^2 \tilde{n}_i(t) = 4p^2 n_i(0) + 2m^2(1 + n_i(0) - n_j(0)), \tag{12.74}$$

where $\omega_n = 2\sqrt{p^2 + m^2}$. Notice that ω_n, which is the frequency with which both $\tilde{n}_i(t)$ and $\tilde{n}_j(t)$ oscillate, is twice the eigenfrequency in Eq. (12.66).

Eq. (12.74) is a simple second order differential equation. There are two boundary conditions, which are

$$\tilde{n}_i(0) = n_i(0) \text{ and } \left. \frac{d\tilde{n}_i(t)}{dt} \right|_{t=0} = 0,$$

where the second condition above comes from Eq. (12.68). These conditions lead to a solution of the form

$$\tilde{n}_i(t) = a_i + b_i \cos(\omega_n t),$$

where a_i and b_i are constants. There is a particularly interesting solution for the case when $p = 0$. Then we find

$$\tilde{n}_i(t) = \frac{1}{2}(n_i(0) - n_j(0) + 1 + (n_i(0) + n_j(0) - 1)\cos(2mt) \tag{12.75}$$

and

$$\tilde{n}_j(t) = \frac{1}{2}(n_j(0) - n_i(0) + 1 + (n_i(0) + n_j(0) - 1)\cos(2mt)). \qquad (12.76)$$

Here we need to recall that we are dealing with fermions, so we should choose $n_i(0) = 0$ or 1 and $n_j(0) = 0$ or 1. If we choose the Fock state $|0, 1\rangle$, then we find $\tilde{n}_i(t) = 0$ and $\tilde{n}_j(t) = 1$. This is because there is no interaction, and $\tilde{n}_i(t)$ remains zero and $\tilde{n}_j(t)$ remains 1. Similarly, if we choose a Fock state of $|1, 0\rangle$, then $\tilde{n}_i(t)$ remains 1 and $\tilde{n}_j(t)$ remains zero. However, if we choose a Fock state, $|1, 1\rangle$, where both states are initially occupied, we find

$$\tilde{n}_i(t) = \tilde{n}_j(t) = \frac{1}{2}(1 + \cos(2mt)) \qquad (12.77)$$

and the mean values of both occupation numbers oscillate in phase, between 0 and 1, because then there is an interaction between the pair of fermions. This result is not too surprising. However, something really surprising occurs when we begin with a vacuum, i.e., with a Fock state where both particle states are initially empty, so with $|0, 0\rangle$. Then we get

$$\tilde{n}_i(t) = \tilde{n}_j(t) = \frac{1}{2}(1 - \cos(2mt)). \qquad (12.78)$$

This is a truly revealing result. Even though we begin with the vacuum state, $|0, 0\rangle$, the coupling between the creation of a particle in one state and the creation of an anti-particle (or equivalently the loss of a particle) in the other, means that the particle/anti-particle pair emerge from the vacuum. Their mean occupation numbers subsequently oscillate together, i.e., in phase, between values of 0 and 1.

The above results may also be readily be understood in terms of the conservation rule in Eq. (12.70). If we begin with a state $|0, 1\rangle$, then $\hat{N}_i - \hat{N}_j$ has an eigenvalue of -1. There is no other eigenstate with this eigenvalue for the system to move to, so the system with this starting state cannot change to another state. Similarly with a starting state of $|1, 0\rangle$. Here $\hat{N}_i - \hat{N}_j$ has an eigenvalue of 1. No other eigenstate has this eigenvalue and again the system has no alternative state to move to. However, with the states, $|0, 0\rangle$ and $|1, 1\rangle$, then the eigenvalue of $\hat{N}_i - \hat{N}_j$ is zero in both cases, so the system can flip between these two states. What is more, the oscillation frequency, from Eqs. (12.77) and (12.78), that is associated with the states with equal occupation numbers is equal to $2m$. In conventional frequency units, this frequency is $\frac{2mc^2}{\hbar}$, where c is the speed of light in a vacuum and m is the fermionic rest mass. This is the *zitterbewegung* frequency [52].

So we have a picture of the *zitterbewegung* phenomenon, not as a rotation, nor as a configuration space oscillation about an equilibrium point, but as the rate at which a particle/anti-particle pair comes into and out of existence from the vacuum. This picture of the electron and positron is very different from that of permanent microscopic lumps of matter. Here the electron is seen as a manifestation of natural number dynamics; a flickering presence rather than a material object. This is quite a remarkable result, coming, as it does, ultimately from the universal quantum equation, Eq. (1.1).

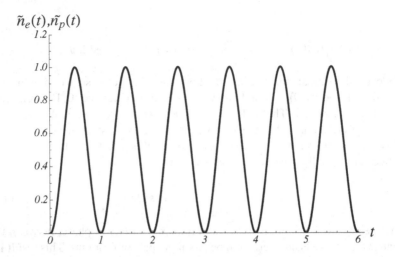

Figure 12.12 The time dependent occupation number expectation values for the case when the initial Fock state is $|0,0\rangle$, from Eq. (12.78). Here we have redefined $\tilde{n}_i(t)$ as the mean electron occupation number, $\tilde{n}_e(t)$, and $\tilde{n}_j(t)$ as the mean positron occupation number, $\tilde{n}_p(t)$. Because $\tilde{n}_e(t) = \tilde{n}_p(t)$, only a single curve can be seen. The time scale is the number of *zitterbewegung* periods.

Fig. 12.12 is an illustration of the time dependent occupation number expectation values for the case when the initial Fock state is $|0,0\rangle$, where we have redefined $\tilde{n}_i(t)$ for the mean electron occupation number, $\tilde{n}_e(t)$, and $\tilde{n}_j(t)$ as the mean positron occupation number, $\tilde{n}_p(t)$. Then p and the m in Eqs. (12.64) and (12.66) represent respectively, the electron (positron) momentum and electron (positron) rest mass. In the case shown in Fig. 12.12, $p = 0$. Because $\tilde{n}_e(t) = \tilde{n}_p(t)$, the two curves coincide and only a single curve can be seen. Fig. 12.13 is a schematic representation of the *zitterbewegung* phenomenon, where a electron in the positive energy state $E = m$, the solid circle, is excited together with a positron, that corresponds to the loss of an electron from the state $E = -m$, the open circle.

These results are a strong indication that the pairing of fermions with their antiparticles is a key element of their fundamental character. As we saw above, in the case of the initial Fock states with single particles, nothing happens; there is no excitation or zitterbewegung process. Incidentally, the same paired particle process is a key element of the BCS theory of superconductors that we met in Section 9.3. There the pairing was between an electron in a state above the Fermi level and a hole in state below the Fermi level. In that case the *zitterbewegung* frequency is replaced by the frequency associated with the scaled superconducting energy gap, 2Δ, as in Fig. 9.2.

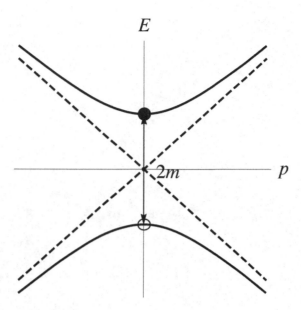

Figure 12.13   A schematic representation of the *zitterbewegung* phenomenon. The curve are plots of the eigenvalues $E = \pm \sqrt{p^2 + m^2}$ as functions of p. An electron in the positive energy state $E = m$, solid circle, is excited, together with a positron, that corresponds to the loss of an electron from the state $E = -m$, the open circle.

13 Epilogue

13.1 EMERGENT PHYSICS

This book is based on a single conceptual conjecture, that the simplest description of any system is the number of items it contains and that we can learn all we need to know about this system by counting these items. The mathematical realization of this basic idea takes the form of the universal quantum equation, Eq. (1.1)

$$\hat{N}\Psi_n = n\Psi_n.$$

Almost everything else that has been discussed in this book follows from this decidedly parsimonious starting point. This primitive universe is devoid of space, devoid of time, devoid of structure. All we have is the natural numbers that are needed for counting items. From these humble beginnings were derived natural number dynamics and we learned how counting leads to a knowledge of the behaviour of the natural world, a world that is, by its very nature, a quantum world. An important consideration in this endeavour has been not to have any preconceived ideas about what constitutes the physical contents of the universe. So, in some ways, this has been a journey of discovery, or at least, rediscovery, of what rules the items in our universe have to obey. Maybe, that is a key result of this journey; we have discovered rules rather than *things*. We certainly have not discovered all of the rules yet, but we have learned enough to have a good idea how the Universe works, at least that part of it that we already know about.

On this journey have emerged canonical quantization and standard quantum mechanics in both Schrödinger and Heisenberg forms, including a general theory of potentials and stationary energy states. By generalizing to multiple categories, we found that we needed time. Also, bosonic and fermionic behaviour emerged, including a closed group of three Hermitian fermionic operators that could be recognized as components of spin, without any preconceived notions involving tiny spinning particles. This generalization, coupled with symmetry associated with degeneracy, also lead to three components of configuration space, of linear momentum, and of angular momentum, via a Goldilocks principle that implies that two components are too few, but four components are too many.

Electromagnetic fields and photons emerged from a generalization of the three components of canonical quantization, involving the components of configuration space paired with their corresponding components of linear momentum. These fields could support propagating waves, from which phase invariance lead naturally to the theory of special relativity and relativistic wave mechanics, without the need to invoke the principle of equivalence nor the quantization of classical relativistic particle motion. Also from multi-category systems emerges standard many-body physics and an explanation of potentials through particle exchange and scattering. The special

DOI: 10.1201/9781003377504-13

case of fermionic scattering via the exchange of bosons provides an explanation for fermionic mass.

We also explored the application of natural number dynamics to systems outside the field of physics. One of the most interesting examples of such an application was to cell division. Natural number dynamics was able to identify processes that can be categorized as normal cell division that leads to stable cell populations, and onco-genic cell division, which leads to unstable growth that can be identified as the onset of cancer. Applying similar natural number dynamics to a two-category fermionic system in which the creation operator of one category is coupled to the annihilation operator of the other yields a remarkably revealing picture of *zitterbewegung* and the excitation of particle/anti-particle pairs from the vacuum, where the *zitterbewegung* frequency is identified with the time dependence of the particle numbers, without any need to resort to configuration space oscillations or localized rotations.

That such a meagre starting point as counting, which entered the human mind so long ago, could lead, as has been shown in this exploration, to so much familiar and fundamental physics might seem somewhat surprising. That view notwithstanding, it arguably offers some new insights into the nature of a physical universe that can be discovered from an examination of mere numbers. One could speculate why starting with the natural numbers works at all. Maybe it is because the set, $\mathbb{N}_0$, is the only relevant, unambiguously defined, unscalable, dimensionless and invariant quantity that there is. It is both abstract and at the same time concrete. It is a kind of absolute. This may not be a bad property for a universe to have.

13.2 WHAT NEXT?

The results in this book can be developed in a variety of ways. It is beyond its present scope to give an exhaustive list and to go into great detail in this regard. Here we just very briefly touch on how we may work our way into just three important examples of what our next steps might be. The three areas to mention are path integral methods, cosmology and thermodynamics.

The construction of quantum theory via *path integral* methods [33, 71, 115, 135] is usually attributed to Feynman and was one of his early breakthroughs in quantum electrodynamics that lead to his winning the Nobel prize for physics [32]. His starting point was the Hamiltonian of a single non-relativistic particle

$$\hat{H} = \frac{\hat{p}^2}{2m} + V(\hat{x}),$$

together with the unitary factor

$$\exp(-it\hbar^{-1}\hat{H}),$$

that enables the time evolution of a system to be evaluated. These two elements, which are both derivable from the universal quantum equation, Eq. (1.1), lead to the *phase integral* [135]

$$\exp(i\hbar^{-1}\int (\frac{1}{2}m\left(\frac{d\hat{x}}{dt}\right)^2 - V(\hat{x}))dt),$$

where we recognize

$$\hat{L} = \frac{1}{2}m\left(\frac{d\hat{x}}{dt}\right)^2 - V(\hat{x})$$

as the Lagrangian. This opens up the whole apparatus of Lagrangian mechanics, which, together with *Noether's theorem* [109] takes us to the gauge theories of modern physics, that are the conventional basis for quantum field theory (QFT) [109, 115, 125, 135]. These results are similar to those found in Section 11.8, but path integral methods are considered more general and are widely used in quantum field theory.

With regard to phenomena on a large scale that are important in cosmology, Kauffman [68, 69] has argued that the commutation relations in H-type equations, that play such a key role in the theory developed in this book, can be viewed as discrete derivatives that are indicative of the discreteness of space itself. Kauffman has shown that this is related to gauge theories, to the curvature of space and ultimately to gravity. We saw in Chapter 8 that the non-commutation of components of linear momentum lead to electromagnetic fields. In a similar way, the curvature of space that underpins Einstein's interpretation of gravity may be associated with the non-commutation of configuration space components.

A further great area of physics that has not been touched on at all so far is thermodynamics. However, this can be addressed from the standpoint of natural number dynamics by extending it to *open systems* [14]. This involves populations of items like those in Chapter 12, being allowed to interact with external populations that act as reservoirs or sinks [7]. Then thermodynamic phenomena like dissipation and time irreversibility enter into consideration.

The results in this book, which can all be traced back to the universal quantum equation, Eq. (1.1), are a testament maybe to the unifying effect of the natural numbers in our understanding of the natural world. It clearly puts number at the centre of physics. This may be contrasted with much of modern physical thinking which has geometry at its heart. The advantage of number as a fundamental conceptual basis is that it links physics to information and, in particular, to the increasingly important field of quantum information [85]. The approach adopted here can provide insights into the physical consequences of the idea of a universe of bits and qubits. Pythagorus is supposed to have said, 'all things are numbers', although no-one seems to be quite sure what he meant by it [113]. Maybe we can now see that there might be some truth in what he said after all.

13.3 FORMS AND SHADOWS: DE-QUANTIZATION AND CLASSICAL PHYSICS

Having traced the origins of the quantum nature of things to the concept of itemization of the natural world through counting, it is time, at the end of this exploration, to consider how classical physics comes into the picture. It is appropriate that we think of classical physics last, since it really is an afterthought and an epilogue to quantum physics. In Platonic terms, we have discovered the *quantum forms* of reality and now it is time to see what *classical shadows* are cast on the walls of our cave.

We have seen that quantum theory comes about because we need operators to describe the natural world. The fundamental operator, is the natural number operator in the universal quantum equation, Eq. (1.1). It is needed in order to give a mathematical framework with which to deal with the natural numbers that turn itemization into a quantitative form. This is the natural route by which operators enter the theory. The quantum properties then follow from the non-commutation of the natural number amplitude operators, $\hat{A}$ and its adjoint, that can be seen in the fundamental commutation relation, Eq. (3.17), i.e.

$$[\hat{A},\hat{A}^{\dagger}] = 1.$$

We can argue that quantum behaviour will disappear if we can neglect the non-commutation property above. This is what we will refer to as *de-quantization*. The question then becomes: compared to what can we neglect $[\hat{A},\hat{A}^{\dagger}]$? Since the commutator is just $\hat{A}\hat{A}^{\dagger} - \hat{A}^{\dagger}\hat{A}$, then we should be able to neglect it if it is small compared to the product of the sizes of $\hat{A}$ and $\hat{A}^{\dagger}$. Expectation values are a good way to estimate the size of an operator, and the simplest eigenstates to use for this purpose are the Fock states, $|n\rangle$. Now of course, as we saw in Chapter 3, the expectation values of $\hat{A}$ and $\hat{A}^{\dagger}$ are zero in this case. However, we can use rms values. Since $\hat{A}$ and $\hat{A}^{\dagger}$ are not Hermitian, we can define the rms size of $\hat{A}$ as $\sqrt{\langle \hat{A}^{\dagger}\hat{A}\rangle} = \sqrt{\langle \hat{N}\rangle} = \sqrt{n}$, and that of $\hat{A}^{\dagger}$ as $\sqrt{\langle \hat{A}\hat{A}^{\dagger}\rangle} = \sqrt{\langle \hat{N}+1\rangle} = \sqrt{n+1}$. Then

$$\frac{\langle[\hat{A},\hat{A}^{\dagger}]\rangle}{\sqrt{\langle \hat{A}^{\dagger}\hat{A}\rangle}\sqrt{\langle \hat{A}\hat{A}^{\dagger}\rangle}} = \frac{1}{\sqrt{n(n+1)}}.$$

So, in the limit of very large n, i.e., $n \to \infty$, the non-commutation of $\hat{A}$ and $\hat{A}^{\dagger}$ becomes negligible and hence quantum effects disappear. $n \to \infty$ [131] can refer either to the number of particles as in a *bao board representation*, in Fig. 0.2, or the quantum number associated with an energy level, as in an *Ishango representation*, in Fig. 0.1.

As an example of the former condition we can note, somewhat naively perhaps, that a macroscopic body with a mass of 1 kg will contain of the order of $n = 10^{26}$ atomic particles, which would certainly satisfy the condition for n being large compared to one.

An example of the classical limit corresponding to an energy level having a large quantum number is shown in Fig. 13.1 for the harmonic oscillator. The solid curve shows the modulus squared of the wave function of a harmonic oscillator with $n = 20$. This represents the probability density for the value of x, which is the location of the oscillating particle in a one-dimensional quadratic potential. The grey curve represents the purely classical calculation of the probability density [22]. The classical calculation is close to the mean of the quantum calculation. As n increases, the classical result and the mean quantum result become indistinguishable, showing that the quantum prediction tends to the classical one as $n \to \infty$.

The equivalent canonical quantization relation in standard units is

$$[\hat{p},\hat{x}] = -i\hbar,$$

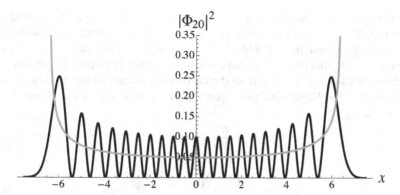

Figure 13.1 The dark curve illustrates the squared modulus of the eigenfunction of the natural number operator for $n = 20$. The grey curve illustrates the probability density for the location of a classical particle, subject to a quadratic potential.

where $\hat{p}$ is the linear momentum operator and $\hat{x}$ is the one-dimensional position operator. It is sometimes argued that the commutator in this case can be neglected under some circumstances, because the value of the (reduced) Planck constant, $\hbar = \frac{h}{2\pi} = 1.05457266 \times 10^{-34}$ Js is so small anyway. However, as in the case with the non-commutation of the natural number amplitudes, we must ask, compared to what can $\hbar$ be considered small; certainly not compared to zero. Bearing in mind that $\hat{p}$ and $\hat{x}$ are both Hermitian, but that their commutator bracket is imaginary, we can follow a similar line to that taken with the natural number amplitudes and compare $\| \langle [\hat{p}, \hat{x}] \rangle \|$ to $\sqrt{\langle \hat{p}^2 \rangle \langle \hat{x}^2 \rangle}$. In the case of the harmonic oscillator we can again use the Fock state as an eigenstate. Then we can utilize Eqs. (5.44) and (5.45) to get

$$\frac{\| \langle [\hat{p}, \hat{x}] \rangle \|}{\sqrt{\langle \hat{p}^2 \rangle \langle \hat{x}^2 \rangle}} = \frac{\hbar}{\hbar(n + \frac{1}{2})} = \frac{1}{n + \frac{1}{2}}.$$

Thus, as expected, the commutator bracket is negligible for $n \to \infty$ and so we get the same condition for de-quantization as before. It is not difficult to show we get a similar result for a particle trapped in an infinite square well.

We can generalize the above arguments to treat macroscopic bodies like canon balls, if we adopt heuristic arguments and treat $\sqrt{\langle \hat{p}^2 \rangle}$ as the typical momentum of a macroscopic body and $\sqrt{\langle \hat{x}^2 \rangle}$ as a measure of its linear size. Then, if we consider a typical macroscopic body with a mass of 1 kg, moving with a speed of $1 \mathrm{ms}^{-1}$ and having a linear size of 1m, we get

$$\frac{\| \langle [\hat{p}, \hat{x}] \rangle \|}{\sqrt{\langle \hat{p}^2 \rangle \langle \hat{x}^2 \rangle}} \sim 1.0 \times 10^{-34},$$

from which we can conclude that non-commutation is negligible in this case. Compare the macroscopic case with an electron of mass, 10^{-30}kg, moving at $10^6 \mathrm{ms}^{-1}$ and confined to an atom of radius 10^{-10}m, then the ratio is about 1, and we have a result that needs quantum mechanics, as expected.

Once we have de-quantized the theory, we can take the linear momentum as a scalar variable, p, and the position as a scalar variable x, since these no longer need to be non-commuting variables. However, since we have carried out the de-quantization of the theory *after* we have constructed the Hamiltonian operator and the operator equations of motion, from the universal quantum equation, then, for the scalar version of the Hamiltonian of a particle in a potential in a one-dimensional system, we just get

$$H = \frac{p^2}{2m} + V(x),$$

$$p = m\frac{\mathrm{d}x}{\mathrm{d}t},$$

and

$$\frac{\mathrm{d}p}{\mathrm{d}t} = -\frac{\mathrm{d}V}{\mathrm{d}x}.$$

The above equations are just classical Newtonian mechanics. The electromagnetic results from Chapter 8 are also easily rendered classical in a similar manner.

Now, of course, Newtonian dynamic is non-relativistic, so what about classical relativistic equations of motion? We cannot turn the Dirac equation directly into a classical equation by turning operators into scalars, because it contains the spin matrices, which have no classical equivalent. However, the eigenvalues of the Dirac equation, such as Eq. (11.92), which we have already pointed out is a perfectly good, energy-momentum equation, and because it is between scalar eigenvalues, is also a perfectly good classical equation too. Equations of motion can be obtained from the Lagrangian equations, Eq. (11.136), together with the scalar form of the Lagrangian in Eq. (11.137).

The de-quantization process, above, makes the correspondence principle redundant, since there is no need to argue that classical physics of itself has any basis in reality. It is simply an approximation to quantum reality, which arises from the discreteness of nature via the universal quantum equation. To paraphrase Lucretius' *De rerum natura* (The nature of things) [76], after 100 years of quantum theory it is about time that we really did accept *the quantum nature of things*.

A Peano axioms

We briefly outline the axioms that were devised at the end of the nineteenth century by the Italian mathematician, Giuseppe Peano, to give a firm mathematical foundation to natural numbers [117, 18]. Let, S be a function that maps the set of natural numbers, $\mathbb{N}_0$ (including 0), into itself, i.e.,

$$S : \mathbb{N}_0 \to \mathbb{N}_0.$$

The definition of S is that it maps a natural number $n \in \mathbb{N}_0$ to a successor, $S(n)$. The Peano axioms that ensure that $\mathbb{N}_0$ exist are

1. There exists $0 \in \mathbb{N}_0$ for which $S(n) \neq 0$ for any $n \in \mathbb{N}_0$.
2. If $S(n) = S(m)$, then $n = m$.
3. If $\Sigma \subseteq \mathbb{N}_0$ is such that $0 \in \Sigma$ and $n \in \Sigma$ implies $S(n) \in \Sigma$ for all $n \in \mathbb{N}_0$, then $\Sigma = \mathbb{N}_0$.
4. There exists a set, $\mathbb{N}_0$, and a function, $S : \mathbb{N}_0 \to \mathbb{N}_0$, satisfying 1 to 3.

These axioms allow the properties of the set of natural numbers to be deduced. The successor function has a role similar to a raising operator, although the definition of $S(n) = n + 1$ is not strictly part of the formal structure of the Peano axioms, since the symbol '$+$' is not specifically defined. However, the idea behind the successor function is that $S(1) = 2$ and $S(2) = 3$, and so on. Particularly relevant here is the role of the natural number 0. By Peano's axiom (1), 0 is by definition, not the successor of any natural number, but it has a successor, i.e., $S(0) = 1$. 0 clearly plays a role similar to the Fock state, $|0\rangle$, that constitutes the essential backstop condition that prevents negative eigenvalues for the number operator. However, the Peano axioms lack the rich algebraic structure of operators that have such a powerful influence on the mathematical content of quantum mechanics. It is arguable that the bosonic natural number operator and its companion creation and annihilation operators are an alternative way of generating the set of natural numbers.

DOI: 10.1201/9781003377504-A

B Virial theorem and discrete energy levels

The virial theorem in quantum mechanics takes the general form [42]

$$\langle T \rangle = \frac{1}{2} \langle r \frac{dV(r)}{dr} \rangle, \tag{B.1}$$

where T is the kinetic energy, $V(r)$ is the potential energy, r is the displacement from a mean position and the angle brackets denote averages obtained from inner products of operators on a Hilbert space. We assume that the potential in scaled form depends on r as r^q, where q is a real number. Also, if the motion takes place along a single co-ordinate, x, we can replace r by $|x|$, then

$$T \sim \frac{q}{2} |x|^q. \tag{B.2}$$

In quantum mechanics, $T = \frac{\hbar^2 k^2}{2}$, where k is a wavenumber. Then in scaled form T is just λ^{-2}, where λ is a wavelength. We now make the assumption that $V(|x|)$ is a trapping potential and a number of wavelengths fit approximately into the scale length $|x|$, so that $|x| \sim (n+\delta)\lambda$, where n is the number of wavelengths and δ allows for the fact that there may not be an exact number of wavelengths in the distance $|x|$. Then Eq. (B.2) becomes

$$\lambda^{-2} \sim \frac{q}{2} ((n+\delta)\lambda)^q \tag{B.3}$$

and hence

$$\lambda(n) \sim (\frac{2}{q})^{\frac{1}{(q+2)}} (n+\delta)^{(-\frac{q}{(q+2)})}. \tag{B.4}$$

Substituting Eq. (B.4) into $E(n) = T + V$, where $E(n)$ is the nth energy level, then yields

$$E(n) \sim \varepsilon (n+\delta)^{(\frac{2q}{q+2})}, \tag{B.5}$$

where ε is a constant factor.

Although, Eq. (B.5) has been obtained by relatively crude methods, it is remarkably accurate at predicting the n-dependence of energy levels in some well-known cases. For example, in the case of the harmonic oscillator, $q = 2$ and we get, $E_n \sim \varepsilon (n+\delta)$, which is correct with $\delta = \frac{1}{2}$. For $q \to \infty$, then $V(|x|) = 0$ for $|x| \le 1$ and $V(|x|) = \infty$, for $|x| > 1$. This is just an infinite square well between $x = \pm 1$ and Eq. (B.5) gives $E_n \sim \varepsilon (n+\delta)^2$, which again is correct with $\delta = 1$.

We also note that for $q = 0$, V is independent of x and E does not depend on n, so the particle is not trapped. Remarkably the method also works for the

DOI: 10.1201/9781003377504-B

Coulomb potential ($q = -1$). Then $E_n = \varepsilon(n + \delta)^{-2}$, precisely as predicted by standard methods of quantum mechanics, with $\delta = 1$. A rather less well-known example is that of a simplified model for quark interactions in a nucleus, where $q = 1$. Then $E_n = \varepsilon(n + \delta)^{\frac{2}{3}}$, which agrees with the result reported in ref. [118].

The simple analysis above also allows us relate the functional dependence of the number operator to the Hamiltonian in Chapter 5. Then Eq. (B.5) is roughly equivalent to

$$\hat{H} = \hat{B}^\dagger \hat{B} \sim \hat{N}^{\left(\frac{2q}{q+2}\right)}. \tag{B.6}$$

So, given a relationship of the form, $\hat{H} = \hat{N}^p$, then this is expected to come from a potential of the form $V \sim |x|^{\frac{2p}{(2-p)}}$.

There are alternatives ways of obtaining the above results from the virial theorem. See for example, ref. [97].

C Wave kinematics

Consider a wave field, $\phi(x,t)$, that is a Fourier superposition of plane waves given by [82]

$$\phi(x,t) = \frac{1}{\sqrt{2\pi}} \int_{-\infty}^{\infty} F(k)\exp(i(kx - \omega(k)t)\mathrm{d}k. \tag{C.1}$$

The scaled energy in the wave field is then, by Parsival's theorem

$$E = \int_{-\infty}^{\infty} \phi^*(x,t)\phi(x,t)\mathrm{d}x = \int_{-\infty}^{\infty} F^*(k)F(k)\mathrm{d}k. \tag{C.2}$$

The inverse transform to Eq. (C.1) is

$$F(k)\exp(-i\omega(k)t) = \frac{1}{\sqrt{2\pi}} \int_{-\infty}^{\infty} \phi(x,t)\exp(-ikx)\mathrm{d}x, \tag{C.3}$$

so

$$\frac{\partial(F(k)\exp(-i\omega(k)t))}{\partial k} = -\frac{i}{\sqrt{2\pi}} \int_{-\infty}^{\infty} x\phi(x,t)\exp(-ikx)\mathrm{d}x. \tag{C.4}$$

The lhs of Eq. (C.4) is

$$\frac{\partial(F(k)\exp(-i\omega(k)t))}{\partial k} = \frac{\partial F(k)}{\partial k}\exp(-i\omega(k)t) - iF(k)t\frac{\partial\omega(k)}{\partial k}. \tag{C.5}$$

Then

$$\int_{-\infty}^{\infty} x\phi^*(x,t)\phi(x,t)\mathrm{d}x = \int_{-\infty}^{\infty} F^*(k)\frac{\partial F(k)}{\partial k}\mathrm{d}k + t\int_{-\infty}^{\infty} \frac{\partial\omega(k)}{\partial k}F^*(k)F(k)\mathrm{d}k. \tag{C.6}$$

Thus, the mean speed of the field, $\bar{v}$ is given by

$$\bar{v} = \frac{\bar{x}(t) - \bar{x}(0)}{t} = \langle\frac{\partial\omega(k)}{\partial k}\rangle_k, \tag{C.7}$$

where

$$\bar{x}(t) = \frac{1}{E} \int_{-\infty}^{\infty} x\phi^*(x,t)\phi(x,t)\mathrm{d}x, \tag{C.8}$$

$$\bar{x}(0) = \frac{1}{E} \int_{-\infty}^{\infty} F^*(k)F'(k)\mathrm{d}k, \tag{C.9}$$

DOI: 10.1201/9781003377504-C

and

$$\langle \frac{\partial \omega(k)}{\partial k} \rangle_k = \frac{1}{E} \int_{-\infty}^{\infty} \frac{\partial \omega(k)}{\partial k} F^*(k) F(k) \mathrm{d}k. \qquad (C.10)$$

Thus the mean speed of the field is the same as the group speed averaged over the spectrum of plane waves.

D Lie groups

Consider a set of three Hermitian operators, $\hat{F}_1$, $\hat{F}_2$ and $\hat{F}_3$ that obey the commutation rule

$$[\hat{F}_i, \hat{F}_j] = i\hat{F}_k, \tag{D.1}$$

for $i = 1$, $j = 2$, $k = 3$ and further two cyclic combinations of the subscripts. This implies that the set forms the 3 elements of a closed Lie group[1][39], with the added symmetry due to their cyclic property. If we construct a vector, $\hat{\mathbf{F}} = (\hat{F}_1, \hat{F}_2, \hat{F}_3)$, then one can further define a square modulus, $\hat{\mathbf{F}}^2 = \hat{F}_1^2 + \hat{F}_2^2 + \hat{F}_3^2$, which must also be Hermitian, and then it is easy to prove from Eq. (D.1) that

$$[\hat{\mathbf{F}}^2, \hat{F}_i] = 0, \tag{D.2}$$

for $i = 1, 2$ and 3. Thus $\hat{\mathbf{F}}^2$ and $\hat{F}_i$ have a common eigenvector, although the three components, the $\hat{F}_i$, do not commute among themselves and so do not share a common eigenvector with each other. To this extent the vector $\hat{\mathbf{F}}$ is in a sense fictitious, since its components do not share a common eigenstate, a state of affairs which means the vector cannot actually be constructed although we know its length. So $\hat{\mathbf{F}}$ represents a spherical surface rather than a directed length. However, we can define that length as we shall show next.

Let us take the common eigenvector of $\hat{\mathbf{F}}^2$ and $\hat{F}_3$ as $|\lambda, f\rangle$ such that

$$\hat{\mathbf{F}}^2 |\lambda, f\rangle = \lambda |\lambda, f\rangle \quad \text{and} \quad \hat{F}_3 |\lambda, f\rangle = f |\lambda, f\rangle, \tag{D.3}$$

where λ and f are real eigenvalues, since both $\hat{\mathbf{F}}^2$ and $\hat{F}_3$ are Hermitian. Let $\hat{G} = \hat{F}_1 - i\hat{F}_2$, then it is easy to show that

$$[\hat{F}_3, \hat{G}] = -\hat{G} \tag{D.4}$$

and

$$[\hat{\mathbf{F}}^2, \hat{G}] = 0. \tag{D.5}$$

Then we find

$$\hat{F}_3 \hat{G} |\lambda, f\rangle = \hat{G}\hat{F}_3 |\lambda, f\rangle - \hat{G}|\lambda, f\rangle = (f-1)\hat{G}|\lambda, f\rangle, \tag{D.6}$$

which means that $\hat{G}|\lambda, f\rangle = \alpha(\lambda, f)|\lambda, f-1\rangle$, where $\alpha(\lambda, f)$ is a scalar factor that may depend on λ and f. Notice that $\hat{G}$ does not affect the value of λ because of Eq. (D.5).

---

[1] In general, a Lie group of elements, which need not be Hermitian, obey the Lie product rule, $[\hat{F}_i, \hat{F}_j] = i\varepsilon_{ijk}\hat{F}_k$, where the ε_{ijk} are termed the structure constants. There can be any number of elements, $\hat{F}_i$, but it is essential that the Lie group is closed under the Lie product, but not all Lie groups have the cyclic property of Eq. (D.1).

DOI: 10.1201/9781003377504-D

In a similar manner we can show that $\hat{G}^\dagger|\lambda,f\rangle = \beta(\lambda,f)|\lambda,f+1\rangle$, where $\beta(\lambda,f)$ is a scalar factor that may depend on λ and f. So, $\hat{G}^\dagger$ and $\hat{G}$ serve as a pair of raising and lowering operators that change the eigenvalue f by ± 1. It can also be seen that

$$[\hat{G},\hat{G}^\dagger] = [\hat{F}_2 - i\hat{F}_1, \hat{F}_2 + i\hat{F}_1] = -2\hat{F}_3, \tag{D.7}$$

$$\hat{G}^\dagger\hat{G} = \hat{F}_1^2 + \hat{F}_2^2 + \hat{F}_3 = \hat{\mathbf{F}}^2 - \hat{F}_3^2 + \hat{F}_3, \tag{D.8}$$

and

$$\hat{G}\hat{G}^\dagger = \hat{F}_1^2 + \hat{F}_2^2 - \hat{F}_3 = \hat{\mathbf{F}}^2 - \hat{F}_3^2 - \hat{F}_3. \tag{D.9}$$

Hence

$$\hat{G}^\dagger\hat{G}|\lambda,f\rangle = (\lambda - f(f-1))|\lambda,f\rangle. \tag{D.10}$$

and so

$$\langle\lambda,f|\hat{G}^\dagger\hat{G}|\lambda,f\rangle =\|\,\hat{G}|\lambda,f\rangle\,\|^2= \alpha^2(\lambda,f)$$
$$= \lambda - f(f-1), \tag{D.11}$$

so $\alpha(\lambda,f) = \sqrt{\lambda - f(f-1)}$. Furthermore

$$\hat{G}^\dagger\hat{G}|\lambda,f\rangle = \beta(\lambda,f-1)\alpha(\lambda,f)|\lambda,f\rangle, \tag{D.12}$$

which makes $\beta(\lambda,f) = \sqrt{\lambda - f(f+1)} = \alpha(\lambda,f+1)$.

To see that the eigenvalues of $\hat{\mathbf{F}}^2$ must be non-negative we note that, from Eqs. (D.8) and (D.9) that

$$\hat{F}_1^2 + \hat{F}_2^2 = \frac{1}{2}(\hat{G}^\dagger\hat{G} + \hat{G}\hat{G}^\dagger). \tag{D.13}$$

Now both terms on the rhs of Eq. (D.13) are moduli and so have non-negative eigenvalues. So since the eigenvalues, f^2, of $\hat{F}_3^2$, must also be non-negative then the eigenvalues of $\hat{\mathbf{F}}^2$ must also be non-negative and what is more, they must be greater than or equal to the eigenvalue of $\hat{F}_3^2$, i.e., $\lambda \geq f^2$.

Now the raising and lowering operators $\hat{G}^\dagger$ and $\hat{G}$ can change f by ± 1 and so f can be either positive or negative. This means that, for a given λ, there must be a cutoff for the largest positive value that f can take, f_u, say, and for its lowest negative value, f_v say. These conditions imply

$$\hat{G}^\dagger|\lambda,f_u\rangle = \sqrt{\lambda - f_u(f_u+1)}|\lambda,f_u+1\rangle = 0 \tag{D.14}$$

and

$$\hat{G}|\lambda,f_v\rangle = \sqrt{\lambda - f_v(f_v-1)}|\lambda,f_v-1\rangle = 0, \tag{D.15}$$

so, $\lambda = f_u(f_u+1) = f_v(f_v-1)$, from which we can infer that either, $f_v = -f_u$, or, $f_v = f_u + 1$. Now the second of these conditions can be eliminated since obviously, f_u must be greater than f_v, so we must have $f_v = -f_u$. This implies $\lambda = f_u(f_u+1)$.

Finally, we know that, raising the $\hat{F}_3$ quantum number from f_v to f_u via the raising operator, then f_v and f_u can only differ by an integer, so $f_u - f_v = 2f_u = n$, where n

is a natural number. Then $f_u = \frac{n}{2}$ and so we have $\lambda = j(j+1)$ where $j = \frac{n}{2}$, and we can write

$$\hat{\mathbf{F}}^2|j,f\rangle = j(j+1)|j,f\rangle \text{ and } \hat{F}_3|j,f\rangle = f|j,f\rangle, \tag{D.16}$$

which are derivable from Eq. (D.1) without any further assumptions. We have also seen that for a given value of j, f can take on any integer value between $\pm j$ which means that for a given value of j there are $2j+1$ possible values of f.

The raising and lowering operators then obey

$$\hat{G}^\dagger|j,f\rangle = \sqrt{j(j+1) - f(f+1)}|j,f+1\rangle \tag{D.17}$$

and

$$\hat{G}|j,f\rangle = \sqrt{j(j+1) - f(f-1)}|j,f-1\rangle. \tag{D.18}$$

E Differential operators

Differential operators have some rather subtle properties that can lead to misunderstandings if care is not taken. This is partly due to the fact that there are really two rather different meanings that can be attributed to symbols of the type

$$\frac{\mathrm{d}}{\mathrm{d}u},$$

that we associate with differentiation, which in this case is with respect to a real scalar variable, u. The form of the above symbol arises from the way differentiation was originally developed by Leibniz and Newton in the seventeenth century. The symbol is associated with a limiting ratio of differences as those differences tend to zero. This is the way calculus is still traditionally introduced. It is not the only way of dealing with differential calculus. It can be developed entirely algebraically, without any reference to limits, as is outlined in Appendix F. However, the differential with respect to u above has to some extent lost its meaning as a ratio. So when it appears in the form applied to a function of u and we get say

$$\frac{\mathrm{d}f(u)}{\mathrm{d}u},$$

we really mean by this a function that is equal to the result of differentiating the function $f(u)$ with respect to u. In this case it is often replaced by the symbol $f'(u)$. This then is one meaning of the differential ratio symbol, i.e., it can be treated as a kind of hieroglyph to represent $f'(u)$, which is after all simply another function of u. The other use of the differential symbol is as a stand alone operator. We will make the distinction between the two situations by keeping $\frac{\mathrm{d}}{\mathrm{d}u}$ distinct from $f(u)$ when we are treating $\frac{\mathrm{d}}{\mathrm{d}u}$ as a stand alone operator. Then we note that

$$\frac{\mathrm{d}}{\mathrm{d}u}f(u) \not\equiv \frac{\mathrm{d}f(u)}{\mathrm{d}u}. \tag{E.1}$$

The reason for this is as a consequence of the Leibniz rule for the differentiation of a product, and we should really write

$$\frac{\mathrm{d}}{\mathrm{d}u}f(u) = \frac{\mathrm{d}f(u)}{\mathrm{d}u} + f(u)\frac{\mathrm{d}}{\mathrm{d}u} = f'(u) + f(u)\frac{\mathrm{d}}{\mathrm{d}u}. \tag{E.2}$$

Notice the difference between

$$\frac{\mathrm{d}f(u)g(u)}{\mathrm{d}u} = \frac{\mathrm{d}f(u)}{\mathrm{d}u}g(u) + f(u)\frac{\mathrm{d}g(u)}{\mathrm{d}u} = f'(u)g(u) + f(u)g'(u) \tag{E.3}$$

and

$$\frac{\mathrm{d}}{\mathrm{d}u}f(u)g(u) = \frac{\mathrm{d}f(u)}{\mathrm{d}u}g(u) + f(u)\frac{\mathrm{d}g(u)}{\mathrm{d}u} + f(u)g(u)\frac{\mathrm{d}}{\mathrm{d}u}. \tag{E.4}$$

DOI: 10.1201/9781003377504-E

The point here is that if we multiply $\Psi(u)$ by the lhs of Eq. (E.4) from the left, we get

$$\frac{df(u)g(u)}{du}\Psi(u) = \frac{df(u)}{du}g(u)\Psi(u) + f(u)\frac{dg(u)}{du}\Psi(u)$$

$$= f'(u)g(u)\Psi(u) + f(u)g'(u)\Psi(u), \qquad (E.5)$$

which is still just a function of u, whereas

$$\frac{d}{du}f(u)g(u)\Psi(u) = \frac{df(u)}{du}g(u)\Psi(u) + f(u)\frac{dg(u)}{du}\Psi(u) + f(u)g(u)\frac{d}{du}\Psi(u)$$

$$= f'(u)g(u)\Psi(u) + f(u)g'(u)\Psi(u) + f(u)g(u)\Psi'(u)$$

$$+ f(u)g(u)\Psi(u)\frac{d}{du}, \qquad (E.6)$$

and so on, which still contains the stand alone operator, $\frac{d}{du}$. The last term on the rhs of Eq. (E.6) indicates that we can continue to multiply further functions to the right of the whole sequence of symbols, on which $\frac{d}{du}$ can continue to operate. However, it should be remembered that all operators representing a system, operate on a Hilbert space. If Ψ is a vector in the Hilbert space then it is permissible to drop the last term on the rhs of Eq. (E.6) and the result will be a mapping to a new vector in the Hilbert space.

The properties of stand alone differential operators are best explored in the context of commutator brackets, as follows. Rearranging Eq. (E.2) we get

$$\frac{d}{du}f(u) - f(u)\frac{d}{du} = \left[\frac{d}{du}, f(u)\right] = \frac{df(u)}{du} = f'(u). \qquad (E.7)$$

Thus the commutator shows clearly how the stand alone differential operator is related to the result of differentiating with respect to u. Two important consequences of Eq. (E.7) are, first, letting $f(u) \to f(u)g(u)$, then

$$\left[\frac{d}{du}, f(u)g(u)\right] = f(u)g'(u) + f(u)g'(u) = f(u)\left[\frac{d}{du}, g(u)\right] + \left[\frac{d}{du}, f(u)\right]g(u), \qquad (E.8)$$

which is just the expansion rule for commutators in Eq. (2.21). So the commutation rule itself obeys the Leibniz rule. This is why commutation is so closely related to differentiation. The second important consequence of Eq. (E.7) can be seen by letting $f(u) \to u$, then we get the well known result

$$\left[\frac{d}{du}, u\right] = 1, \qquad (E.9)$$

showing directly that u and $\frac{d}{du}$ do not commute, which is a result that is fundamental to quantum mechanics. We note that a certain class of Hermitian operators can be treated as real scalar variables like u. Then we can ask what is the Hermiticity of their corresponding differential operators. As we pointed out in Section 2.4, the commutator of a pair of Hermitian operators is anti-Hermitian, so it cannot be equal to a real number. This implies that $\frac{d}{du}$ cannot be Hermitian.

To emphasize that we are treating u as an Hermitian operator, we can represent u by the operator symbol, $\hat{U}$. Because u is real and $\hat{U}$ is Hermitian, then, $\hat{U}^\dagger = \hat{U}$ and $u^\dagger = u$. Using Eq. (2.32) then

$$\left[\frac{d}{du}, u\right]^\dagger = \left[u, (\frac{d}{du})^\dagger\right] = 1. \tag{E.10}$$

Writing

$$\left[u, (\frac{d}{du})^\dagger\right] = \left[-(\frac{d}{du})^\dagger, u\right] = 1 \tag{E.11}$$

and comparing Eqs. (E.10) and (E.11), then we must have

$$\left(\frac{d}{du}\right)^\dagger = -\frac{d}{du} \tag{E.12}$$

and we can conclude that the stand-alone differential operator $\frac{d}{du}$ must be anti-Hermitian, as long as u is real.

Now we can demonstrate the distinction between $\frac{d}{du} f$ where $\frac{d}{du}$ is a stand-alone operator and what we have previously referred to a a hieroglyph, $\frac{df}{du}$ the traditional differential of f. We note that $u^\dagger = u$ implies that $f^\dagger(u) = f(u^\dagger) = f(u)$, so, $(f'(u))^\dagger = f'(u)$ and hence [116]

$$\left(\frac{df}{du}\right)^\dagger = \frac{df}{du}. \tag{E.13}$$

On the other hand,

$$\left(\left(\frac{d}{du}\right)f\right)^\dagger = f\left(\frac{d}{du}\right)^\dagger = -f\left(\frac{d}{du}\right), \tag{E.14}$$

which is completely different from the result in Eq. (E.13).

We can now check the meaning of the differential operator in the H-type equation, Eq. (2.63). Let's look at its adjoint form

$$(i\frac{d\hat{G}}{d\xi})^\dagger = [\hat{G}, \hat{K}]^\dagger = [\hat{K}, \hat{G}^\dagger], \tag{E.15}$$

where the final term follows from the rule for the adjoint of a product and $\hat{K}$ being Hermitian. Recall that $\hat{G}$ in the H-type equation, Eq. (2.63) is a function of ξ and $\hat{G}(\xi) = \exp(i\xi\hat{K})\hat{G}(0)\exp(-i\xi\hat{K})$, so

$$\hat{G}(\xi)^\dagger = (\exp(i\xi\hat{K})\hat{G}(0)\exp(-i\xi\hat{K}))^\dagger = (\exp(-i\xi\hat{K}))^\dagger\hat{G}^\dagger(0)(\exp(i\xi\hat{K})^\dagger$$
$$= \exp(i\xi\hat{K})\hat{G}^\dagger(0)\exp(-i\xi\hat{K}). \tag{E.16}$$

It is straightforward to show that $(\exp(-i\xi\hat{K}))^\dagger = \exp(i\xi\hat{K})$ by first expanding the function of $\hat{K}$ as a power series (see Section 2.6). This is actually an important result since it shows that the adjoint of an H-type operator is itself an H-type operator as it must be if the H-type equation is to be universally applied to operators that are functions of the same continuous real scalar variable like ξ. As a consequence,

$$i\frac{d\hat{G}^\dagger}{d\xi} = [\hat{G}^\dagger, \hat{K}]. \tag{E.17}$$

Comparing Eqs. (E.15) and (E.17) implies that

$$\left(\frac{d\hat{G}}{d\xi}\right)^\dagger = \frac{d\hat{G}^\dagger}{d\xi}, \tag{E.18}$$

so we can see that the differentials of the operator function behave simply as an operator function and we must not treat the differential symbol as a stand-alone operator. This means we could write Eq. (E.18) as

$$(\hat{G}'(\xi))^\dagger = \hat{G}^{\dagger\prime}(\xi). \tag{E.19}$$

This result is interesting from the point of view of the definition of the adjoint operator. If we consider the inner product $\langle\Phi|\hat{G}\Psi\rangle$, where Ψ and Φ are vectors in a Hilbert space in H-representation, i.e., independent of ξ, then by definition, $\langle\Phi|\hat{G}\Psi\rangle^* = \langle\hat{G}\Psi|\Phi\rangle = \langle\Phi|\hat{G}^\dagger\Psi\rangle$. Now evaluating differentials of the inner products we have

$$\langle\hat{G}\Psi|\Phi\rangle' = \langle\hat{G}'\Psi|\Phi\rangle \tag{E.20}$$

and

$$\langle\Phi|\hat{G}^\dagger\Psi\rangle' = \langle\Phi|\hat{G}^{\dagger\prime}\Psi\rangle. \tag{E.21}$$

Noting that by definition

$$\langle\hat{G}'\Psi|\Phi\rangle = \langle\Phi|\hat{G}'^\dagger\Psi\rangle, \tag{E.22}$$

then comparing Eqs. (E.21) and (E.22) confirms Eq. (E.19).

There is an important corollary to this discussion about the nature of the differential. Let us write the full functional form of the Eq. (2.63) as

$$i\frac{d\hat{G}(\xi)}{d\xi} = i\hat{G}'(\xi) = [\hat{G}(\xi), \hat{K}], \tag{E.23}$$

where $\hat{G}'(\xi)$ is just an operator valued function that is equal to $\hat{G}(\xi)$ differentiated with respect to ξ. Eq. (E.23) implies

$$i\hat{G}'(0) = [\hat{G}(0), \hat{K}]. \tag{E.24}$$

Now operators of the type $\hat{G}(0)$ and $\hat{G}'(0)$ are independent of ξ and can be considered S-type operators. These can thus be used to operate on S-type state functions that are functions of ξ. Although this result seems rather trivial, it turns out to be useful when dealing with representations of the number operator in Chapter 3.

We end this appendix by noting that the differential is not confined to real variables and Hermitian operators. Consider a pair of non-Hermitian operators, $\hat{A}$ and $\hat{B}$ such that

$$[\hat{A}, \hat{B}] = 1.$$

Iterating the expansion rule for the above commutator one finds

$$[\hat{A}, \hat{B}^p] = p\hat{B}^{p-1},$$

from which we can infer using the same arguments as above that

$$[\hat{A}, f(\hat{B})] = f'(\hat{B}^{p-1}) = \frac{df(\hat{B})}{d\hat{B}},$$

and we can infer that

$$\hat{A} = \frac{d}{d\hat{B}}.$$

Notice also that

$$[-\hat{B}, \hat{A}] = 1,$$

so by the same token we could write

$$\hat{B} = -\frac{d}{d\hat{A}}.$$

Also

$$[\hat{A}, \hat{B}]^\dagger = [\hat{B}^\dagger, \hat{A}^\dagger] = [-\hat{A}^\dagger, \hat{B}^\dagger] = 1,$$

so

$$[-(\frac{d}{d\hat{B}})^\dagger, \hat{B}^\dagger] = 1,$$

and hence

$$(\frac{d}{d\hat{B}})^\dagger = -\frac{d}{d\hat{B}^\dagger},$$

but this does not allow us to infer any particular relationship between $\frac{d}{d\hat{B}}$ and $\frac{d}{d\hat{B}^\dagger}$.

F Calculus without limits

Differential calculus is almost invariably introduced to students in the context of finding the tangent to a curve by using the limiting process associated with the methods originally devised by Newton and Leibniz in the seventeenth century. In this appendix it is shown how differential calculus can be reconstructed entirely algebraically in a manner that does not rely at all on taking limits. All that is needed is to use a non-associative algebra that obeys the Leibniz rule. The mathematical objects that obey this rule are called *derivations* [36]. We begin by defining a linear operator, $\hat{\mathfrak{L}}$, that operates on functions, which it transforms into new functions. The linear property means that

$$\hat{\mathfrak{L}}(f_1 + f_2) = \hat{\mathfrak{L}} f_1 + \hat{\mathfrak{L}} f_2, \tag{F.1}$$

where f_1 and f_2 are functions. Now we define $\hat{\mathfrak{L}}$ as a Leibnizian operator that obeys Leibniz rule for the differentiation of a product, i.e.

$$\hat{\mathfrak{L}}(fg) = (\hat{\mathfrak{L}} f)g + f(\hat{\mathfrak{L}} g), \tag{F.2}$$

where f and g are functions. The meaning of the lhs of Eq. (F.2) is that the two functions are first multiplied together to produce a new function and $\hat{\mathfrak{L}}$ then operates on the product function. The first term on the rhs of Eq. (F.2) means that $\hat{\mathfrak{L}}$ first operates on f and the resulting function is then multiplied by g. The second term on the rhs of Eq. (F.2) means $\hat{\mathfrak{L}}$ first operates on g and then the resulting function is multiplied by f. This is the perfectly familiar Leibniz rule of differential calculus. Notice that it means that the operator $\hat{\mathfrak{L}}$ does not obey the associative rule when otherwise, the lhs of Eq. (F.2) would simply be equal to the first term on the rhs of Eq. (F.2).

Now let us choose a specific Leibnizian, such that

$$(\hat{\mathfrak{L}}_x x) = 1, \tag{F.3}$$

where x is a function that is simply equal to the variable, x and $\hat{\mathfrak{L}}_x$ is a Leibnizian. We can use this definition of $\hat{\mathfrak{L}}_x$ to generate differential calculus without the need to know anything about its origins in infinitesimal calculus and its association with limits. First note that $x = 1x$, so, according to Eq. (F.2)

$$\hat{\mathfrak{L}}_x(1x) = (\hat{\mathfrak{L}}_x 1)x + 1(\hat{\mathfrak{L}}_x x). \tag{F.4}$$

However, $\hat{\mathfrak{L}}_x(1x) = \hat{\mathfrak{L}}_x x = 1$, and so $\hat{\mathfrak{L}}_x(1) = 0$. We can easily show that $\hat{\mathfrak{L}}_x 0 = 0$, since $\hat{\mathfrak{L}}_x(1 + 0) = \hat{\mathfrak{L}}_x 1 + \hat{\mathfrak{L}}_x 0 = 0$. Furthermore, because of the linearity property, $\hat{\mathfrak{L}}_x 2 = \hat{\mathfrak{L}}_x(1 + 1) = \hat{\mathfrak{L}}_x 1 + \hat{\mathfrak{L}}_x 1 = 0$. This result can clearly be extended to, for example, $3 = 2 + 1$ etc., and so $\hat{\mathfrak{L}}_x n = 0$ for any natural number, n. All of the integers may also be included by noting first that $\hat{\mathfrak{L}}_x 0 = \hat{\mathfrak{L}}_x(1 - 1) = \hat{\mathfrak{L}}_x 1 + \hat{\mathfrak{L}}_x(-1) = 0$, so $\hat{\mathfrak{L}}_x(-1) = 0$, and then $\hat{\mathfrak{L}}_x(-n) = (\hat{\mathfrak{L}}_x(-1))n - 1\hat{\mathfrak{L}}_x n = 0$.

DOI: 10.1201/9781003377504-F

Next we can show that $\hat{\mathfrak{L}}_x(n^{-1}) = 0$ from $\hat{\mathfrak{L}}_x(n \times n^{-1}) = \hat{\mathfrak{L}}_x 1 = 0$ and again applying Eq. (F.2). This is then easily extended to showing that $\hat{\mathfrak{L}}_x$ operating on any rational number gives zero. From this, noting that any number can be constructed from an infinite series of rational numbers, means that $\hat{\mathfrak{L}}_x q = 0$, where q is any number whether, it be irrational or not. Thus we have shown that $\hat{\mathfrak{L}}_x(qf(x)) = q\hat{\mathfrak{L}}_x f(x)$ quite generally.

Now we note that $\hat{\mathfrak{L}}_x x^2 = (\hat{\mathfrak{L}}_x x)x + x\hat{\mathfrak{L}}_x x = 2x$. This result may be iterated to give

$$\hat{\mathfrak{L}}_x x^p = px^{p-1}, \tag{F.5}$$

where p is any integer. Notice that Eq. (F.5) shows that $\hat{\mathfrak{L}}_x$ maps any power x^p to x^{p-1} apart from when $p = 0$, since $x^0 = 1$, which does not map to x^{-1}. The reason is easy to see; it is because the coefficient of the result is 0, so x^0 maps to $0x^{-1} = 0$, in this case. The function with a differential of x^{-1} will be dealt with later. For now let us see that we can also deal with roots of x. First $\hat{\mathfrak{L}}_x(\sqrt{x} \times \sqrt{x}) = (\hat{\mathfrak{L}}_x \sqrt{x})\sqrt{x} + \sqrt{x}\hat{\mathfrak{L}}_x \sqrt{x} = 1$ from which we get $\hat{\mathfrak{L}}_x \sqrt{x} = -1/(2\sqrt{x})$, which shows that Eq. (F.5) works perfectly well for $p = 1/2$. Iterating this result by writing, for example, $x = x^{1/3} \times x^{1/3} \times x^{1/3}$ shows that Eq. (F.5) works for any rational number. Later we will show how Eq. (F.5) may be generalized to any irrational exponent of x.

Since we can express any analytic function of x as a polynomial in x of infinite order, i.e.

$$f(x) = \sum_{n=0}^{\infty} \alpha_n x^n, \tag{F.6}$$

where α_n is a constant coefficient, then we can see that

$$\hat{\mathfrak{L}}_x f(x) = \sum_{n=0}^{\infty} n\alpha_n x^{n-1} = f'(x), \tag{F.7}$$

where $f'(x)$ is the derivative of $f(x)$ familiar from infintessimal calculus. Take for example the exponential function, $\exp x$, defined by $\hat{\mathfrak{L}}_x \exp x = \exp x$. Let

$$\exp x = \sum_{n=0}^{\infty} \alpha_n x^n, \tag{F.8}$$

then from the definition of $\exp x$ above

$$\hat{\mathfrak{L}}_x \exp x = \sum_{n=0}^{\infty} n\alpha_n x^{n-1} = \sum_{n=0}^{\infty} \alpha_n x^n. \tag{F.9}$$

Equating coefficients with the same exponents on each side of Eq. (F.9) yields $\alpha_n = n\alpha_{n-1}$. Since $\exp x$ is only defined to within a constant factor, we can let $\alpha_0 = 1$ and then we get the usual infinite series for $\exp x$,

$$\exp x = \sum_{n=0}^{\infty} \frac{x^n}{n!}. \tag{F.10}$$

We can then use $\exp(ix) = C(x) + iS(x)$ together with $x \to ix$ in Eq. (F.10) to get $\hat{\mathfrak{L}}_x S(x) = C(x)$ and $\hat{\mathfrak{L}}_x C(x) = -S(x)$ together with other properties of $S(x)$ and $C(x)$, which are obviously the well-known trigonometrical functions, $\sin x$ and $\cos x$ respectively.

Another useful result is

$$\hat{\mathfrak{L}}_x(f(x)\frac{1}{f(x)}) = \hat{\mathfrak{L}}_x 1 = 0, \tag{F.11}$$

which leads to $\hat{\mathfrak{L}}_x(1/f(x)) = -f'(x)/f(x)^2$.

Finally, we can show that

$$\hat{\mathfrak{L}}_x(f(g(x))) = \hat{\mathfrak{L}}_g f(g) \hat{\mathfrak{L}}_x g(x), \tag{F.12}$$

where, by definition $\hat{\mathfrak{L}}_g g = 1$, by noting that $f(g)$ can be written as a polynomial in g. So $f(g)$ will simply be a polynomial in powers of g of the form

$$f(g(x)) = \sum_{n=0}^{\infty} \alpha_n g(x)^n, \tag{F.13}$$

where α_n is a constant coefficient. Take for example the second order term, then $\hat{\mathfrak{L}}_x g(x)^2 = \hat{\mathfrak{L}}_x (g(x) \times g(x)) = 2g(x)g'(x)$. The nth order term will give $\hat{\mathfrak{L}}_x g(x)^n = ng(x)^{n-1} g'(x)$ and so on. Thus $\hat{\mathfrak{L}}_x g(x) = g'(x)$ will be a common factor to all of the terms of the polynomial in g and we will get as a result

$$\hat{\mathfrak{L}}_x(f(g(x))) = (\sum_{n=1}^{\infty} n\alpha_n g(x)^{n-1})\hat{\mathfrak{L}}_x g(x) = (\hat{\mathfrak{L}}_g f(g))\hat{\mathfrak{L}}_x g(x), \tag{F.14}$$

which can be recognized as the correct form for the differential of a function of a function. Notice that the $n = 0$ term has disappeared from the sum in Eq. (F.14) since $\hat{\mathfrak{L}}_x g(x)^0 = 0$.

Recalling that differentiating any power, p of x, leads to a power of $p-1$, apart from $p = 0$, we are now in a position to find the function $g(x)$ for which $\hat{\mathfrak{L}}_x g(x) = x^{-1}$. First we note that $\ln(\exp x) = x$. Let $\exp x = g$, then from Eq. (F.14)

$$\hat{\mathfrak{L}}_x(\ln(\exp(x))) = \hat{\mathfrak{L}}_g \ln(g) \hat{\mathfrak{L}}_x \exp(x) = (\hat{\mathfrak{L}}_g \ln(g)) \exp(x) = (\hat{\mathfrak{L}}_g \ln(g))g \tag{F.15}$$

So, we can conclude, since $\hat{\mathfrak{L}}_x(\ln(\exp(x))) = \hat{\mathfrak{L}}_x x = 1$, that $\hat{\mathfrak{L}}_g \ln(g) = g^{-1}$, from which, $\hat{\mathfrak{L}}_x \ln(x) = x^{-1}$.

Eq. (F.14) may also be used to show that Eq. (F.5) can be generalized to any irrational exponent of x, as follows. By Eq. (F.14)

$$\hat{\mathfrak{L}}_x(\ln(x^q)) = x^{-q}\hat{\mathfrak{L}}_x(x^q), \tag{F.16}$$

where q is an irrational number. However, $\ln(x^q) = q \ln x$, so the lhs of Eq. (F.16) is $q\hat{\mathfrak{L}}_x(\ln x) = qx^{-1}$, and then we get $\hat{\mathfrak{L}}_x(x^q) = qx^{q-1}$, as required.

We have now succeeded in reconstructing the key results of differential calculus entirely by algebraic means without recourse to taking limits and in so doing shown the power of algebraic rules. Clearly, it is important to know that we can get the same results by considering limiting chords on curves as tangents, as Newton and Leibniz did, in the original formulation of what we now call differential calculus, that gives an interpretation of calculus in terms of the geometrical gradients of those curves and hence giving a connection to rates of change in dynamical systems. However, there is also something remarkable about being able to get these powerful results simply by examining the logical consequences that follow from an operator that does not obey the associative rule for multiplication.

As we have seen, the operator $\hat{\mathcal{L}}_x$ has the property of a derivation. It is clear that it does not obey the associative rule. However, from the earliest mathematical formulation of quantum mechanics [50, 24], operators were represented by matrices. Indeed, quantum mechanics was often referred to as *matrix mechanics*. However, matrix algebra is associative and thus $\hat{\mathcal{L}}_x$ cannot be represented by a matrix. Dirac, in his very first paper on the subject [24] showed how matrices could be used to represent the differentials of operators. It is actually a great advantage to be able to represent differentials of operators by an associative non-commutative algebra since this allows operators to be represented, for explicit calculations, by matrices, which of course obey the associative rule under multiplication. Dirac [24] actually started his argument with the two rules in Eqs. (F.1) and (F.2). Let us write the differential of an operator $\hat{A}$ as $\hat{A}'$, then Eqs. (F.1) and (F.2) translate into

$$(\hat{A}+\hat{B})' = \hat{A}' + \hat{B}' \tag{F.17}$$

and

$$(\hat{A}\hat{B})' = \hat{A}'\hat{B} + \hat{A}\hat{B}'. \tag{F.18}$$

Dirac argued that Eq. (F.17) implied that $\hat{A}'$ must be linear in $\hat{A}$. This actually agrees with our result above since we can interpret $\hat{\mathcal{L}}_x$ as $x' = \hat{\mathcal{L}}_x x$, and more generally

$$f'(x) = \hat{\mathcal{L}}_x f(x), \tag{F.19}$$

where $f(x)$ is any analytic function of x.

Now suppose $\hat{C}$ is a linear operator coefficient in the linear relationship between $\hat{A}'$ and $\hat{A}$ that obeys the associative rule $\hat{C}(\hat{A}\hat{B}) = (\hat{C}\hat{A})\hat{B}$. Then clearly $\hat{A}' \neq \hat{C}\hat{A}$, since $(\hat{A}\hat{B})' \neq \hat{C}\hat{A}\hat{B}$. There is another linear relationship, if we allow for non-commutation between operators, i.e., we could consider $\hat{A}' = \hat{A}\hat{D}$, where $\hat{D}$ is a second linear coefficient, which also obeys the associative rule. Clearly, this also fails for the same reason as $\hat{C}\hat{A}$ did. However, we can consider, $\hat{A}' = \hat{C}\hat{A} + \hat{A}\hat{D}$, which is also a linear relation. Then

$$(\hat{A}\hat{B})' = \hat{C}\hat{A}\hat{B} + \hat{A}\hat{B}\hat{D}. \tag{F.20}$$

We also have

$$(\hat{A}\hat{B})' = \hat{A}'\hat{B} + \hat{A}\hat{B}' = \hat{C}\hat{A}\hat{B} + \hat{A}\hat{D}\hat{B} + \hat{A}\hat{C}\hat{B} + \hat{A}\hat{B}\hat{D}, \tag{F.21}$$

which implies $\hat{A}\hat{D}\hat{B} + \hat{A}\hat{C}\hat{B} = \hat{A}(\hat{D}+\hat{C})\hat{B} = 0$ for all $\hat{A}$ and $\hat{B}$. The only way this can be true in general is if $\hat{D}+\hat{C} = 0$. So we get

$$\hat{A}' = -\hat{D}\hat{A} + \hat{A}\hat{D} = [\hat{D}, \hat{A}], \tag{F.22}$$

which is Dirac's result [24], although he used a somewhat different argument involving explicit matrix components.

Assuming that $\hat{A}$ has an adjoint, $\hat{A}^{\dagger}$, then

$$\hat{A}^{\dagger\prime} = [\hat{D}, \hat{A}^{\dagger}], \tag{F.23}$$

but if $\hat{A}^{\dagger\prime} = \hat{A}^{\prime\dagger}$, then

$$[\hat{D}, \hat{A}^{\dagger}] = [\hat{D}, \hat{A}]^{\dagger} = [\hat{A}^{\dagger}, \hat{D}^{\dagger}] = -[\hat{D}^{\dagger}, \hat{A}^{\dagger}], \tag{F.24}$$

which implies $\hat{D}^{\dagger} = -\hat{D}$ and so $\hat{D}$ must be anti-Hermitian. We can replace $\hat{D}$ by $\pm i\hat{K}$, where $\hat{K}$ is an Hermitian operator. Choosing the positive alternative we get

$$i\hat{A}' = [\hat{A}, \hat{K}], \tag{F.25}$$

which is identical to the H-type equation of motion for an operator that was derived in Chapter 2. The negative alternative is

$$i\hat{A}' = [\hat{K}, \hat{A}], \tag{F.26}$$

which is the form of the von Neumann equation that is important in quantum statistical mechanics [110]. The key point about both Eqs. (F.25) and (F.26) is that it is the product $[\hat{K}, \hat{A}]$ that has the derivation property, i.e.

$$[\hat{K}, \hat{A}\hat{B}] = [\hat{K}, \hat{A}]\hat{B} + \hat{A}[\hat{K}, \hat{B}],$$

despite the fact that $(\hat{K}\hat{A})\hat{B} = \hat{K}(\hat{A}\hat{B})$, etc. So just as we have shown in this appendix that calculus can be constructed algebraically, without limits, from the Leibniz rule, then so can the H-type equation of motion. As we have seen in Section 2.9, it is the derivation property of commutator algebra that underlies these results.

G Mean-field approximation

The mean-field approximation allows a simplification of the many-body interaction Hamiltonian, Eq. (9.25), that was derived in Section 9.2. That equation is reproduced here as

$$\hat{\Omega} = \sum_i W_i \hat{C}_i^\dagger \hat{C}_i + \sum_{jklm} V_{jklm} \hat{C}_j^\dagger \hat{C}_k^\dagger \hat{C}_l \hat{C}_m. \tag{G.1}$$

We are interested in a single term in the sum in the interaction part of the Hamiltonian in Eq. (G.1), but also recognize that included in the sum is an adjoint term which ensures that the Hamiltonian as a whole is Hermitian. So let us examine one such term and its adjoint, i.e.

$$V_{jklm} \hat{C}_j^\dagger \hat{C}_k^\dagger \hat{C}_l \hat{C}_m + V_{jklm}^* \hat{C}_m^\dagger \hat{C}_l^\dagger \hat{C}_k \hat{C}_j.$$

The mean-field approximation then involves treating the combination like $\hat{C}_l \hat{C}_m$ as equal to a large mean $\langle \hat{C}_l \hat{C}_m \rangle$ plus a small time varying operator so we replace each pair, $\hat{C}_l \hat{C}_m$, by

$$\langle \hat{C}_l \hat{C}_m \rangle + \hat{C}_l \hat{C}_m$$

where $\hat{C}_l \hat{C}_m$ is small compared to $\langle \hat{C}_l \hat{C}_m \rangle$, so that we can neglect products of such terms. Similarly, we can replace the original $\hat{C}_m^\dagger \hat{C}_l^\dagger$ by $\langle \hat{C}_m^\dagger \hat{C}_l^\dagger \rangle + \hat{C}_m^\dagger \hat{C}_l^\dagger$. Then we get a contribution to the interaction Hamiltonian of

$$V_{jklm} \hat{C}_j^\dagger \hat{C}_k^\dagger \langle \hat{C}_l \hat{C}_m \rangle + V_{jklm}^* \langle \hat{C}_m^\dagger \hat{C}_l^\dagger \rangle \hat{C}_k \hat{C}_j = \Delta_{jk} \hat{C}_j^\dagger \hat{C}_k^\dagger + \Delta_{jk}^* \hat{C}_k \hat{C}_j, \tag{G.2}$$

where $\Delta_{jk} = V_{jklm} \langle \hat{C}_l \hat{C}_m \rangle$ is a complex number. There is a corresponding pair of terms to those in Eq. (G.2),

$$\Delta_{lm} \hat{C}_m^\dagger \hat{C}_l^\dagger + \Delta_{lm}^* \hat{C}_l \hat{C}_m,$$

but these are essentially Eq. (G.2), with a different pair of dummy indices, so the Eq. (G.1), in the mean field approximation becomes

$$\hat{\Omega} = \sum_i W_i \hat{C}_i^\dagger \hat{C}_i + \sum_{jk} (\Delta_{jk} \hat{C}_j^\dagger \hat{C}_k^\dagger + \Delta_{jk}^* \hat{C}_k \hat{C}_j). \tag{G.3}$$

There will also be constant terms of the form

$$\langle \hat{C}_j^\dagger \hat{C}_k^\dagger \rangle \langle \hat{C}_l \hat{C}_m \rangle,$$

but these can be neglected because they have no effect on the Heisenberg equation of motion for the operators of the system.

DOI: 10.1201/9781003377504-G

Finally, we note in the case of the BCS Hamiltonian, attention is concentrated on coupled pairs, all of which lie close to the Fermi energy, and then, without loss of generality, we can take $\Delta_{jk} = \Delta_{jk}^* = \Delta$, so

$$\hat{\Omega} = \sum_i W_i \hat{C}_i^\dagger \hat{C}_i + \Delta \sum_{jk} (\hat{C}_j^\dagger \hat{C}_k^\dagger + \hat{C}_k \hat{C}_j). \tag{G.4}$$

Recalling that, in the case of the BCS Hamiltonian, the interaction term is negative, so we have $\Delta \rightarrow -\Delta$, which leads to Eq. (9.27).

H | Legendre transformation

We begin with the two functions,

$$\lambda = \lambda(v_x, v_y, v_z, x, y, z, t) \text{ and } \omega = f_\omega(k_x, k_y, k_z, x, y, z, t), \tag{H.1}$$

together with the relation between them

$$\lambda = \sum_i k_i v_i - \omega, \tag{H.2}$$

where i indicates the three components of the vectors $\mathbf{k} = (k_x, k_y, k_z)$ and $\mathbf{v} = (v_x, v_y, v_z)$. Eq. (H.2) is referred to as a Legendre transformation. Then

$$d\lambda = \sum_i k_i dv_i + \sum_i (v_i - \frac{\partial f_\omega}{\partial k_i}) dk_i - \sum_i \frac{\partial f_\omega}{\partial x_i} dx_i - \frac{\partial f_\omega}{\partial t} dt. \tag{H.3}$$

Since λ does not depend on k_i, we must have

$$v_i - \frac{\partial f_\omega}{\partial k_i} = 0,$$

and so

$$\frac{\partial \omega}{\partial \mathbf{k}} = \mathbf{v} = \frac{d\mathbf{x}}{dt}, \tag{H.4}$$

which confirms that the group velocity is equal to the frame velocity in Section 11.8. From Eq. (H.3) we also get

$$\frac{\partial \lambda}{\partial \mathbf{v}} = \mathbf{k}, \tag{H.5}$$

which is the definition of a canonical momentum in standard mechanics. In addition, from Eq. (H.3) we also get

$$\frac{\partial \lambda}{\partial \mathbf{x}} = -\frac{\partial \omega}{\partial \mathbf{x}} \text{ and } \frac{\partial \lambda}{\partial t} = -\frac{\partial \omega}{\partial t}. \tag{H.6}$$

DOI: 10.1201/9781003377504-H

Bibliography

1. Adler, S. L.: Quaternionic quantum mechanics and quantum fields, Oxford University Press, Oxford (1995)

2. Ali, S. T., Antoine, J.-P., Gazeau, J.-P.: Coherent states, Wavelets and their generalization, Springer, New York (2000)

3. Auletta, G., Fortunato, M., Paris, G.: Quantum mechanics, Cambridge University Press, Cambridge (2009)

4. Bacaër, N.: A short history of mathematical population dynamics, Springer-Verlag, London (2011)

5. Bagarello, F.: An operational approach to stock markets, J. Phys. A **39**, 6823–6840 (2006)

6. Bagarello, F., Oliveri, F.: An operator-like description of love affairs, SIAM J. Appl. Math. **70**, 3235–3251 (2010)

7. Bagarello, F.: Quantum dynamics for classical systems, Wiley, New York (2013)

8. Bagarello, F., Oliveri, F.: Dynamics of closed ecosystems described by operators, Ecological Modelling **275**, 89–99 (2014)

9. Bagarello, F., Gargano, F.: Non-Hermitian operator modelling of basic cancer cell dynamics, Entropy **20**, 270 (2018)

10. Barnett, S. M., Pegg, D. T.: Phase in quantum optics, J. Phys. A: Math. Gen. **19**, 3849–3862 (1986)

11. Becker, A.: What is real?, John Murray, London (2018)

12. Bender, C. M., Brody, D. C., Jones, H. F.: Must Hamiltonians be Hermitian, Am. J. Phys. **71**, 1095–1102 (2003)

13. Born, M., Wolf, E.: Principles of optics (6e), Pergamon Press, Oxford (1980)

14. Breuer, H-P., Petruccione, F.: The theory of open quantum systems, Oxford University Press, Oxford (2002)

15. Bruus, H., Flensberg, K.: Many-body quantum theory in condensed matter physics, Oxford University Press, Oxford (2004)

16. Calvocoressi, R.: Magritte, Phaidon Press, London (1984)

17. Carruthers, P., Nieto, M. N.: Phase and angle variables in quantum mechanics, Rev. Mod. Phys. **40**, 411–440 (1968)

18. Cockshott, P., Mackenzie, L. M., Michaelson, G.: Computation and its limits, Oxford University Press, Oxford (2012)

19. Cohen-Tannoudji, C., Diu, B., Laloë, F.: Quantum mechanics (2e), Wiley-VCH (2020)

20. Daukste, L., Basse, B., Baguley, B.C., Wall, D. J. N., Mathematical determination of cell population doubling times for multiple cell lines, Bull. Math. Biol. **74**, 2510–2534 (2012)

21. DeAngelis, D. L.: Mathematical modelling relevant to closed artificial ecosystems, Adv. Space Res. **31**, 1657–1665 (2003)

22. Dicke, R. H., Wittke, J. P.: Introduction to quantum mechanics, Addison-Wesley Publishing Co., Reading, Massachusetts (1960)

23. Dickson, L. E.: Linear algebras, Cambridge University Press, Cambridge (1914)

24. Dirac, P. A. M.: The fundamental equations of quantum mechanics, Proc. Roy. Soc. Lond. A **109**, 642–653 (1925)

25. Dirac, P. A. M.: Lectures on quantum mechanics, Belfer Graduate School of Science, Yeshiva University, New York (1964)

26. Dong, S.-H.: Factorization method in quantum mechanics, Springer, Dordrecht (2010)

27. Duane, W.: The transfer in quanta of radiation to matter, Proc. Nat. Acad. Sci. Wash. **9**, 158–164 (1923)

28. Dyson, F. J.: Feynman's proof of the Maxwell equations, Am. J. Phys. **58**, 209–211 (1990)

29. Eddington, A. S.: Fundamental theory, Cambridge University Press, Cambridge (1946)

30. Einstein, A.: The meaning of relativity, Chapman and Hall, London (1967)

31. Elkins, J.: Six stories from the end of representation, Stanford University Press, Stanford, California (2008)

32. Feynman, R. P.: Nobel lecture: The development of the space-time view of quantum electrodynamics (1965) https://www.nobelprize.org/prizes/physics/1965/feynman/biographical/

33. Feynman, R. P., Hibbs, A. R.: Quantum mechanics and path integrals, McGraw-Hill Co. Ltd., New York (1965)

34. Feynman, R. P.: The character of physical law, Penguin books, London (1992)

35. Fraser, D., Koberinski, A.: The Higgs mechanism and superconductivity, Stud. Hist. Philos. Sci. **55**, 72–91 (2016)

36. Freudenburg, G.: Algebraic theory of locally nilpotent derivations (2e), Springer, Berlin (2017)

37. Gamow, G.: Thirty years that shook physics, Dover Publications Inc, New York (1985)

38. Gerry, C. C., Knight, P. L.: Introductory quantum optics, Cambridge University Press, Cambridge (2005)

39. Gilmore, R.: Lie groups, physics and geometry, Cambridge University Press, Cambridge (2008)

40. Giordiano, G., Blannchini, F., Bruno, R., Colaneri, P., Di Filippo, A., Di Matteo, A., Colaneri, M.: Modelling the COVID-19 epidemic and implementation of population-wide interventions in Italy, Nat. Med. **26**, 855–860, https://doi.org/10.1038/s41591-020-0883-7 (2020)

41. Goldstein, H.: Classical mechanics (2e), Addison-Wesley Publishing Company, Reading, Massachusetts (1980)

42. Gottfried, K., Yan, T.-M.: Quantum mechanics: Fundamentals (2e), Springer-Verlag, New York (2003)

43. Haken, H.: Quantum field theory of solids, North-Holland, Amsterdam (1976)

44. Haken, H.: Light (Vol. 1), North-Holland, Amsterdam (1981)

45. Hall, B. C.: Quantum theory for mathematicians, Springer, New York (2013)

46. Hanahan, D., Weinberg, R. A.: The hallmark of cancer, Cell **100**, 57–70 (2000)

47. Harrison, P.: Angels on pinheads and needles' points, Notes Queries **63**, 45–47 (2016)

48. Healey, R.: The quantum revolution in philosophy, Oxford University Press, Oxford (2017)

49. Heinosaari, T., Ziman, M.: The mathematical language of quantum theory, Cambridge University Press, Cambridge (2012)

50. Heisenberg, W.: Über die quantentheoretische Umdeutung kinematischer und mechanischer Beziehungen, Z. Phys. **33**, 879–893 (1925)

51. Heisenberg, W.: Physics and philosophy, Penguin books, London (1989)

52. Hestenes, D., Weingartshofer, A. (Eds.): The electron, Kluwer Academic Publishers, Dordrecht (1991)

53. Horn, R. A., Johnson, C. R.: Matrix analysis (2e), Cambridge University Press, Cambridge (2013)

54. Horowitz, J., Normand, M. D., Corradini, M. G., Peleg, M.: Probabilistic model of microbial cell growth, division, and mortality, Appl. Environ. Microbiol. **76**, 230–242 (2010)

55. Hughes, R.: The shock of the new (2e), Thames and Hudson, London (1991)

56. Hunter, J. K., Nachtergaele, B.: Applied analysis, World Scientific, Singapore (2001)

57. Huylebrouck, D.: Africa and mathematics, Springer Nature Switzerland AG, Cham, Switzerland (2019)

58. Isea, R., Lonngren, K. E.: A mathematical model of cancer under radiotherapy, Int. J. Public Health Res. **3**, 340–344 (2014)

59. Isea, R., Lonngren, K. E.: A preliminary mathematical model for the dynamic transmission of Dengue, Chikungunya and Zika, Am. J. Mod. Phys. Applic. **3**, 11–15 (2016)

60. Jackson, J. D.: Classical electrodynamics (3e), John Wiley and Sons, Inc., New York (1999)

61. Jacoby, J. A., Curran, M., Wolf, D. R., Freericks, J. K.: Proving the existance of bound states for attractive potentials in on-dimension and two-dimensions without calculus, Eur. J. Phys. **40**, 045404 (2019)

62. Johnson, F.: A standard Swahili-English dictionary, Oxford University Press, Nairobi (1939)

63. Johnson, R. J.: The Deleuze-Lucretius encounter, Edinburgh University Press, Edinburgh (2017)

64. Joy, L. S.: Gassendi the atomist, Cambridge University Press, Cambridge (1987)

65. Kant, I.: Critique of pure reason (trans. Guyer, P., Wood, A. W.), Cambridge University Press, Cambridge (1998)

66. Kaplan, I. G., Navarro, O., Sanchez, J. A.: Exact commutation relations for Cooper pair operators and the problem of two interacting Cooper pairs, Physica C **419**, 13–17 (2005)

67. Kauffman, L. H., Noyes, H. P.: Discrete physics and the Dirac equation, Phys. Lett. A **218**, 139–146 (1996)

68. Kauffman, L. H.: Non-commutative worlds, New J. Phys. **6**, 173 (2004)

69. Kauffman, L. H.: Non-commutative worlds and classical constraints, Entropy **20**, 483 (2018)

70. Kittel, C.: Quantum theory of solids (2e), John Wiley and Sons, New York (1987)

71. Kleinert, H.: Path integrals in quantum mechanics, statistics, polymer physics and financial markets (3e), World Scientific, Singapore (2004)

72. Kuhn, T. S.: Black-body theory and the quantum discontinuity, 1894-1912, The University of Chicago Press, Chicago (1987)

73. Landau, L. D., Lifshitz, E. M.: The classical theory of fields (3e), Pergamon Press, Oxford (1971)

74. Landé, A.: New foundations of quantum mechanics, Cambridge University Press, London (1965)

75. Lotka, A. J.: Elements of physical biology, Williams and Wilkins, Baltimore, Maryland (1925)

76. Lucretius (Titus Lucretius Carus): *De rerum natura* (On the nature of things) (2e), English translation by W. H. D. Rouse, revised by M. F. Smith, Harvard University Press, Cambridge, Massachusetts (1982)

77. Mallat, S.: A wavelet tour of signal processing, Academic Press, San Diego (1998)

78. Mandel, L., Wolf, E.: Optical coherence and quantum optics, Cambridge University Press, Cambridge (1995)

79. Mason, A. S.: Plato, Acumen Publishing Ltd, Durham (2010)

80. Merzbacher, E.: Quantum mechanics, John Wiley and Sons Inc., New York (1961)

81. Messiah, A.: Quantum mechanics, North Holland Publishing Co., Amsterdam (1965)

82. Morse, P. M., Feshbach, H.: Methods of theoretical physics, McGraw-Hill Book Company, New York (1953)

83. Nambu, Y.: Nobel lecture: Spontaneous symmetry breaking in particle physics: A case of cross fertilization, Rev. Mod. Phys. **81**, 1015–1018 (2009)

84. Neumann, J. von: Mathematical foundations of quantum mechanics, Princeton University Press, Princeton (1955)

85. Nielsen, M. A., Chuang, I. L.: Quantum computation and quantum information, Cambridge University Press, Cambridge (2000)

86. Ohnuki, Y., Kamefuchi, S.: Quantum field theory and parastatistics, Springer-Verlag, New York (1982)

87. Olsen, M. K., Horowicz, R. J., Plimak, L. I., Treps, N., Fabre, C.: Quantum-noise-induced macroscopic revivals in second-harmonic generation, Phys. Rev. A **61**, 021803(R) (2000)

88. Olsen, M. K., Plimak, L. I., Khoury, A. Z.: Quantum analysis of the nondegenerate optical parametric amplifier with injected signal, Opt. Commun. **215**, 101–111 (2003)

89. Orszag, M.: Quantum optics (3e), Springer International Publishing, Switzerland (2016)

90. Partington, J. R.: The origins of atomic theory, Ann. Sci. **4**, 245–282 (1939)

91. Pauli, W.: General principles of quantum mechanics, Springer, Berlin (1980)

92. Pike, E. R., Sarkar, S.: The quantum theory of radiation, Oxford University Press, Oxford (1995)

93. Pletser, V., Huylebrouck, D.: The Ishango artefact: The missing base 12 link, Forma **14**, 339–346 (1999)

94. Prykarpatsky, A. K., Taneri, U., Bogolubov, Jr, N. N.: Quantum field theory with applications to quantum nonlinear optics, World Scientific, New Jersey (2002)

95. Rae, A. I. M., Napolitano, J.: Quantum mechanics (6e), Taylor and Francis, Boca Raton, Florida (2016)

96. Rahimi, I., Chen, F., Gandomi, A. H.: A review on COVID-19 forecasting models, Neural. Comput. Appl., https://doi.org/10.1007/s00521-020-05626-8 (2020)

97. Razavy, M.: Heisenberg's quantum mechanics, World Scientific Publishing Co. Pte. Ltd., Singapore (2011)

98. Reed, M., Simon, B.: Methods of modern mathematical physics, vol. 1, Functional analysis, Academic Press, San Diego, California (1980)

99. Rickless, S. C.: Plato's forms in transition, Cambridge University Press, Cambridge (2007)

100. Robinson, T. R.: The heating of the high latitude ionosphere by high power radio waves, Phys. Rep. **179**, 79–209 (1989)

101. Robinson, T. R.: A note on the Klein-Gordon equation and its solutions with applications to certain boundary value problems involving waves in plasma and in the atmosphere, Ann. Geophysicae **12**, 220–225 (1994)

102. Robinson, T. R.: Mass and charge distributions of the classical electron, Phys. Lett. A **200**, 335–339 (1995)

103. Robinson, T. R.: On Klein tunneling in graphene, Am. J. Phys. **80**, 141–147 (2012)

104. Robinson, T. R., Haven, E.: Quantization and quantum-like phenomena: A number amplitude approach, Int. J. Theor. Phys. **54**, 4576–4590 (2015)

105. Robinson, T. R.: The equally spaced energy levels of the quantum harmonic oscillator revisited: a back-to-front reconstruction of an n-body Hamiltonian, Eur. J. Phys. **38**, 065403 (2017)

106. Robinson, T. R., Haven, E., Fry, A. M.: Quantum counting: Operator methods for discrete population dynamics with application to cell division, Prog. Biophys. Mol. Biol. **130**, 106–119 (2017)

107. Robinson, T. R.: Natural number dynamics: reconstructing physics from generalized atomicity, Eur. J. Phys. **42**, 025402 (2021)

108. Rowlands, P.: The foundations of physical law, World Scientific Publishing Co. Pte. Ltd, Singapore (2015)

109. Ryder, L.: The quantum field theory (2e), Cambridge University Press, Cambridge (1996)

110. Schieve, W. C., Horwitz, L. P.: Quantum statistical mechanics, Cambridge University Press, Cambridge (2009)

111. Schrieffer, J. R.: Theory of superconductivity, CRC Press, Boca Raton, Florida (2018)

112. Schrödinger, E.: A method of determining quantum-mechanical eigenvalues and eigenfunctions, Proc. Roy. Irish Acad. **46**, 9–16 (1940)

113. Schrödinger, E.: Nature and the Greeks, Cambridge University Press, Cambridge (1954)

114. Seife, C.: Zero, Souvenir Press, London (2000)

115. Srednicki, M.: Quantum field theory, Cambridge University Press, Cambridge (2007)

116. Stakgold, I.: Green's functions and boundary value problems, John Wiley and Sons, New York (1979)

117. Stewart, I., Tall, D.: The foundations of mathematics (2e), Oxford University Press, Oxford (2015)

118. Swanson, D. G.: Quantum mechanics: Foundations and applications, Taylor and Francis, New York (2007)

119. Szymanski, T., Freericks, J. K.: Algebraic derivation of Kramers-Pasternack relations based on the Schrödinger factorization method, Eur. J. Phys. **42**, 025409 (2021)

120. Taylor, C. C. W.: The atomists: Leucippus and Democritus-Fragments, University of Toronto Press, Toronto (2011)

121. Taylor, P. L., Heinonen, O.: A quantum approach to condensed matter physics, Cambridge University Press, Cambridge (2002)

122. Tipler, P. A.: Physics of scientists and engineers (6e), W. H. Freeman and Co., New York (2007)

123. Volterra, V.: Fluctuations in the abundance of categories considered mathematically, Nature **118**, 558–560 (1926)

124. Waerden, B. L. van der: Sources of quantum mechanics, Dover Publications, New York (1967)

125. Weinberg, S.: The quantum theory of fields (Vol. 1), Cambridge University Press, Cambridge (1995)

126. Weinberg, S.: Lectures in quantum mechanics, Cambridge University Press, Cambridge (2013)

127. Weinstein, I. B., Joe, A.: Oncogene addiction, Cancer Res. **68**, 3077–3080 (2008)

128. Whitham, G. B.: Linear and nonlinear waves, John Wiley and Sons, New York (1974)

129. Whyte, L. L.: Essay on atomism: From Democritus to 1960, Thomas Nelson and Sons, London (1961)

130. Wilcox, R. M.: Exponential operators and parameter differentiation in quantum physics, J. Math. Phys. **8**, 962–982 (1967)

131. Yaffe, L.G.: Large N limits as classical mechanics, Rev. Mod. Phys. **54**, 407–436 (1982)

132. Zagoskin, A. M.: Quantum theory of many-body systems, Springer-Verlag, New York (1998)

133. Zagoskin, A. M., Sowa, A. P.: Can we see or compute our way out of Plato's cave?, J. Appl. Computat. Math. **3**, 192, doi:10.4172/2168-9679.1000192 (2014)

134. Zaslavsky, C.: Africa counts (3e), Lawrence Hill Books, Chicago (1999)

135. Zee, A.: Quantum field theory in a nutshell, Princeton University Press, Princeton (2003)

Index

Printed in the United States
by Baker & Taylor Publisher Services